The Minerals, Metals & Materials Series

Paul Mason · Charles R. Fisher · Ryan Glamm
Michele V. Manuel · Georg J. Schmitz
Amarendra K. Singh · Alejandro Strachan
Editors

Proceedings of the 4th World Congress on Integrated Computational Materials Engineering (ICME 2017)

Editors
Paul Mason
Thermo Calc Software Inc.
McMurray, PA
USA

Charles R. Fisher
Naval Surface Warfare Center
Fairfax, VA
USA

Ryan Glamm
Boeing
Seattle, WA
USA

Michele V. Manuel
University of Florida
Gainesville, FL
USA

Georg J. Schmitz
MICRESS Group at ACCESS e.V.
Aachen
Germany

Amarendra K. Singh
Kanpur
India

Alejandro Strachan
Purdue University
West Lafayette, IN
USA

ISSN 2367-1181   ISSN 2367-1696 (electronic)
The Minerals, Metals & Materials Series
ISBN 978-3-319-57863-7   ISBN 978-3-319-57864-4 (eBook)
DOI 10.1007/978-3-319-57864-4

TMS owns copyright; Springer has full publishing rights

Library of Congress Control Number: 2017938620

© The Minerals, Metals & Materials Society 2017
This work is subject to copyright. All rights are reserved by the Publisher, whether the whole or part of the material is concerned, specifically the rights of translation, reprinting, reuse of illustrations, recitation, broadcasting, reproduction on microfilms or in any other physical way, and transmission or information storage and retrieval, electronic adaptation, computer software, or by similar or dissimilar methodology now known or hereafter developed.
The use of general descriptive names, registered names, trademarks, service marks, etc. in this publication does not imply, even in the absence of a specific statement, that such names are exempt from the relevant protective laws and regulations and therefore free for general use.
The publisher, the authors and the editors are safe to assume that the advice and information in this book are believed to be true and accurate at the date of publication. Neither the publisher nor the authors or the editors give a warranty, express or implied, with respect to the material contained herein or for any errors or omissions that may have been made. The publisher remains neutral with regard to jurisdictional claims in published maps and institutional affiliations.

Printed on acid-free paper

This Springer imprint is published by Springer Nature
The registered company is Springer International Publishing AG
The registered company address is: Gewerbestrasse 11, 6330 Cham, Switzerland

Preface

This is a collection of manuscripts presented at the 4th World Congress on Integrated Computational Materials Engineering, a specialty conference organized by The Minerals, Metals & Materials Society (TMS) and the seven conference organizers, and held in Ypsilanti, Michigan, USA, on May 21–25, 2017.

Integrated computational materials engineering (ICME) has received international attention as it has been proven to shorten product and process development time, while lowering cost and improving outcomes. Building on the great success of the first three World Congresses on Integrated Computational Materials Engineering, which started in 2011, the 4th World Congress convened researchers, educators, and engineers to assess the state-of-the-art ICME and determine paths to further the global advancement of ICME. More than 200 authors and attendees from all over the world contributed to this conference in the form of presentations, lively discussions, and manuscripts presented in this volume. The international advisory committee members representing 11 different countries actively participated and promoted the conference.

The specific topics highlighted during this conference included integration framework and usage, ICME design tools and application, microstructure evolution and phase field modeling, mechanical performance using multiscale modeling, ICME success stories and applications, and a special focus on additive manufacturing. The conference consisted of integrated all-conference plenary talks, invited talks, contributed presentations, and a number of excellent poster presentations.

The 34 papers presented in this volume represent a cross section of the presentations and discussions from this congress. It is our hope that the 4th World Congress on ICME and these proceedings will further the global implementation of ICME, broaden the variety of applications to which ICME is applied, and ultimately help industry design and produce new materials more efficiently and effectively.

Paul Mason
Charles R. Fisher
Ryan Glamm
Michele V. Manuel
Georg J. Schmitz
Amarendra K. Singh
Alejandro Strachan

Acknowledgements

The organizers/editors would like to acknowledge the contributions of a number of people without whom this 4th World Congress, and the proceedings, would not have been possible.

First, we would like to offer many thanks to the TMS staff who worked tirelessly to make this an outstanding conference and an excellent proceedings publication.

Second, we want to thank the international advisory committee for their input in the planning of the conference, the promotion of the conference, and their participation in the conference. This international committee included the following: Dipankar Banerjee, Indian Institute of Science, India; Annika Borgenstam, KTH—Royal Institute of Technology, Sweden; Masahiko Demura, University of Tokyo, Japan; Dennis Dimiduk, BlueQuartz, LLC, USA; B.P. Gautham, Tata Consulting Services, India; Liang Jiang, Central South University, China; Kai-Friedrich Karhausen, Hydoaluminium, Germany; Peter Lee, Imperial College, UK; Mei Li, Ford Motor Company, USA; Baicheng Liu, Tsinghua University, China; Javier Llorca, IMDEA, Spain; Jiangfeng Nie, Monash University, Australia; Warren Poole, University of British Columbia, Canada; Antonio J. Ramirez, The Ohio State University, USA; James Warren, National Institute of Standards and Technology, USA; and Erich Wimmer, Materials Design, France.

Finally, we would especially like to acknowledge the financial support of all our sponsors. We are also grateful for the participation and contributions of all of the attendees.

Contents

Part I Integration Framework and Usage

An Attempt to Integrate Software Tools at Microscale and Above Towards an ICME Approach for Heat Treatment of a DP Steel Gear with Reduced Distortion 3
Deepu Mathew John, Hamidreza Farivar, Gerald Rothenbucher,
Ranjeet Kumar, Pramod Zagade, Danish Khan, Aravind Babu,
B.P. Gautham, Ralph Bernhardt, G. Phanikumar and Ulrich Prahl

Integrated Microstructure Based Modelling of Process-Chain for Cold Rolled Dual Phase Steels 15
Danish Khan, Ayush Suhane, P. Srimannarayana, Akash Bhattacharjee,
Gerald Tennyson, Pramod Zagade and B.P. Gautham

Improving Manufacturing Quality Using Integrated Computational Materials Engineering 23
Dana Frankel, Nicholas Hatcher, David Snyder, Jason Sebastian,
Gregory B. Olson, Greg Vernon, Wes Everhart and Lance Carroll

ICME Based Hierarchical Design Using Composite Materials for Automotive Structures 33
Azeez Shaik, Yagnik Kalariya, Rizwan Pathan and Amit Salvi

Towards Bridging the Data Exchange Gap Between Atomistic Simulation and Larger Scale Models 45
David Reith, Mikael Christensen, Walter Wolf, Erich Wimmer
and Georg J. Schmitz

A Flowchart Scheme for Information Retrieval in ICME Settings 57
Georg J. Schmitz

An Ontological Framework for Integrated Computational Materials Engineering......... 69
Sreedhar Reddy, B.P. Gautham, Prasenjit Das,
Raghavendra Reddy Yeddula, Sushant Vale and Chetan Malhotra

European Materials Modelling Council......... 79
Nadja Adamovic, Pietro Asinari, Gerhard Goldbeck, Adham Hashibon,
Kersti Hermansson, Denka Hristova-Bogaerds, Rudolf Koopmans,
Tom Verbrugge and Erich Wimmer

Facilitating ICME Through Platformization......... 93
B.P. Gautham, Sreedhar Reddy, Prasenjit Das and Chetan Malhotra

Bridging the Gap Between Bulk Properties and Confined Behavior Using Finite Element Analysis......... 103
David Linder, John Ågren and Annika Borgenstam

Ontology Dedicated to Knowledge-Driven Optimization for ICME Approach......... 113
Piotr Macioł, Andrzej Macioł and Łukasz Rauch

Integration of Experiments and Simulations to Build Material Big-Data......... 123
Gun Jin Yun

Part II ICME Design Tools and Application

ICME-Based Process and Alloy Design for Vacuum Carburized Steel Components with High Potential of Reduced Distortion......... 133
H. Farivar, G. Rothenbucher, U. Prahl and R. Bernhardt

Study of Transient Behavior of Slag Layer in Bottom Purged Ladle: A CFD Approach......... 145
Vishnu Teja Mantripragada and Sabita Sarkar

Developing Cemented Carbides Through ICME......... 155
Yong Du, Yingbiao Peng, Peng Zhou, Yafei Pan, Weibin Zhang,
Cong Zhang, Kaiming Cheng, Kai Li, Han Li, Haixia Tian, Yue Qiu,
Peng Deng, Na Li, Chong Chen, Yaru Wang, Yi Kong, Li Chen,
Jianzhan Long, Wen Xie, Guanghua Wen, Shequan Wang,
Zhongjian Zhang and Tao Xu

CSUDDCC2: An Updated Diffusion Database for Cemented Carbides......... 169
Peng Deng, Yong Du, Weibin Zhang, Cong Chen, Cong Zhang,
Jinfeng Zhang, Yingbiao Peng, Peng Zhou and Weimin Chen

Part III Microstructure Evolution

Multi-scale Modeling of Quasi-directional Solidification of a Cast Si-Rich Eutectic Alloy 183
Chang Kai Wu, Kwan Skinner, Andres E. Becerra, Vasgen A. Shamamian and Salem Mosbah

Numerical Simulation of Macrosegregation in a 535 Tons Steel Ingot with a Multicomponent-Multiphase Model 193
Kangxin Chen, Wutao Tu and Houfa Shen

Validation of CAFE Model with Experimental Macroscopic Grain Structures in a 36-Ton Steel Ingot 203
Jing'an Yang, Zhenhu Duan, Houfa Shen and Baicheng Liu

Analysis of Localized Plastic Strain in Heterogeneous Cast Iron Microstructures Using 3D Finite Element Simulations 217
Kent Salomonsson and Jakob Olofsson

An Integrated Solidification and Heat Treatment Model for Predicting Mechanical Properties of Cast Aluminum Alloy Component 227
Chang Kai Wu and Salem Mosbah

Linked Heat Treatment and Bending Simulation of Aluminium Tailored Heat Treated Profiles 237
Hannes Fröck, Matthias Graser, Michael Reich, Michael Lechner, Marion Merklein and Olaf Kessler

Numerical Simulation of Meso-Micro Structure in Ni-Based Superalloy During Liquid Metal Cooling Process 249
Xuewei Yan, Wei Li, Lei Yao, Xin Xue, Yanbin Wang, Gang Zhao, Juntao Li, Qingyan Xu and Baicheng Liu

Part IV Phase Field Modeling

Multiscale Simulation of α-Mg Dendrite Growth via 3D Phase Field Modeling and Ab Initio First Principle Calculations 263
Jinglian Du, Zhipeng Guo, Manhong Yang and Shoumei Xiong

Macro- and Micro-Simulation and Experiment Study on Microstructure and Mechanical Properties of Squeeze Casting Wheel of Magnesium Alloy 273
Shan Shang, Bin Hu, Zhiqiang Han, Weihua Sun and Alan A. Luo

Solidification Simulation of Fe–Cr–Ni–Mo–C Duplex Stainless Steel Using CALPHAD-Coupled Multi-phase Field Model with Finite Interface Dissipation 283
Sukeharu Nomoto, Kazuki Mori, Masahito Segawa and Akinori Yamanaka

Phase-Field Modeling of θ′ Precipitation Kinetics in W319 Alloys 293
Yanzhou Ji, Bita Ghaffari, Mei Li and Long-Qing Chen

Part V Mechanical Performance Using Multi-scale Modeling

Hybrid Hierarchical Model for Damage and Fracture Analysis in Heterogeneous Material 307
Alex V. Vasenkov

Fatigue Performance Prediction of Structural Materials by Multi-scale Modeling and Machine Learning 317
Takayuki Shiraiwa, Fabien Briffod, Yuto Miyazawa and Manabu Enoki

Nano Simulation Study of Mechanical Property Parameter for Microstructure-Based Multiscale Simulation 327
K. Mori, M. Oba, S. Nomoto and A. Yamanaka

Part VI ICME Success Stories and Applications

Multiscale, Coupled Chemo-mechanical Modeling of Bainitic Transformation During Press Hardening 335
Ulrich Prahl, Mingxuan Lin, Marc Weikamp, Claas Hueter, Diego Schicchi, Martin Hunkel and Robert Spatschek

Development of Microstructure-Based Multiscale Simulation Process for Hot Rolling of Duplex Stainless Steel 345
Mototeru Oba, Sukeharu Nomoto, Kazuki Mori and Akinori Yamanaka

A Decision-Based Design Method to Explore the Solution Space for Microstructure After Cooling Stage to Realize the End Mechanical Properties of Hot Rolled Product 353
Anand Balu Nellippallil, Vignesh Rangaraj, Janet K. Allen, Farrokh Mistree, B.P. Gautham and Amarendra K. Singh

Influence of Computational Grid and Deposit Volume on Residual Stress and Distortion Prediction Accuracy for Additive Manufacturing Modeling 365
O. Desmaison, P.-A. Pires, G. Levesque, A. Peralta, S. Sundarraj, A. Makinde, V. Jagdale and M. Megahed

Author Index 375

Subject Index 379

About the Editors/Organizers

Paul Mason graduated in 1989 from South Bank University in London, UK, with an Hons degree in physical sciences and scientific computing. On graduation, he joined the Atomic Energy Research Establishment at Harwell and worked in the area of civil nuclear power for 14 years focusing on materials R&D issues, particularly at high temperatures. Paul began his career mostly involved in experimental work and then moved into the modeling realm. In 2004, Paul was appointed President of Thermo-Calc Software Inc when the Swedish-based Thermo-Calc Software AB started a US subsidiary. Since that time Paul has been responsible for the marketing and sales, technical support, training, and customer relations for the North American market which includes the USA, Canada, and Mexico. Paul has been actively involved in the TMS ICME Committee since its formation and was Chair of the Committee from 2014 to 2016.

Charles R. Fisher is a Materials Scientist in the Welding, Processing, and Nondestructive Evaluation (NDE) Branch at the Naval Surface Warfare Center, Carderock Division (NSWCCD) near Bethesda, Maryland. Charles has been with the US Navy for 4 years working in computational weld mechanics (CWM) and ICME-related programs, with a focus on residual stress and distortion analysis in welded structures. He holds a B.S. in materials engineering from Iowa State University and a M.S. and Ph.D. in materials science and engineering from the University of Florida. He first became involved with TMS as a student member of Material Advantage in 2005, and has been a continuous and active member since.

Ryan Glamm is a Materials and Processes Engineer for Boeing Research and Technology focused on developing novel alloys and processes for structural airframe applications. He has a B.S. in materials science and engineering from The Ohio State University and a Ph.D. in materials science and engineering from Northwestern University. He has lead the introduction of integrated computational materials engineering for metals within Boeing and has authored numerous internal and four external publications and has two patents under review. Additionally, he has had assignments supporting KC-46, 777X, and the qualification of fastening assembly automation. He has interest in utilizing computational methods for accelerating technology implementation and in solving in-production engineering challenges. Ryan is active in the Puget Sound ASM International Chapter as the Education Chair.

Michele V. Manuel is a Professor and Chair in the Department of Materials Science and Engineering at the University of Florida. She received her Ph.D. in materials science and engineering at Northwestern University in 2007 and her B.S. in materials science and engineering at the University of Florida. She is the recipient of the 2013 Presidential Early Career Awards for Scientists and Engineers (PECASE), NSF CAREER, NASA Early Career Faculty, ASM Bradley Stoughton Award for Young Teachers, AVS Recognition for Excellence in Leadership, TMS Early Career Faculty, TMS Young Leaders Professional Development, and TMS/JIM International Scholar Awards. Her research lies in the basic understanding of the relationship among processing, structure, properties, and performance. She uses a system-based materials design approach that couples experimental research with theory and mechanistic modeling for the accelerated development of materials. Her current research is focused on the use of system-level design methods to advance the development of new materials through microstructure optimization. Of specific interest are lightweight alloys, self-healing metals, computational thermodynamics and kinetics, shape-memory alloys, and materials in extreme environments—specifically under high magnetic fields and irradiation.

Georg J. Schmitz obtained his Ph.D. in materials science in 1991 from RWTH Aachen University in the area of microstructure control in high-temperature superconductors. At present, he is senior scientist at ACCESS e.V., a private, nonprofit research center at the RWTH Aachen University. His research interests comprise microstructure formation in multicomponent alloys, modeling of solidification phenomena, phase field models, and thermodynamics. He is the official agent for Thermo-Calc Software AB in Germany and provides global support for MICRESS®. He has been appointed as expert by several institutions and is an active member of the committee on "Digital Transformation in Materials Engineering" of the Association of German Engineers (VDI), member of the TMS Integrated Computational Materials Engineering (ICME) Committee, and member of the European Materials Modelling Council (EMMC). He is editor and reviewer for a number of journals and has published more than 150 scientific articles, a book on a platform concept for ICME, and a recent Handbook of Software Solutions for ICME.

Amarendra K. Singh received his B.Tech. degree in metallurgical engineering in 1987 and M.Tech. and Ph.D. degrees in metallurgical engineering and materials science from the Indian Institute of Technology, Kanpur. He is currently a Professor with the Department of Materials Science and Engineering at the Indian Institute of Technology, Kanpur. Prior to joining IIT Kanpur in April 2015, he was with TCS Innovations Lab-TRDDC, Pune, for 23 years where he had successfully demonstrated the power and utility of mathematical modeling in optimization manufacturing operations leading to significant productivity gains, quality improvements, energy saving, and/or environment compliance. Dr. Singh has actively carried out research in number of areas including liquid metal processing (steelmaking, copper, and aluminum), solidification processing (ingot and continuous casting of steel/aluminum), and combustion and multiphase reacting flows (power plant boilers and metallurgical furnaces). Professor Singh's current research focus is on the areas of mathematical modeling of metallurgical operations with special focus on steelmaking and processing operations and integrated computational materials engineering (ICME). Dr. Singh has received many awards and honors, including being selected for the AICTE-INAE Distinguished Visiting Professorship at IIT Bombay from 2009 to 2011, TCS Distinguished Scientist Award in 2011, and the Metallurgist of the Year Award from the Ministry of Steel, Govt. of India, in 2014.

Alejandro Strachan is a Professor of materials engineering at Purdue University and the Deputy Director of the Purdue's Center for Predictive Materials and Devices (c-PRIMED) and of NSF's Network for Computational Nanotechnology. Before joining Purdue, he was a Staff Member in the Theoretical Division of Los Alamos National Laboratory and worked as a Postdoctoral Scholar and Scientist at Caltech. He received a Ph.D. in Physics from the University of Buenos Aires, Argentina, in 1999. Among other recognitions, Prof. Strachan was named a Purdue University Faculty Scholar (2012–2017), received the Early Career Faculty Fellow Award from TMS in 2009 and the Schuhmann Best Undergraduate Teacher Award from the School of Materials Engineering, Purdue University, in 2007. Professor Strachan's research focuses on the development of predictive atomistic and molecular simulation methodologies to describe materials from first principles, their application to problems of technological importance, and quantification of associated uncertainties. Application areas of interest include coupled electronic, chemical and thermo-mechanical processes in devices of interest for nanoelectronics and energy as well as polymers and their composites, molecular solids and active materials, including shape memory and high energy density materials. He has published over 120 articles in peer-reviewed scientific literature.

Part I
Integration Framework and Usage

An Attempt to Integrate Software Tools at Microscale and Above Towards an ICME Approach for Heat Treatment of a DP Steel Gear with Reduced Distortion

Deepu Mathew John, Hamidreza Farivar, Gerald Rothenbucher, Ranjeet Kumar, Pramod Zagade, Danish Khan, Aravind Babu, B.P. Gautham, Ralph Bernhardt, G. Phanikumar and Ulrich Prahl

Abstract Finite element simulation of heat treatment cycles in steel could be challenging when it involves phase transformation at the microscale. An ICME approach that can take into account the microstructure changes during the heat treatment and the corresponding changes in the macroscale properties could greatly help these simulations. Dual phase steel (DP steel) are potential alternate materials for gears with reduced distortion. Inter-critical annealing in DP steel involves phase transformation at the microscale and the finite element simulation of this heat treatment could be greatly improved by such an ICME approach. In the present work, phase field modeling implemented in the software package Micress is used to simulate the microstructure evolution during inter-critical annealing. Asymptotic Homogenization is used to predict the effective macroscale thermoelastic properties from the simulated microstructure. The macroscale effective flow curves are

---

D.M. John (✉) · A. Babu · G. Phanikumar
Integrated Computational Materials Engineering Laboratory, Department of Metallurgical and Materials Engineering, Indian Institute of Technology Madras, Chennai 600036, India
e-mail: deepumaj@gmail.com

H. Farivar · U. Prahl
Integrated Computational Materials Engineering, Institut für Eisenhüttenkunde (IEHK), der RWTH Aachen, Intzestraße 1, 52072 Aachen, Germany

G. Rothenbucher · R. Kumar · R. Bernhardt
Simufact Engineering GmbH, 21079 Hamburg, Germany

P. Zagade · D. Khan · B.P. Gautham
TRDDC, TCS Research, Tata Consultancy Services, Pune, India

© The Minerals, Metals & Materials Society 2017
P. Mason et al. (eds.), *Proceedings of the 4th World Congress on Integrated Computational Materials Engineering (ICME 2017)*,
The Minerals, Metals & Materials Series, DOI 10.1007/978-3-319-57864-4_1

obtained by performing Virtual Testing on the phase field simulated microstructure using Finite Element Method. All the predicted effective properties are then passed on to the macro scale Finite Element simulation software Simufact Forming, where the heat treatment cycle for the inter-critical annealing is simulated. The thermal profiles from this simulation are extracted and passed on to microscale to repeat the process chain. All the simulation softwares are integrated together to implement a multi-scale simulation, aiming towards ICME approach.

Keywords DP steel · Multi phase field modeling · Homogenization · Inter-critical annealing · Gear · Finite element method · Micress · Simufact

Introduction

Production of tailored components with improved properties is one of the primary aims of the industry at present. This requires materials with complex microstructures and strategic process design and control. Integrated computational materials engineering (ICME) is one of the present areas of interest for both academic and industrial research, as it uses physics based models, empirical models and human expertise in an integrated manner to significantly reduce the time and cost of development of new materials and their manufacturing processes. ICME had been successfully used for materials design, development and rapid qualification [1, 2]. Dual phase steel (DP steel) is one of the potential alternate materials for gears as it shows improved fatigue life and decrease in heat treatment distortion [3]. Low pressure vacuum carburizing (LPC), along with high pressure gas quenching (HPGQ) can be used to produce carburized components with less distortion, compared to other heat treating methods. ICME tools can help in controlling the gear distortion in such a heat treatment [4]. ICME approach had also been applied to optimize the metallurgy and improve the performance of carburizable ferrium steels, which are now commercially available for gear and bearing applications [5]. Multiscale modeling can be used to achieve ICME in order to assess the effects of constituent properties and processing on the performance of materials [6]. The present work aims towards an ICME approach for the design of a DP steel gear with reduced distortion. Vertical integration (multiscale modeling) is one of the aspects of an ICME approach. The present work addresses this aspect for the micro and macro scales. The data input-output of all the simulation tools was modeled on an ICME platform. LPC with HPGQ was used to carburize and heat treat the DP steel gear. A chemical composition selected using Calphad tools for carburized DP steel gear with maximum hardenability and maximum difference between Ae_3 and Ae_m was used for the simulations.

Simulations at Different Length Scales

Microscale: Phase Field Simulation

Phase field modeling is one of the widely used technique to predict the microstructure evolution during diffusional phase transformations. It had been used successfully to simulate the austenite to ferrite transformation in DP steels [7–9]. In the present work, phase field modeling implemented in the commercial software package Micress® was used to simulate the microstructure evolution during phase transformation. The austenite to ferrite transformation during inter-critical annealing in DP steel was simulated at the microscale. In order to simulate this phase transformation, a two dimensional multi-phase field simulation was performed on a multi-component system with chemical composition Fe-0.35Cr-0.75Mn-0.5Mo-0.4Si-0.1Ni-0.18C. An initial synthetic microstructure was created using voronoi tessellation, using the initial austenite grain size obtained by averaging the grain size from several experimental micrographs. Figure 1a shows the initial austenite microstructure with an average grain size of 10 µm. For simulating the nucleation of ferrite, seeds were defined at the triple junctions. The thermodynamic and the kinetic data required for the phase field simulation was obtained from ThermoCalc database, using the TQ coupling feature of Micress. Periodic boundary conditions were defined in all directions and the local equilibrium negligible partitioning (nple) approach was used to simulate the redistribution of the alloying elements. Simulations were run with 12 threads on an Intel Xeon E5-2630 processor and the average simulation time was around 9 h. The other simulation parameters used are reported in Table 1. Some of these parameters are taken from literature [10].

The simulation was performed for the heat treatment cycle consisting of cooling and holding (inter-critical annealing) as shown in Fig. 1c. The inter-critical annealing (IC annealing) time chosen for the simulation was 30 min. Figure 1b shows the final microstructure after the 30 min of holding. The white colour represents the ferrite phase and the remaining region represents the austenite phase.

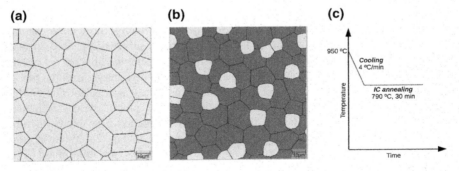

Fig. 1 a Initial austenite microstructure. b Final microstructure. c Heat treatment cycle used for simulation

Table 1 Micress simulation parameters

Simulation parameter		Value
Interfacial mobility: cm^4/Js	$\gamma-\gamma, \alpha-\alpha$	1×10^{-5}
	$\gamma-\alpha$	2×10^{-5}
Interfacial energy: J/cm^4	$\gamma-\gamma, \alpha-\alpha$	2×10^{-5}
	$\gamma-\alpha$	4×10^{-5}
Domain size		100×100 μm
Grid size		0.25 μm
Time step		Automatic (based on stability criterion)

Effective Properties: Asymptotic Homogenization

Asymptotic homogenization could be used to predict the effective thermo-elastic properties of an RVE (Representative Volume Element). The method uses assumption of microstructure periodicity and uniformity of the macroscopic fields within a unit cell domain [11]. In the present work, in order to predict the effective thermo-elastic properties from the phase field simulated microstructure, asymptotic homogenization implemented in the commercial software tool Homat® was used. The simulated microstructure in Micress (RVE) was meshed with C3D8 (8 node linear) hexahedral elements using Mesh2Homat tool. The meshed microstructure was passed on to Homat tool for performing the asymptotic homogenization. In order to perform the homogenization, Homat requires the geometric description and the properties of the individual phases. Geometric description (microstructure) was obtained from Micress whereas the properties of individual phases were obtained from JMatPro® database. Thermo-elastic homogenization was performed at 790 °C to obtain the effective macroscale elastic modulus, poissons ratio, density, specific heat capacity, thermal expansion coefficient and thermal conductivity.

Virtual Testing: Effective Flow Curve

Virtual testing on an RVE can be used to predict the effective flow curve of the macroscopic material [12]. Ramazani et al. [13] have successfully used virtual tensile testing on RVE in two and three dimensions to predict the flow curves of DP steel. In the present work, in order to predict the effective flow curve from the phase field simulated microstructure (RVE), a uniaxial tension test was performed on the microstructure using finite element method. The phase field simulated microstructure was meshed with C3D8 hexahedral elements. The microstructure consisted of two phases, austenite and ferrite as shown in Fig. 2a. White region represents ferrite phase and the remaining portion represents austenite phase. The properties of ferrite and austenite phases obtained from JMatPro® database were used for performing the finite element simulations. The meshed microstructure was

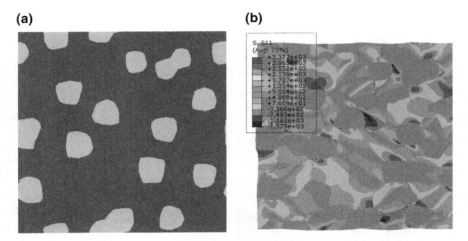

Fig. 2 a Initial microstructure for virtual testing. **b** Microstructure with 7% strain

loaded uniaxially in x direction to 7% strain and the average stress-strain response from the entire microstructure was recorded. Figure 2b shows the distribution of stresses in x direction for a strain of 7%. All operations were performed using commercial finite element tools. The average stress-strain response of all the elements in the RVE was used to obtain the effective flow curve of the macroscopic material. This flow curve is passed on to the software Simufact Forming® to start the macroscale simulation.

Macroscale Simulation of Heat Treatment

The inter-critical annealing heat treatment was simulated at the macroscale using finite element method, implemented in the commercial software package Simufact Forming®. Two dimensional axisymmetric simulation was performed on a cylindrical layered geometry. The geometry is assumed to be extracted from one tooth of a carburized gear as shown in Fig. 3a, b. The outer layer corresponds to the carburized layer (case of the gear). It is assumed that this layer has a uniform carbon composition of 0.7% throughout. The inner layer corresponds to the core of the gear with 0.18% carbon. The austenite to ferrite phase transformation is assumed to happen only in the core region. For the case region, no phase transformation happens during the inter-critical annealing. A fully austenite microstructure as shown in Fig. 1a is used to assign the effective macroscale material property for this layer. The material properties are assigned separately for the core and the case layers. The macroscale heat treatment cycle shown in Fig. 1c is simulated on the layered geometry. 2D quadrilateral elements were used for the simulation. Figure 3c shows the temperature distribution at an intermediate stage of the simulation

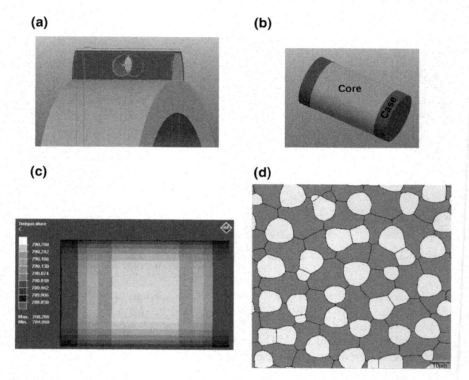

Fig. 3 a One tooth of a carburized gear. **b** Selected layered geometry. **c** Temperature distribution at an intermediate stage in Simufact. **d** Phase field simulated microstructure using the data from one of the tracking point in Simufact

in Simufact. Tracking points were used in the simulation to extract the thermal profile across the geometry during the simulation. The data from these tracking points were used to start a phase field simulation to obtain the microstructure evolution across the macroscale geometry. Figure 3d shows the phase field simulated microstructure using the data from one of the tracking points. The phase fraction from this simulated microstructure was compared with experimental data and was found to be matching well.

Multiscale Simulation Chain

A multi-scale simulation chain as shown in Fig. 4 was completed. For implementing this simulation chain, homogenization and virtual testing was performed for the phase field simulated microstructure at various instants along the heat treatment cycle. The effective macroscale properties calculated were fed to the macroscale heat treatment simulation in Simufact. The tracking point data in

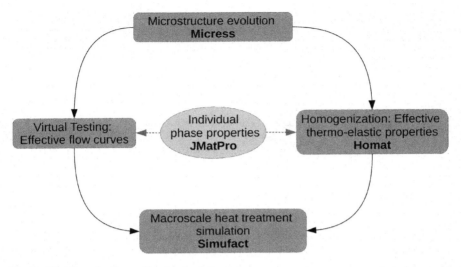

Fig. 4 Workflow for the multiscale simulation

Simufact was used to start a phase field simulation to obtain the microstructure evolution at the tracking point during the macroscale heat treatment process. In this way, the microstructure evolution at various points across the geometry was calculated and the corresponding effective macroscale material properties were obtained using homogenization and virtual testing. These properties were in turn fed to macroscale heat treatment simulation in Simufact to repeat the multiscale simulation chain.

ICME Platform

ICME approach requires an efficient information exchange between the simulation tools at various length scales. An ICME platform can greatly help in facilitating this information exchange and can also help in tracking the results along the process chain. This in turn eases the effort required and also gives a better understanding of the mechanisms by tracking the simulation results along the production chain [14]. In the present work, the data input-output of all the simulation tools was modeled on the ICME platform PREMAP (Platform for Realization of Engineered Materials and Products). PREMAP is an IT platform from Tata Consultancy Services (TCS) that facilitates integration of models, knowledge, and data for designing both the material and the product [15]. PREMAP requires ontological definitions of various entities including for product with requirements, manufacturing processes and material description at different scales. It essentially provides semantic bases which can be used for working with different tools through a unified semantic

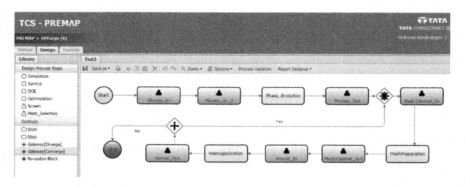

Fig. 5 Workflow implemented in PREMAP for micress and homat

language. This enables extensibility as well as use of different tools within an engineering workflow. This ontology can be used to express various forms of knowledge in forms or rules, expressions, etc. to help take engineering decisions. In the present work, the platform is used to help in repeating the multi scale simulation chain for various conditions, across the macroscale geometry, in order to arrive at the best possible process conditions. Figure 5 shows a snapshot of the simulation chain implemented in PREMAP for micress and homat. With the help of the ICME platform, several runs with different conditions could be made across the simulation chain to arrive at microstructure which gives minimum distortion and best possible combination of mechanical properties at the macroscale.

ICME Implementation Strategy

In order to start the multi-scale simulation chain to implement ICME approach, a starting temperature profile was required. In order to obtain this profile across the macroscale geometry, a Simufact simulation was performed using the material property data obtained from JMatPro database. The temperature-time data from the tracking points in this simulation was used to start a phase field simulation in Micress, as shown in Fig. 6. The simulated microstructure was then used to perform homogenization and virtual testing to obtain the macroscale effective properties. These properties were then fed to Simufact to start the macroscale simulation of heat treatment. This process chain was repeated until the target values for the desired properties were achieved. Once all the simulations are calibrated with experiments, this strategy could be used to obtain the microstructure and the process conditions corresponding to minimum distortion and best possible combination of mechanical properties at the macroscale.

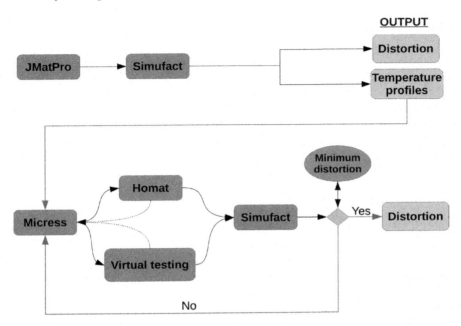

Fig. 6 Workflow for the implementation of a part of an ICME approach

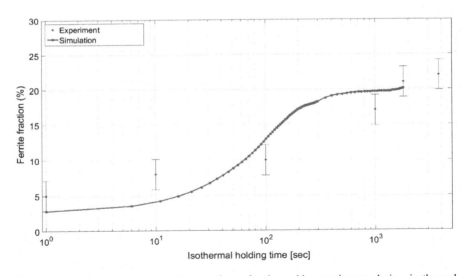

Fig. 7 Comparison of simulated ferrite volume fraction with experiments during isothermal holding at 790 °C

Experimental Validation

In order to validate the Micress phase field simulations, the macroscale heat treatment process was physically simulated on cylindrical samples in Baehr® Dilatometer and the phase fractions were obtained from the optical micrographs. The phase fraction evolution during the inter-critical annealing, obtained from these experiments were compared with that simulated in Micress, as shown in Fig. 7. The carburizing process and the final distortion at macroscale are simulated on Navy C-Ring specimens and experimentally validated in another work by the authors [16].

Conclusion

The present work implements the vertical integration aspect (multi-scale modeling), aiming towards an ICME approach for the microstructure and process design of a DP steel gear with reduced distortion. The microstructure and property evolution during inter-critical annealing heat treatment in DP steels were simulated using commercial software tools. The macroscale heat treatment was simulated on a layered cylindrical geometry to mimic the heat transfer across the cross-section of one tooth of a carburized gear. The data input-output of all the software tools was modeled on an ICME platform. The simulation chain implemented in this work could be used to obtain the microstructure with minimum distortion and best possible combination of macroscale material properties. In this way, the number of experiments required for such a microstructure and process design could be reduced.

Acknowledgements The authors would like to acknowledge the funding from IGSTC (Indo-German Science and Technology Centre) for funding through the project 'Combined Process and Alloy Design of a microalloyed DP Forging Steel based on Integrative Computational Material Engineering (DP-Forge)'.

References

1. J.T. Sebastian, G.B. Olson, Examples of QuesTek innovations' application of ICME to materials design, development, and rapid qualification, in *55th AIAA/ASMe/ASCE/AHS/SC Structures, Structural Dynamics, and Materials Conference* (2014)
2. O. Guvenc, F. Roters, T. Hickel, M. Bambach, ICME for crashworthiness of TWIP steels: from ab initio to the crash performance. JOM **67**, 120–128 (2015)
3. F. Kazuaki, T. Kunikazu, S. Tetsuo, Examination of surface hardening process for dual phase steel and improvement of gear properties. JFE Tech. Rep. **15**, 17–23 (2010)
4. J. Wang, X. Su, M. Li, R. Lucas, W. Dowling, ICME tools can help control gear distortion from heat treating. Adv. Mater. Process. **171**, 55–58 (2013)

5. J. Grabowski, J. Sebastian, A. Asphahani, C. Houser, K. Taskin, D. Snyder, Application of ICME to optimize metallurgy and improve performance of carburizable steels, in *American Gear Manufacturers Association Fall Technical Meeting 2014 (FTM 2014)* (2014), pp. 216–225
6. E. Pineda, B. Bednarcyk, S. Arnold, Achieving ICME with multiscale modeling: the effects of constituent properties and processing on the performance of laminated polymer matrix composite structures, in *55th AIAA/ASMe/ASCE/AHS/SC Structures, Structural Dynamics, and Materials Conference* (2014)
7. B. Zhu, M. Militzer, Phase-field modeling for intercritical annealing of a dual-phase steel. Metall. Mater. Trans. A **46**, 1073–1084 (2014)
8. J. Rudnizki, U. Prahl, W. Bleck, Phase-field modelling of microstructure evolution during processing of cold-rolled dual phase steels. Integr. Mater. Manuf. Innov. **1**, 3 (2012)
9. M.G. Mecozzi, Phase field modelling of the austenite to ferrite transformation in steels. Ph.D. thesis, TU Delft (2007)
10. J. Rudnizki, B. Bottger, U. Prahl, W. Bleck, Phase-field modeling of austenite formation from a ferrite plus pearlite microstructure during annealing of cold-rolled dual-phase steel. Metall. Mater. Trans. A **42**, 2516–2525 (2011)
11. G. Laschet, Homogenization of the thermal properties of transpiration cooled multi-layer plates. Comput. Methods Appl. Mech. Eng. **191**, 4535–4554 (2002)
12. G. Schmitz, U. Prahl, *Integrative Computational Materials Engineering: Concepts and Applications of a Modular Simulation Platform* (Wiley-VCH Verlag GmbH, 2012)
13. A. Ramazani, K. Mukherjee, H. Quade, U. Prahl, W. Bleck, Correlation between 2D and 3D flow curve modelling of DP steels using a microstructure-based RVE approach. Mater. Sci. Eng. A **560**, 129–139 (2013)
14. G. Schmitz, S. Benke, G. Laschet, M. Apel, U. Prahl, P. Fayek, S. Konovalov, J. Rudnizki, H. Quade, S. Freyberger, T. Henke, M. Bambach, E. Rossiter, U. Jansen, U. Eppelt, Towards integrative computational materials engineering of steel components. Prod. Eng. Res. Devel. **5**, 373–382 (2011)
15. B.P. Gautham, A.K. Singh, S.S. Ghaisas, S.S Reddy, F. Mistree, PREMAP: a platform for the realization of engineered materials and products, in *ICoRD'13*. Lecture Notes in Mechanical Engineering (2013)
16. H. Farivar, G. Rothenbucher, U. Prahl, R. Bernhardt, *ICME-based process and alloy design for carburized steel components with high potential of reduced distortion* (ibid)

Integrated Microstructure Based Modelling of Process-Chain for Cold Rolled Dual Phase Steels

Danish Khan, Ayush Suhane, P. Srimannarayana,
Akash Bhattacharjee, Gerald Tennyson, Pramod Zagade
and B.P. Gautham

Abstract The properties of dual phase (DP) steels are governed by the underlying microstructure, the evolution of which is determined by the processing route. In order to design a dual phase steel with tailored properties, it is therefore important to model and design each of the process involved at the microstructure level in an integrated fashion. In this work, an integrated approach is used to predict the final microstructure and mechanical properties of dual phase steels through microstructure based modelling of cold rolling, intercritical annealing and quenching processes. Starting with a representative volume element (RVE) of initial ferrite-pearlite microstructure, cold-reduction during rolling is simulated in a FEM based micromechanics approach under appropriate boundary conditions. The deformed microstructure with plastic strain energy distribution after cold-reduction serves as input for modelling static recrystallization and ferrite/pearlite to austenite transformation during intercritical annealing using a phase-field approach. A micromechanics based quenching simulation is then used to model austenite to martensite transformation, related volume expansion and evolution of transformational stress/strain fields. The resultant microstructure with its complete state is used to evaluate the flow behavior under uniaxial loading conditions in a FEM based micromechanics approach under periodic boundary conditions. Property variation for different initial microstructure, composition and processing conditions are studied and discussed.

Keywords Micromechanics · Phase-field · Intercritical annealing · Process integration · Microstructure modelling · Phase transformation · Property prediction

---

D. Khan (✉) · A. Suhane · P. Srimannarayana · A. Bhattacharjee · G. Tennyson ·
P. Zagade · B.P. Gautham
TRDDC, TCS Research, Tata Consultancy Services, Pune, India
e-mail: d.khan2@tcs.com

Introduction

Cold rolled dual phase steels are known for their high strength, high toughness and high formability. This makes them a suitable candidate for producing strength-relevant and crash-relevant body-in-white components having complex geometries such as cross-beams, pillars and other reinforcements [1]. This has been possible due to the multiphase nature of these steels wherein the different microstructure constituents impart varied properties to the steels. The constant pursuit of the auto-makers to cater to the increasing fuel-efficiency demands and safety regulations have led to the necessity of continual improvement and optimization of the properties and hence the microstructure of these steels.

Researchers have continuously tried to optimize the properties of the dual phase steels with new processing routes and processing conditions [2, 3]. The final properties of dual phase steels not only depend upon the individual properties of these micro-constituents but also on their morphology and distribution. On the other hand, the properties of the individual phases, their morphology and distribution depend upon the processing history of the steel. Therefore, in order to optimize the final properties of steels in a systematic way, it is not only important to optimize the individual processes involved but also the entire process-chain with explicit tracking of the microstructure evolution. With the advent of microstructure based process-structure [4, 5] and structure-property [6, 7] modelling techniques, it is now possible to model the evolution of microstructure and properties with the processing conditions. Apart from that, with more efforts being put towards solving problem through ICME route, optimizing the entire process-chains of the products in a closed-loop with systematic decision making at each decision point is the need of the hour.

The final steps of a typical processing route for the production of cold-rolled dual phase steels is shown in Fig. 1. There has been number of attempts in past for sequential integration of microstructure based process models as well as integration of process-property models. Madej et al. [8] carried out microstructure based modelling of cold-rolling of ferritic-pearlitic steels using FEM and used its plastic energy distribution output for modelling static recrystallization (SRX) during inter-critical annealing (ICA) using cellular automata in a digital material

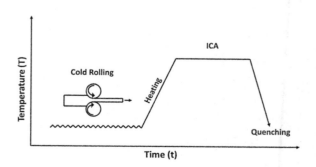

Fig. 1 A typical processing route for cold-rolled dual phase steels

representation framework. Rudinizki [9] on the other hand studied the through process modelling of production of dual phase steels by modelling ICA using phase-field approach followed by property prediction of the microstructure thus obtained using FEM based micromechanics approach. Ramazani et al. [10] modelled the process chain for dual-phase by integrating a similar phase-field approach based ICA model with FEM based micromechanics model of property prediction that took into account the effect of geometrically necessary dislocation formed during quenching on the final properties. However, none of these efforts attempt the integration of process-chain right from the cold-rolling till the final property-predictions in an ICME framework.

The present work involves the integration of microstructure based models of cold-deformation, inter-critical annealing and quenching processes to take into account the effect of each of them on the final microstructure and properties prediction of a cold-rolled dual phase steel. The focus is on the integration of models on an ICME-enabling platform that allows running the process-chain simulations in a loop with decision-making at each stage, thereby opening up the opportunity for optimizing the process-chain in a closed-loop.

Integrated Numerical Models

Figure 2 shows the integration of various micro-scale process models used in this study along with relevant phenomena modelled and related information exchange. The process-chain simulation starts with a 2D RVE of ferritic-pearlitic microstructure having certain statistics defined in terms of ferrite grain size, pearlite colony size etc. This RVE represents a typical ferritic-pearlitic microstructure obtained at the end of runout table (ROT). A typical RVE used in this study is shown in Fig. 3a. The microstructure RVE was subjected to different mechanical and thermal boundary conditions of the subsequent processes and the essential physics involved was modelled to keep track of the evolution of microstructure along with its state of stress and strain. Following sections describe the details of the various microstructure-scale process and property models used.

Cold-Rolling

Cold reduction was modelled as a plain-strain compression of the RVE under homogenous boundary conditions using a FEM model in ABAQUS^{TM}. Chemical composition based flow curves for ferrite and pearlite [11], were used as input for the model. Based on the stress and strain partitioning between ferrite and pearlite phases, plastic strain energy was calculated at each material point of the RVE. A typical plastic strain energy distribution in a ferritic-pearlitic RVE having 14% pearlite, deformed to 50% cold reduction is shown in Fig. 3b.

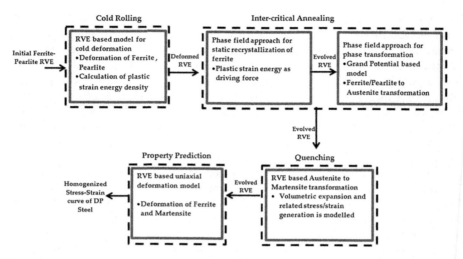

Fig. 2 Integration of various micro-scale process models along with information exchange

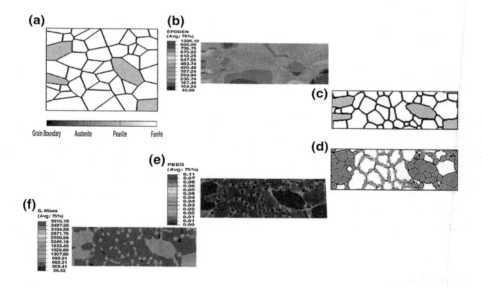

Fig. 3 Microstructures involved at different stages of micro-scale process-chain simulations. **a** Initial ferritic-pearlitic RVE. **b** Plastic strain energy density (J/mm^3) distribution in deformed RVE at the end of cold-reduction. **c** Microstructure obtained at the end of ferrite recrystallization during ICA. **d** Final microstructure obtained at the end of ICA simulation. **e** Plastic strain distribution in the microstructure at the end of quenching simulation. **f** Stress (MPa) distribution in the final ferritic-martensitic microstructure at the end of uniaxial loading

Inter-critical Annealing (ICA)

ICA involves two key microstructural phenomena viz., static recrystallization of ferrite and phase transformation from ferrite/pearlite to austenite. Two different phase-field approaches were used to model recrystallization [4] and phase transformation [12]. Stored plastic strain energy within the deformed grains from the cold-rolling model was used as the driving force for recrystallization and a grand potential difference was used as the driving force for phase transformation. Binary Fe–C was considered for the simulations assuming no recrystallization in pearlite phase which was assumed to be homogenous and hard [4]. Periodic boundary conditions were considered throughout the model. All grain boundaries were assumed to be high angle boundaries and respective orientation effects were neglected. Grain boundary properties were taken from Raabe and Hantcherli [13]. Typical microstructures obtained during ICA, isothermally held at 760 °C for 3 min are shown in Fig. 3c, d.

Quenching

In order to model the effect of volume expansion and transformation strain associated with austenite to martensite ($\gamma \rightarrow \alpha'$) transformation, and calculation of resultant residual stresses, a microstructure based quenching simulation was set-up as a FEM micromechanics model in ABAQUS^{TM}. The microstructure obtained from the phase field models of ICA serves as input for the model. Temperature profile was imposed on the microstructure as thermal load under periodic boundary conditions and the volume expansion associated with $\gamma \rightarrow \alpha'$ transformation was modelled using different temperature dependent thermal expansion coefficients [10] for ferrite and austenite/martensite. High temperature flow curves for ferrite and austenite/martensite were modelled as ratios of room-temperature flow curves, as used by Ramazani et al. [10]. Calculated plastic strain distribution at α/α' interface, due to $\gamma \rightarrow \alpha'$ transformation, at the end of quenching simulation is shown in Fig. 3e.

Property Prediction

The microstructure obtained at the end of quenching simulation, with its complete state of stress and strain, was subjected to uniaxial loading conditions in a FEM micromechanics model under periodic boundary conditions. Chemical composition dependent phenomenological work hardening models for individual phases, as developed by Rodriguez and Gutierrez [14], were used as input for the model. First order volumetric homogenisation of the calculated stress and plastic strain values

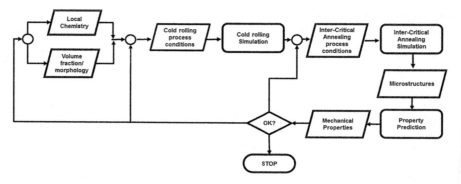

Fig. 4 Integrated work-flow of all the micro-scale models on the ICME platform, TCS PREMAP

was carried out over the entire RVE, at each time-step of the analysis, in order to calculate the uniaxial flow curve for the steel. Stress distribution in the deformed microstructure at the end of uniaxial loading is shown in Fig. 3f.

Figure 4 shows the integrated workflow of the various models on the ICME enabling platform, TCS PREMAP [15]. Such a workflow enables a systematic optimisation of the dual phase steels properties by running the process-chain simulations in a loop and enabling the user to make decisions at different stages of the process-flow in order to arrive at the most-suitable process-conditions.

Figure 5 shows the comparison of final flow curves of dual phase steels obtained after running the process-chain simulations for 50% cold-reduction followed by 3 min Inter-critical annealing at 760 and 780 °C.

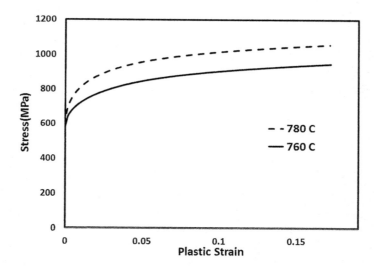

Fig. 5 Flow curves for two dual phase steels obtained after 50% cold-reduction followed by ICA at 760 and 780 °C

Summary

Cold-rolled dual phase steels are an important class of AHSS steels that are finding increasing usage in the automotive bodies in order to achieve desired targets of mass saving and safety regulations. Based on the increasing requirements of the industries, optimising the properties and hence the microstructure of these steels, in a systematic way, to find user-specific suitable combination of composition and processing conditions, is very much needed. In order to achieve this, microstructure based models for different processes and property prediction were implemented and integrated in a work-flow that enables the user to explore various processing scenarios. The idea has been demonstrated by running the process-chain models for different processing scenarios and the results have been reported.

References

1. Steel Market Development Institute, FutureSteelVehicle Final Engineering Report (WordAutoSteel, 2011)
2. X.-L. Cai, A.J. Garratt-Reed, W.S. Owen, The development of some dual phase steel structures from different starting microstructures. Metallur. Trans. A **16A**, 543–557 (1985)
3. K. Park et al., Effect of the martensite distribution on the strain hardening and ductile fracture behaviors in dual-phase steel. Mater. Sci. Eng. A **604**, 135–141 (2014)
4. B. Zhu, M. Militzer, Phase-field modeling for intercritical annealing of a dual-phase steel. Metallur. Mater. Trans. A **46**(3), 1073–1084 (2015)
5. C. Zheng, Dierk Raabe, Interaction between recrystallization and phase transformation during intercritical annealing in a cold-rolled dual-phase steel: A cellular automaton model. Acta Mater. **61**(14), 5504–5517 (2013)
6. S. Sodjit, V. Uthaisangsuk, Microstructure based prediction of strain hardening behavior of dual phase steels. Mater. Des. **41**, 370–379 (2012)
7. A. Ramazani et al., Modelling the effect of microstructural banding on the flow curve behaviour of dual-phase (DP) steels. Comput. Mater. Sci. **52**, 46–54 (2012)
8. L. Madej et al., Multi scale cellular automata and finite element based model for cold deformation and annealing of a ferritic–pearlitic microstructure. Comput. Mater. Sci. **77**, 172–181 (2013)
9. J. Rudnizki, Through-process model for the microstructure of dual-phase steel. Ph.D. thesis (IEHK RWTH Aachen University 2011), pp. 81–132
10. A. Ramazani et al., Quantification of the effect of transformation-induced geometrically necessary dislocations on the flow-curve modelling of dual-phase steels. Int. J. Plast. **43**, 128–152 (2013)
11. N. Ishikawa et al., Micromechanical modeling of ferrite-pearlite steels using finite element unit cell models. ISIJ Int. **40**(11), 1170–1179 (2000)
12. A. Choudhury, B. Nestler, Grand-potential formulation for multicomponent phase transformations combined with thin-interface asymptotics of the double-obstacle potential. Phys. Rev. E **85**, 021602 (2012)
13. D. Raabe, L. Hantcherli, 2D cellular automaton simulation of the recrystallization texture of an IF sheet steel under consideration of Zener pinning. Comput. Mater. Sci. **34**, 299–313 (2005)

14. R.M. Rodriguez, I. Gutierrez, Unified formulation to predict the tensile curves of steels with different microstructures. Mater. Sci. Forum **426–432**, 4525–4530 (2003)
15. B.P. Gautham et al., PREMAP: a platform for the realization of engineered materials and products, in *ICoRD'13*. Lecture Notes in Mechanical Engineering, ed. by A. Chakrabarti, R. V. Prakash (Springer India, 2013), pp. 1301–1313

Improving Manufacturing Quality Using Integrated Computational Materials Engineering

Dana Frankel, Nicholas Hatcher, David Snyder, Jason Sebastian, Gregory B. Olson, Greg Vernon, Wes Everhart and Lance Carroll

Abstract The prediction of materials properties and their variation within a specification or design space is key in ensuring reliable production uniformity. To capture the complex mechanisms that underpin materials' performance, processing-structure-properties links are established using a "systems design" approach. QuesTek Innovations LLC has previously utilized multi-scale ICME modeling methodologies and tools (e.g., CALPHAD thermodynamic and kinetic databases, property models, etc.) and advanced characterization techniques to design advanced materials with improved performance. This work focuses on building an ICME infrastructure to predictively model properties of critical

---

Honeywell Federal Manufacturing & Technologies manages and operates the Kansas City National Security Campus for the United States Department of Energy under contract number DE-NA-0002839.

D. Frankel (✉) · N. Hatcher · D. Snyder · J. Sebastian (✉) · G.B. Olson
QuesTek Innovations LLC, Evanston, IL, USA
e-mail: dfrankel@questek.com

J. Sebastian
e-mail: jsebastian@questek.com

N. Hatcher
e-mail: nhatcher@questek.com

D. Snyder
e-mail: dsnyder@questek.com

G.B. Olson
e-mail: golson@questek.com

G. Vernon (✉) · W. Everhart · L. Carroll
Honeywell Federal Manufacturing & Technologies, LLC, Kansas City, MO, USA
e-mail: gvernon2@kcp.com

W. Everhart
e-mail: weverhart@kcp.com

L. Carroll
e-mail: lcarroll@kcp.com

materials for energy and defense applications by optimizing existing materials, performing calculations to quantify uncertainty in material properties, and defining target specification ranges and processing parameters necessary to ensure design allowables. Focusing on two material case studies, 304L austenitic stainless steel and glass-ceramic-to-metal seals, we show how these ICME techniques can be used to better understand process-structure and structure-property relationships. These efforts provide pathways to novel, fully optimized alloys and production processes using the Accelerated Insertion of Materials (AIM) methodology within ICME. The AIM method is used for probabilistic properties forecasting to enable rapid and cost-efficient process optimization and material qualification.

Keywords ICME · Materials design · Manufacturing · CALPHAD · Uncertainty quantification

Introduction

Integrated Computational Materials Engineering (ICME) has supplanted traditional design of experiments (DOE) approaches to materials development. Beginning with its design of ultra-high performance *Ferrium*™ steels, QuesTek has pioneered the *Materials by Design*® approach to rapidly design, scale up, and qualify new materials. The foundation of QuesTek's *Materials by Design* technology is its systems-based design framework that considers the various interactions between chemistry, processing, microstructure and properties to achieve a new material's desired performance [1]. QuesTek's design process for a new material begins by working with the key stakeholders such as final product users as well as alloy producers to establish specific material property and processing criteria. Honeywell manages and operates the Kansas City National Security Campus for the United States Department of Energy and is a leading producer of specialized engineered materials. Honeywell invents, develops, and characterizes materials and processes for a variety of materials to meet unique requirements for customer components. Honeywell specializes in product enhancement and transitioning those enhancements to manufacturing production.

Here, QuesTek and Honeywell have used the ICME approach to improve reliability and production uniformity for novel processing methods such as Additive Manufacturing (AM) as well as traditional material processing methods for a wide variety of materials. During this program, QuesTek has worked to implement ICME tools and methodologies into the manufacturing process at Honeywell. The ICME modeling includes simulation of material behavior across length scales (from the atomic to continuum level) and will be used to predict properties variations (including minimum engineering properties) given the range of composition and processing conditions within each material's established specifications. QuesTek has pioneered several "Accelerated Insertion of Materials" (AIM) methodologies to enable these predictions including computational model integration, location-specific microstructure and property modeling of components, and uncertainty quantification and management [2]. In the current program, QuesTek has worked with Honeywell to apply various ICME models and AIM methodologies to selected production material systems.

QuesTek and Honeywell have validated and refined ICME models using data including properties measurements, microstructural evaluations, and processing conditions of targeted materials. ICME models were further calibrated and extended using this data. Future efforts in this program will also focus on incorporating ICME models into Honeywell's simulations capabilities. These techniques will be incorporated into Honeywell's manufacturing process to further develop material processing/structure/properties relationships and predict material behavior while reducing uncertainty and manufacturing variability.

Approach

A number of computational modeling tools underlie QuesTek's ICME-based approach to modeling and design of materials. QuesTek utilizes CALPHAD (Calculation of PHAse Diagrams)-based tools such as Thermo-Calc for multi-component thermodynamic and phase diagram calculations, DICTRA (Diffusion Controlled phase TRAnsformations) for diffusion simulations, and TC-PRISMA for simulation of 3D multiparticle diffusive precipitation kinetics. Modules in QuesTek's proprietary Integrated Computation Materials Dynamics (iCMD®) software interact with Thermo-Calc to produce step and contour plots while easily varying input parameters over desired ranges. More than 25 modules exist that can be customized for calculation of specific thermodynamic and physical properties including but not limited to equilibrium phase fraction, driving force, liquidus/solidus/solvus temperature, T_0 temperature, M_s temperature, TTT/CCT diagrams, yield strength, and thermal conductivity. One particular strength of iCMD is the ability to produce and overlay contour plots of calculated parameters or functions.

This work will show how QuesTek and Honeywell (1) applied ICME-based modeling to critical Honeywell materials to allow for better understanding of process-structure-property relationships and quantitative prediction of material properties, (2) expanded the application of ICME modeling from conventional metal alloy systems to complex multi-material and non-metallic material systems, and (3) facilitated efforts to develop fabrication and qualification capabilities for parts produced via additive manufacturing techniques. To build an ICME framework for each material or material system of interest, system design charts (SDCs) were constructed after extensive technical review and feedback from materials users regarding application-specific performance requirements, quantitative property objectives, operating conditions, and processing constraints. This iterative process served to reinforce program goals and identify gaps in the existing modeling capabilities to describe process-structure and structure-property linkages.

Subsequent effort has focused on applying existing ICME tools to better describe material systems and quantify variability while concurrently working to develop additional models to fill identified ICME gaps. Examples of program work presented here include case studies for both a conventional ferrous alloy (SS304L) as well as a complex multi-material, metallic/non-metallic system (glass-ceramic-to-metal seals).

Material Case Study #1: 304L Stainless Steel

304L stainless steel is an austenitic steel with low carbon content (≤ 0.030 wt%) which allows for good weldability and resistance to intergranular corrosion. Microsegregation within the solidification structure can lead to banded compositional variations within the austenitic matrix as a consequence of the forging process, resulting in the formation of delta ferrite stringers which can impact weld integrity and general ductility [3]. Subsequent heat treatments can lead to decomposition of the delta phase and formation of brittle intermetallic phase (i.e. sigma and chi phases) and a corresponding decrease in uniform ductility [4]. Due to the low carbon content and relatively low carbide-forming additions, the microstructure is largely carbide-free.

A systems design chart constructed for SS304L is shown in Fig. 1. The left-hand column represents the processing flow, which includes melting, solidification, hot rolling, forging, annealing, machining, welding, and operation. Relevant structures include the solidification structure, brittle phases, grain structure, and the metallic matrix consisting mostly of austenite but with low levels of delta ferrite as well as possible martensite resulting from stress-induced transformations during working or machining. Certain applications can involve H-rich service environments leading to H embrittlement, and therefore H diffusivity in the matrix as well as the density and potency of H trapping sites are of interest for modeling purposes. Weldability is governed by the Cr and Ni equivalent ratio of the austenite phase, which can be calculated directly from the matrix composition. Structural properties such as strength and ductility are modeled using composition and microstructure-dependent mechanistic contributions, such that given a specification space and computational thermodynamics tools based on Calculation of Phase Diagrams (CALPHAD) methodologies [5], variations in properties can be predicted.

CALPHAD-based computational tools such as Thermo-Calc and DICTRA software packages [6] have been utilized to model solidification (solidification sequence, freezing temperature range, segregation profiles), homogenization, and the resulting microstructure (austenite composition, sigma phase thermal stability). An example of such calculations can be found in Fig. 2. Figure 2a shows a Thermo-Calc equilibrium step diagram with predicted equilibrium phase fraction as a function of temperature, calculated using Thermo-Calc Software's TCFE8 thermodynamic database for Fe-based alloy systems. This calculation is useful for determining the temperature window of sigma phase stability and thus identifying

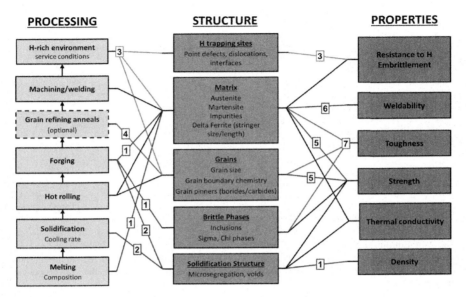

Fig. 1 Systems design chart for SS304L. *Green links* indicate existing modeling capabilities; *purple links* indicate models from the literature that can be validated and implemented; *red links* indicate models that need further development; *black lines* indicate models that are outside of the scope of the project. Linkages represent models including (*1*) Thermo-Calc, (*2*) DICTRA, (*3*) H embrittlement model, (*4*) Grain size model, (*5*) yield strength model, (*6*) Weldability models including Cr/Ni equivalent ratio, (*7*) Toughness model (color online)

critical processing steps that could result in material embrittlement. Figure 2b compares solidification simulations using different cooling rates, plotting the percent solidified against the temperature. The blue curve is a Thermo-Calc equilibrium simulation in which infinitely slow cooling is assumed, allowing for full back-diffusion in the solid and therefore no resulting microsegregation in the solidification structure. The purple curve is a Thermo-Calc Scheil simulation which assumes an infinitely fast cooling rate with no back diffusion in the solid and therefore predicts the worst-case microsegregation. A DICTRA simulation with a finite cooling rate utilizes a multidimensional diffusion model to calculate the solidification and back diffusion rates, resulting in an intermediate cooling rate and level of microsegregation (green/red curve). The DICTRA solidification model, while computationally more expensive than the Thermo-Calc models, provides the most realistic solidification simulation.

In addition to thermodynamic calculations and property prediction functionalities, the *i*CMD platform has two tools to evaluate material properties over a multi-dimensional range of processing variables: Min/Max and Uncertainty Analysis. Both of these tools take as inputs elemental composition, processing temperature, and other model variables. Uncertainty ranges for each desired input value are then specified. Output may be any desired thermodynamic parameters such as

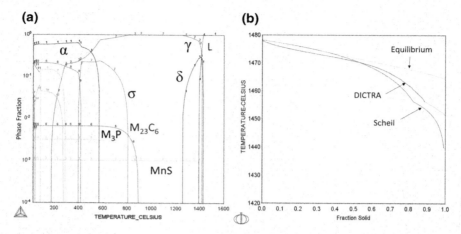

Fig. 2 a Equilibrium step diagram from Thermo-Calc showing predicted phase fractions (log scale) in a SS304L steel versus temperature. **b** Results from equilibrium and Scheil simulations (Thermo-Calc) as well as DICTRA solidification simulation highlighting the differences in predicted solidification ranges

phase fractions of secondary phases, phase compositions, etc. Min/Max evaluates the extrema of these outputs by taking the extrema of each of the varied input parameters. Uncertainty Analysis performs a random statistical evaluation in the specified range to determine the probability of a given property measurement given the target composition range. Quantification of property probability, including 1% minimum properties, is key to understanding and controlling process variation in a manufacturing setting.

In Fig. 3, QuesTek has applied the Uncertainty Analysis tool to forecast the variation in predicted strength values within a known processing variability

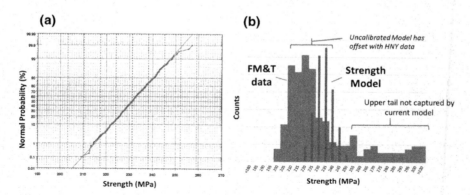

Fig. 3 Uncertainty Analysis for SS304L **a** forecast normal probability across spec range for SS304L yield strength based on *i*CMD strength model for austenitic stainless steel, **b** histogram with experimental data for SS304L yield strength provided by Honeywell (*blue*) with forecast yield strength values (*red*) output from uncertainty analysis in *i*CMD (color online)

window. A yield strength model for austenitic steels has been implemented in *i*CMD which includes contributions from Peierls stress, dislocation strengthening, solid solution strengthening, and grain size strengthening. The normal distribution of the predicted strength values forecast for the entire specification range can be found in Fig. 3a. Figure 3b displays the forecast data in a histogram along alongside measured data for SS304L yield strength provided by Honeywell. The upper tail of the measured data set (blue bars) is uncharacteristic of the expected normal property distribution and likely anomalous. It is suspected that two primary causes for this anomaly are that the data was collected over several decades via various experimental techniques, and that the data was not collected with intent to reuse within an ICME analysis. Due in part to these findings, Honeywell recognizes that careful data collection, enhanced with comprehensive meta-data, and storage of data in accessible, future-proofed formats to be a critical enabler of ICME. Discounting the upper tail reveals a distribution with a similar shape and width to the forecast model (red bars). The offset in the mean yield strength values of the forecast and measured data is likely due to uncalibrated strength model terms. A recalibration of the model to account for unknown inputs (i.e. dislocation density) would allow for better agreement in the predicted mean yield strength models. A key takeaway is that the range of forecast yield strength agrees well between predicted and measured values, and a recalibration of the model would allow for accurate prediction of not only the mean, but also the 1% minimum properties.

Material Case Study #2: Glass-Ceramic-to-Metal Seals

Glass-ceramic (G-C) to metal seals (G-C/M seals) are of key technological importance for providing hermeticity in applications that require insulation of adjacent chemical environments while maintaining electrical isolation of metal pin interconnects across the seal [7]. The Li_2O-Al_2O_3-SiO_2 (LAS) system, such as the commercial Elan 46, is of interest as an industry standard for glass-ceramics used in G-C/M seals due to its easily tunable coefficient of thermal expansion (CTE) and good mechanical properties. Precious metal alloys such as the Pd-Ag-Cu-Au-Pt-Zn Paliney® alloy are used for electrical interconnects due to their good corrosion resistance, high strength, and low electrical contact resistance. The thermal compatibility of the G-C and metal, as well as the character of the bonded interface, are critical factors in determining the reliability and integrity of the G-C/M seal. Models describing this system must address both bulk phenomena of the glass-ceramic and metal contact material as well as heterogeneous interfacial phenomena which are not easily described by bulk thermodynamics.

The systems design chart in Fig. 4 captures the high complexity of the G-C/M system. Glass-ceramics consist of an amorphous glassy matrix phase containing crystalline ceramic grains. In lithia-alumina-silica (LAS)-type glass-ceramics, the ceramic grains are a quartz or cristobalite solid solution which nucleate heterogeneously on crystallized lithium phosphate in the glass melt. The

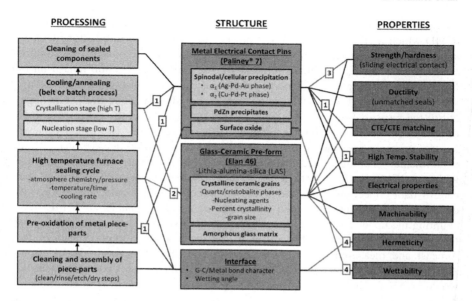

Fig. 4 Systems design chart for glass-ceramic-to-metal (G-C/M) seals. *Green links* indicate existing modeling capabilities; *purple links* indicate models from the literature that will can be validated and implemented; *red links* indicate models that need further development; *black lines* indicate models that are outside of the scope of the project. Linkages represent models including (*1*) Thermo-Calc, (*2*) Glass crystallization theory, (*3*) Yield strength model, (*4*) Interface model (color online)

Pd-Ag-Cu-Au-based alloys are age-hardenable via discontinuous and/or spinodal precipitation due to an fcc miscibility gap [8–9]. Additions of Zn mainly contribute to alloy strength via solid solution hardening, although precipitation of a Zn-rich phase at low temperatures might additionally contribute to the alloy strength [10]. Different phase separation mechanisms dominate at intermediate and high temperatures, resulting in a large dependence of alloy strength on the thermal profile of the belt process [11]. Developing a thorough process-structure understanding for both the Paliney alloy and glass-ceramic seal is integral for reliable property prediction.

The key properties for glass-ceramic seal reliability and hermeticity include good wettability on metal surfaces as well as high strength. In order to maintain good structural integrity, the coefficient of thermal expansion (CTE) of both the glass-ceramic seal and the metal contact surface must be carefully controlled such that a significant CTE mismatch does not occur at operating temperature, which could lead to cracking in the glass ceramic and failure of the seal. The melting temperature and high temperature strength of the contact pin are of importance because of the high temperatures the part is exposed to during sealing and operation.

Thermodynamic modeling of the Pd-Ag-Cu-Au-Pt-Zn Paliney system using Thermo-Calc Software's TCSLD3 solder alloy solutions thermodynamic database

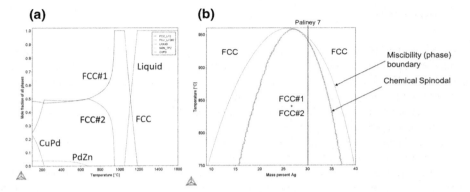

Fig. 5 a Thermo-Calc equilibrium step diagram showing the predicted equilibrium molar phase fraction versus temperature. **b** Thermo-Calc phase diagram showing the miscibility phase boundary between the single phase fcc region and two-phase fcc miscibility gap (*solid line*), overlaid with the calculated chemical spinodal boundary for stable spinodal decomposition

allows for a better understanding of the phase stability in this complex 6-component alloy. Equilibrium map and step diagrams correctly describe the fcc miscibility gap reported in published experimental studies [11], resulting in a 2-phase region with an Ag-rich Au-Pd-Au fcc phase and a Cu-rich Cu-Pd-Pt fcc phase. A step diagram for the Paliney® 7 alloy in Fig. 5a clearly shows the miscibility gap in the fcc phase, as well as the presence of a low-temperature PdZn phase, which has not been experimentally confirmed. Stability calculations allow the chemical spinodal to be plotted in Fig. 5b. The addition of an elastic strain energy term will allow for the plot of the coherent spinodal line, giving further insight into the mechanism of phase separation. A better understanding of the thermodynamics and kinetics of phase separation during the thermal belt process is also necessary for building mechanistic models for key properties such as yield strength, ductility, and CTE.

A gap analysis was performed in order to determine available modeling capabilities in the LAS glass-ceramic system. To assess the phase stability, numerous thermodynamic databases were considered. Thermo-Calc Software's Geochemical/Environmental GCE2 thermodynamic database was found to include relevant elemental components and descriptions for the Quartz polymorphs of interest. However, a full description of devitrification kinetics is necessary to accurately model the most critical features of the glass-ceramic microstructure, and therefore modeling of the equilibrium thermodynamics is insufficient. Future work will focus on modeling of the nucleation and growth kinetics during the thermal belt process.

Another modeling gap identified in the G-C/M system was modeling of the interface itself. Traditional bulk thermodynamic simulations do not capture the chemical and physical phenomena at and near the interface between the glass-ceramic and metal alloy. Future work will address the characterization of the interface in order to better understand reliability issues and failure mechanisms associated with the interfacial region.

Conclusions

QuesTek and Honeywell have utilized an ICME framework to address problems relevant to existing alloys that are considered critical materials for Honeywell's business. The successful completion of the ICME framework for these materials will allow for process modeling, quantitative and mechanistic property forecasting, and uncertainty quantification within a known specification space. Expansion of the ICME methodologies beyond traditional metal alloy systems is valuable for a complex manufacturing environment in which components contain multiple and heterogeneous material systems. The thorough understanding of the process-structure-property relationships enabled by this ICME framework will allow for flexibility in material suppliers, consolidation and simplification of specifications, and eventually design of new alloys for enhanced performance or new processing routes such as additive manufacturing.

References

1. G.B. Olson, Computational design of hierarchically structured materials. Science **277**, 1237–1242 (1997)
2. W. Xiong, G.B. Olson, Cybermaterials: materials by design and accelerated insertion of materials. NPJ Comput. Mater. **2**(August), 1–14 (2015)
3. R. Plaut, C. Herrera, D. Escriba, A short review on wrought austenitic stainless steels at high temperatures: processing, microstructure, properties and performance. Mater. Res. **10**(4), 453–460 (2007)
4. C.C. Tseng et al., Fracture and the formation of sigma phase, M23C6, and austenite from delta-ferrite in an AISl 304L stainless steel. Metall. Mater. Trans. A **25**(6), 1147–1158 (1994)
5. N. Saunders, A.P. Miodownik, *CALPHAD (Calculation of Phase Diagrams): A Comprehensive Guide* (Elsevier Science Ltd., Oxford, UK, 1998)
6. J.O. Andersson et al., Thermo-Calc & DICTRA, computational tools for materials science. Calphad **26**(2), 273–312 (2002)
7. I.W. Donald, Preparation, properties and chemistry of glass- and glass-ceramic-to-metal seals and coatings. J. Mater. Sci. **28**(11), 2841–2886 (1993)
8. M. Ohta, K. Hisatsune, M. Yamane, Age hardening of AgPdCu dental alloy. J. Less-Common Met. **65**(1), P11–P21 (1979)
9. H.J. Seol et al., Age-hardening and related phase transformation in an experimental Ag-Cu-Pd-Au alloy. J. Alloys Compd. **407**(1–2), 182–187 (2006)
10. D.-J. Noh et al., Phase transformation and microstructural changes during aging process in Ag–Pd–Cu–Pt–Zn alloy. Mater. Sci. Technol. **26**(2), 203–209 (2010)
11. D.F. Susan et al., Characterization of continuous and discontinuous precipitation phases in Pd-rich precious metal alloys. Metall. Mater. Trans. A Phys. Metall. Mater. Sci. **45**(9), 3755–3766 (2014)

ICME Based Hierarchical Design Using Composite Materials for Automotive Structures

Azeez Shaik, Yagnik Kalariya, Rizwan Pathan and Amit Salvi

Abstract Composite materials are increasingly being used in transport structures due to their higher specific stiffness and specific strength. They can also be molded relatively easily to achieve aerodynamic shapes. Fiber reinforced composites offer excellent energy absorption under crushing loads and hence are increasingly being used in safety and load bearing applications. Composite material characterization is a complicated task due to micro-scale non-homogeneity and its resulting anisotropy and is generally accomplished with expensive physical tests at coupon level. High fidelity computational models are increasingly being used to accurately establish the elastic as well as inelastic nonlinear behaviour due micro-damage and fracture. The fiber material and its architecture, resin selection and its curing process control the resulting composite properties. Draping of fabrics before resin infusion also leads to geometrical non-linearities in the structure. All these parameters in the above processes need to be tightly coupled and can be altered in turn to provide a maximum performance for a given application under certain loads. A multi-scale methodology to study global-local relations of materials can also be integrated in the entire process. In this paper, Integrated Computational Materials Engineering based hierarchical design process integrated with composite material selection and microstructure based material design is presented. This framework for design decisions is currently being integrated using a TCS PREMAP framework developed in house.

Keywords ICME · Multi scale · Repetitive unit cell · Composite

Introduction

Over the years, the use of composite materials has been increased in many areas such as aerospace, automotive, marine, energy and defense. Most of the structural design in the industry is stiffness driven where composite materials are offering

A. Shaik · Y. Kalariya · R. Pathan · A. Salvi (✉)
TCS Research, TRDDC, Pune, India
e-mail: salvi.amit@tcs.com

excellent performance per unit weight. New class of 2D and 3D fiber composites are pushing the limits of strength based designs. The materials offer enhanced toughness under static as well as crash loading compared to laminated composites. Complex shapes can also be manufactured using weaving operation using fiber bundles (tows). Due to its complicated architecture, these materials accumulate damage though micro-cracks and fiber tow/resin debonding. Thus, the fibre architecture at the local material level is directly connected to the overall structural performance.

The typical structural design cycle consists of appropriate material selection followed by material sizing and shaping (topology optimization). In textile and woven composites, material selection as well as material sizing are embedded at a microstructure level itself. Changes in fiber tow cross-sectional size and weave angles can change the anisotropy of the material as well as their failure mechanism which in turn govern failure strength as well as fracture toughness. Thus, material optimization is not only limited to sizing but also connected to its microstructural design. Traditionally, qualification of materials is done by carrying out expensive and time consuming material property tests to establish stiffness (elastic) as well strength properties in all material coordinates. Thus, to eliminate time and cost associated with the physical testing, "validation through simulation" approach can be adopted. In this approach, the material characterization can be carried out using high fidelity computational models which can be connected to structural computational models.

High fidelity computational models which bridge length scales are increasing being carried out in design of composite structures [1]. In an automotive structure, large assemblies are divided into smaller assemblies and then components. These components are again analysed at ply level and within plies. The analysis typical moves from large length scale to a smaller length scale as shown in Fig. 1. In this approach, local or lower length scale models are used to verify or qualify design strategies at higher length scales, e.g. adhesive bonded joints or rivets can be analysed using fracture analysis to qualify joining strategies at the structural level. The allowables and tolerances on the manufacturing of the joints are thus calculated computationally. This strategy can also be employed to quantify global design by passing material property data such as stiffness and strengths by modelling microstructure at local level. This top down as well as bottom up approach can very well be utilized to design futuristic composite structures. In this study, a methodology is proposed to combine structural as well as material analysis to design composite structures. With the integrated design, introduction of newer materials and their configurations can be readily incorporated without carrying out expensive testing.

Failure analysis of composites materials has been extensively investigated over last two decades [2–4]. Most of the studies link the microstructural damage accumulation and fracture to its material properties at macro level. The performance of the microstructure at macro level (material level) purely depends on the performance of its constitutes such as fibers (fiber tows) and resin as well their interactions. If the constituents and its interaction are modelled correctly, they exhibit various failure modes under variety of loads and boundary conditions. The accuracy

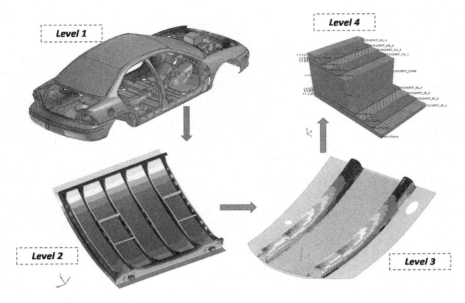

Fig. 1 Hierarchical design of automotive structure

of the properties of constituents and their interactions is extremely important in this multi scale approach.

Textile fiber composite material are made up of fiber bundles (tows) woven in a particular fashion and impregnated with a resin material. These materials exhibit superior out-of-plane stiffness, strength and toughness properties when compared to laminated composites. However, the geometry of this composite class is complex and the choice of possible architectures and constituents is nearly unlimited. Many parameters of woven fiber architectures can be changed, such as fiber tow size, weave type, hybridization or choice of constituents (fiber and resin type) to alter the material properties of the composites. The elastic properties of the complicated microstructures can be predicted using analytical models which are based on its constituent's properties such as tow size, tow undulation and fiber material [5–8].

Non-linear properties of the composites have been investigated using multiscale analysis. In composite materials, geometric repetitive unit cell (RUC) of fiber architecture and resin are designed and analysed for damage and failure [9–16]. The inelastic performance of these complex materials are experimentally validated under varying load conditions, e.g. under crash loads and shock wave. These properties can be effectively used at the macro scale where laws of continuum mechanics are applicable. By connecting these scales, performance of the composite structures can be connected to individual constituents of the micro scale such as fiber, resin and their interface. In this study, the three constituents are fiber tows (fiber bundle impregnated with resin), resin and their interface. The properties of the fiber tows itself can be obtained by using multi-scale analysis.

Design Case Study

A design case study was undertaken to demonstrate proposed integrated material and geometric multi-scale approach. A metallic car door assembly was designed using composite materials as shown in Fig. 2.

The design of this door was carried out using variety of load cases given by the designer. Every component of the assembly is designed using stiffness based deflection criteria. The material selection for every component is carried out for FRC as well as foam or balsa core. Materials selection was also done based on variety of unidirectional (0), bi-directional (0/90) and quad-directional (0/+45/90/−45) fabrics. The fabric selection was done based on the stress (and strain) contours obtained from preliminary run using isotropic materials for design as shown in Fig. 3.

The panel thickness (number of plies) is calculated to match the original deflection criteria of every single component. A failure index based on Tsai-Hill; Tsai-Wu criteria was also checked. Note that further weight reduction can be

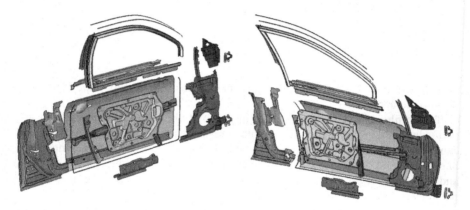

Fig. 2 Exploded view of the automotive door

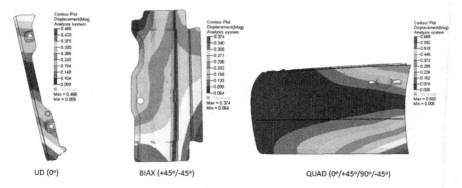

Fig. 3 Selection of directional fabric based on stress/strain contours

obtained by redesigning the whole component and by combing multiple parts but it's out of the scope of this work. Based on above criteria individual component were designed to obtain significant weight reduction with increase safety (failure) margins. This stiffness driven analysis is carried out using ABAQUS standard analysis with continuum shell elements.

Sub Component Design with Oil Canning Loads

The assembly is then subjected to local loads at the subcomponent level to check for local failure. A sub component of the assembly can be subjected to many local load conditions and thus these local loads are applied specifically to a part of the entire assembly. A requirement of oil canning load on the outer panel of the load case is studied to verify material as well as sizing of the plies. The oil canning load or the bump load decides the ability of the panel to deflect or dent under quasi-static load. This load can deform the panel elastically or damage it locally. This problem can be solved by creating the entire assembly of solids and subjecting to the oil canning loads which is computationally expensive as many subcomponents do not bear any loads. A multi-scaling approach can be selected to carry out the study in two steps more quickly and accurately using an ABAQUS sub-modelling technique.

In oil canning, a load of 225 N was applied on the area on the surface of the panel as shown in Fig. 4a with all the four edges constrained as shown. The oil canning analysis is carried out using ABAQUS Explicit analysis. The location of the load is calculated by creating and solving a separate buckling analysis. This location is where the panel is most likely to deform. Time dependent load is applied on the panel and resulting deflections are calculated as shown in Table 1.

To check the failure of this panel, it was then converted to solid elements representing the actual 2 ply layups from continuum shell elements as shown in Fig. 5.

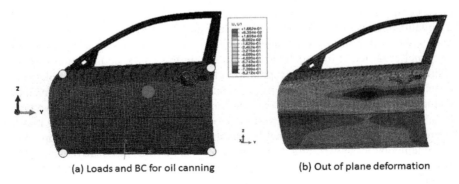

(a) Loads and BC for oil canning   (b) Out of plane deformation

Fig. 4 Oil canning load case

Table 1 Time dependent oil canning load versus displacement

Time (s)	Load (N)	Displacement (mm)
0.01	25	0.01
0.02	60	0.04
0.03	100	0.08
0.04	160	0.15
0.045	225	0.27
0.06	225	0.38

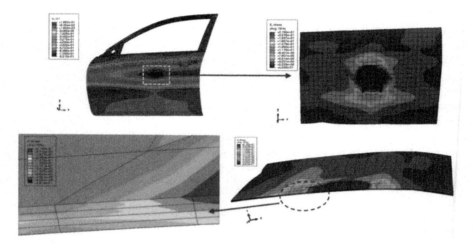

Fig. 5 Sub-model of oil canning load area

It was observed that the delamination between the two plies under the oil canning load was not initiated. For higher oil canning load that ends up in delamination redesign of the panel is warranted. This can either be carried out either by increasing the number of plies to stiffen the panel (geometric sizing), or by changing the material with higher inter-laminar strength (material design) such as woven or textile fabric composite laminate. In both cases the changes are made at the global level and the qualification is carried out again down through the scales. In the geometric sizing option, the weight of the panel increases even though existing in plane stiffness of the panel is sufficient for the deflection criteria. In this study, combination of both approaches are incorporated. Keeping the global stiffness requirement similar, the material is redesigned to increase its strength. This is carried out by using woven or textile fiber weaves. The material can be changed by changing the fiber weave type from 0–90 plain weave to 0–90 textile weave. An analytical model is used to match the in-plane properties of the panel by changing the fiber tow size, their amplitude and undulation as shown for a 3D weave in Fig. 6. Hybrid fiber material tows can also be considered for the design. Slight change in the material properties can be compensated by change in the thickness of the panel.

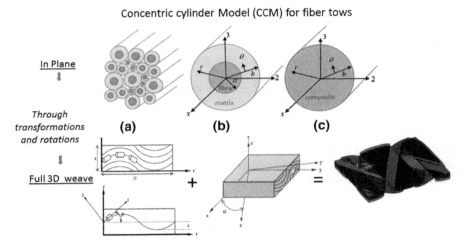

Fig. 6 Calculation of elastic properties through analytical model

Repetitive Unit Cell (RUC) Model

A repetitive unit cell model (RUC) is a geometry model of the microstructure which can be repeated and arranged in a certain order multiple time to obtain a bulk material. Bulk properties (elastic as well as inelastic such as strength and toughness) of the bulk material is averaged resultant of this unit cell. Thus by analysing the RUC, inelastic properties of the bulk can be obtained. Conversely, these models can also be used to design a microstructure that yield desired properties of the bulk through optimization. RUC models of 0–90° plain weave and 0–90° textile weave of carbon fiber tows and epoxy resin matrix are shown in Fig. 7. The RUC consists of fiber tows and the surrounding resin block. Thus by developing exact geometric model of the fiber tows and resin and assigning correct properties to its constituents

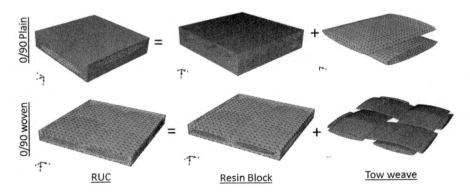

Fig. 7 RUC model of 0/90 plain and 0/90 weave carbon fiber tow with epoxy resin

such as fiber tows, surrounding resin and their interface, non-linear mechanical properties of the RUC can be obtained accurately for variety of load conditions such as tension, compression, shear, fracture toughness etc. This type of "virtual testing" with proper experimental validation can save tremendous amount of time and cost spent on physical testing.

Size independent non-linear response of the bulk material is obtained by virtual testing on representative volume element (RVE) of the composite. The RVE can be made by stacking and repeating RUCs. At some stacking of RUCs or at a given periodicity, the response of the RVE will asymptote, which is size independent material property of the bulk composite. This periodicity will change from fiber architecture, material properties as well as type of the property e.g. R-curve in case of fracture toughness. Figure 8 shows the three 0/90 woven RVE of carbon fiber tows and epoxy resin matrix made up of 1 × 1, 2 × 2 and 3 × 3 periodicity of RUCs.

Figure 9 shows the comparison of the nonlinear tension response of these RVEs. It can be observed that as RVE for the 0/90 woven textile composite can be

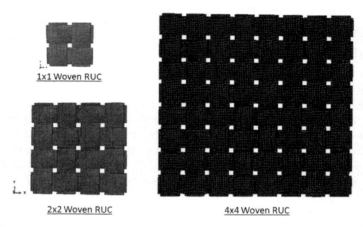

Fig. 8 RVEs made up of 1 × 1, 2 × 2 and 4 × 4 RUCs of carbon fiber tow with epoxy resin

Fig. 9 Tension response of 0/90 woven composite of carbon fiber tow with epoxy resin

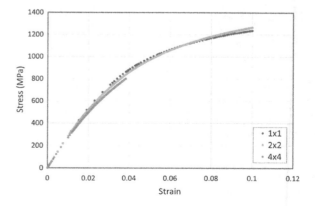

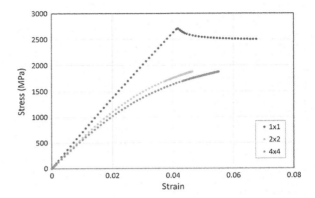

Fig. 10 Tension response of 0/90 plain composite of carbon fiber tow with epoxy resin

established at 1 × 1 and above periodicity, i.e. the asymptotic material properties can be observed at 1 × 1 RUCs.

Figure 10 shows the comparison of the nonlinear tension response of these RVEs. It can be observed that as RVE for the 0/90 plain textile composite can be established at 4 × 4 and above periodicity. This captures phenomenological nature of the material properties of these composites due to strong microstructural dependency.

Automation of the ICME Process

The suggestion of the alternate composite microstructure can be automated using python and fiber weave architecture software. In this study, open source TexGen software is used to generate various fiber architecture RUCs such as plain, woven, braided and 2D/3D textile weaves. These geometries are then converted into Finite Element (FE) input files and relevant material properties are assigned to fiber, resin and the interface. Virtual testing for tension, compression, shear and fracture toughness are automatically carried out and the material properties thus measured are stored and used to qualify for the panel requirements. The process is repeated by changing parameters such as fiber and resin materials, tow sizes, and undulation angles and amplitudes until required properties are met. Figure 11 shows the flowchart of the overall process.

This entire process is being implemented using a TCS PREMΛP platform. TCS PREMΛP is an IT platform for enabling integrated computational materials engineering including design of materials, manufacturing processes and products, by using a combination of physics based simulations, data driven reasoning and guided experiments supported by decision support tools and knowledge engineering systems. The TCS PREMΛP platform is built on TCS's model-driven-engineering architecture making it a highly flexible, configurable and scalable solution that can be applied to a variety of engineering domains and design applications. Figure 12

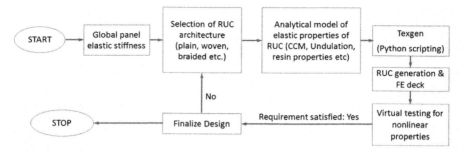

Fig. 11 Flow chart for the material design process

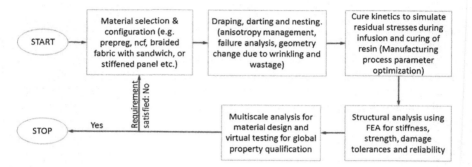

Fig. 12 ICME lead integrated design flow in PREMAP

shows the flowchart of the entire ICME lead design process for the composite structures. Currently, draping analysis and cure kinetics analysis is being incorporated in the platform.

Conclusion

A generalized ICME based design and analysis process to design automotive composite structure is presented in this study. Various analytical and computational multi-scale modelling strategies are used in the process. Material selection and qualification is carried out to maximise structural performance at lower cost. Also, a separate automated material design based on microstructure optimization is carried out. Automation at different length scales is introduced to make the model robust. Structures with large assemblies can be designed efficiently and at lower cost and time using this generalized methodology. Experience based thumb rules in design as well as manufacturing associated with composite materials can be integrated and used effectively. This methodology can be customized by the designer using multiple in-house design platforms.

References

1. M.G. Ostergaard et al., Virtual testing of aircraft structures. CEAS Aeronaut. J. **1**(1–4), 83 (2011)
2. L. Raimondo, M.H. Aliabadi, Multiscale progressive failure analysis of plain-woven composite materials. J. Multiscale Modell. **1**(2), 263–301 (2009)
3. A.C. Orifici, I. Herszberg, R.S. Thomson, Review of methodologies for composite material modelling incorporating failure. Compos. Struct. **86**(1), 194–210 (2008)
4. B. El Said et al., Multi-scale modeling of 3D woven structures for mechanical performance, in *Proceedings of 16th European Conference on Composite Materials (ECCM 2014)*, Seville, Spain, 22–26 June 2014 (2014)
5. S.C. Quek et al., Analysis of 2D triaxial flat braided textile composites. Int. J. Mech. Sci. **45**(6), 1077–1096 (2003)
6. Z. Hashin, B.W. Rosen, The elastic moduli of fiber-reinforced materials. J. Appl. Mech. **31**(2), 223–232 (1964)
7. Z.M. Huang, The mechanical properties of composites reinforced with woven and braided fabrics. Compos. Sci. Technol. **60**(4), 479–498 (2000)
8. R.M. Christensen, F.M. Waals, Effective stiffness of randomly oriented fiber composites. J. Compos. Mater. **6**(3), 518–535 (1972)
9. M. Pankow et al., Split Hopkinson pressure bar testing of 3D woven composites. Compos. Sci. Technol. **71**(9), 1196–1208 (2011)
10. M. Pankow et al., A new lamination theory for layered textile composites that account for manufacturing induced effects. Compos. A Appl. Sci. Manuf. **40**(12), 1991–2003 (2009)
11. S. Song et al., Braided textile composites under compressive loads: modeling the response, strength and degradation. Compos. Sci. Technol. **67**(15), 3059–3070 (2007)
12. D. Zhang et al., Flexural behavior of a layer-to-layer orthogonal interlocked three-dimensional textile composite. J. Eng. Mater. Technol. **134**(3), 031009 (2012)
13. M. Pankow et al., Modeling the response, strength and degradation of 3D woven composites subjected to high rate loading. Compos. Struct. **94**(5), 1590–1604 (2012)
14. M. Pankow et al., Shock loading of 3D woven composites: A validated finite element investigation. Compos. Struct. **93**(5), 1590–1604 (2012)
15. T. Zeng, L.Z. Wu, L.C. Guo, Mechanical analysis of 3D braided composites: a finite element model. Compos. Struct. **64**(3), 399–404 (2004)
16. A. Bogdanovich, M.H. Mohamed, Three-dimensional reinforcements for composites. SAMPE J. **45**(6), 8–28 (2009)

Towards Bridging the Data Exchange Gap Between Atomistic Simulation and Larger Scale Models

David Reith, Mikael Christensen, Walter Wolf, Erich Wimmer and Georg J. Schmitz

Abstract Materials properties are rooted in the atomic scale. Thus, an atomistic understanding of the physics and chemistry is the foundation of computational materials engineering. The MedeA computational environment provides a highly efficient platform for atomistic simulations to predict materials properties from the fundamental interactions effective at the nanoscale. Nevertheless, many interactions and processes occur at much larger time and length scales, that need to be described with microscale and macroscale models, as exemplified by the multiphase field tool MICRESS. The predictive power of these larger scale models can be greatly increased by augmenting them with atomistic simulation data. The notion of per phase-properties including their anisotropies provides e.g., the key for the determination of effective properties of multiphase materials. The key goal of the present work is to generate a common interface between atomistic and larger scale models using a data centric approach, in which the "interface" is provided by means of a standardized data structure based on the hierarchical data format HDF5. The example HDF5 file created by Schmitz et al., Sci. Technol. Adv. Mater. 17 (2016) 411, describing a three phase Al–Cu microstructure, is taken and extended to include atomistic simulation data of the Al–Cu phases, e.g., heats of formation, elastic properties, interfacial energies etc. This is pursued with special attention on using metadata to increase transparency and reproducibility of the data provided by the atomistic simulation tool MedeA.

Keywords Integrated computational materials engineering (ICME) · Materials modeling · Software · Interoperability · Industrial deployment · HDF5 · Multiscale modelling · Metadata · Hierarchy

---

D. Reith (✉) · M. Christensen · W. Wolf · E. Wimmer
Materials Design SARL, 42 avenue Verdier, 92120 Montrouge, France
e-mail: dreith@materialsdesign.com

G.J. Schmitz
MICRESS Group at Access e.V., Intzestr. 5, 52072 Aachen, Germany

Introduction

Integrated Computational Materials Engineering (ICME) offers a unification of various computational approaches and simulation tools addressing multiple phenomena at multiple time and length scales. Such multiscale modeling is vital to accurately describe the processes, structures, properties, and performance of materials [1]. Essentially, all materials properties are rooted in the atomic scale, which defines the shortest length and time scale within the ICME framework. As such, having the ability to compute properties at these scales offers a basis for a fully integrated computational framework able to study properties of materials prior to their actual synthesis. This extends far beyond the currently established practice of many continuum methods, which rely strongly on experimental data.

All atomistic simulation methods use "discrete" models that explicitly relate materials properties to interactions between atoms. At the lowest level, interatomic interactions are defined by the electrons. Describing these requires a quantum mechanical approach based on solving the Schrödinger equation for a many-electron system. In the past decades, density functional theory (DFT) and related methods have established themselves as quantum mechanical modelling workhorses offering a good balance between computational cost and accuracy. Above this quantum mechanical level one can find forcefield methods based on a classical description of the atomic interactions. At this level electrons are not explicitly described and Coulomb interactions are simplified on an abstract atomistic level where charges are attributed to atoms. Besides the atomic charge the interactions between atoms are described by convenient functional forms such as Lennard-Jones potentials, Morse potentials or the forms used by the embedded atom method. Class 2 forcefields include complex coupling terms involving 2-, 3-, and 4-body interactions. Such forcefield approaches are computationally much less demanding than the quantum mechanical approaches, thus enabling atomistic simulations to address and describe the evolution of thousands of atoms during time periods of nanoseconds or the exploration of millions of configurations in Monte Carlo simulations.

A simulation software such as MedeA [2] offers a universal and versatile common data model and a graphical user interface to overcome the inherent methodological differences between these various atomistic approaches, being embedded and integrated within a common, unified computational environment. Figure 1 summarizes the various levels of atomistic simulations. A more detailed insight on status and perspectives of atomistic simulations can be found in the paper of Christensen et al. [3].

The major challenge, which we tackle in the present work, is to bridge the differences between "discrete" atomistic simulation tools and "continuum" based tools. It is important to understand that the issue is not one of mere data exchange, but rather a subtle conceptual difference between the various computational approaches. On a continuum level, the central issue is to accurately describe and classify the state or evolution of a material. Once this has been accomplished within a well-defined classification system every ambiguity has been removed.

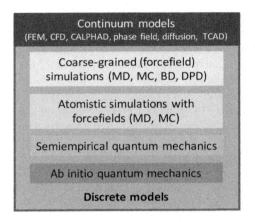

Fig. 1 An overview on various levels of simulations with a focus on discrete atomistic simulation. The abbreviations are: molecular dynamics (*MD*), Monte Carlo (*MC*), Brownian dynamics (*BD*), dissipative particle dynamics (*DPD*), finite element methods (*FEM*), computational fluid dynamics (*CFD*), calculation of phase diagrams (*CALPHAD*), and technology computer aided design (*TCAD*)

Furthermore, many different aspects of the material can be described within a single framework. However, discrete atomistic simulation operates on a calculation centric basis. Each calculation is setup to identify and describe a single property or a class of similar properties. The quality of such a description always depends on the methodological choice and settings with which the calculation has been performed. While the aim is to use sufficiently accurate settings the setup should not be too computationally expensive. Such settings are not only dependent on the applied approach, quantum mechanical or classical, but also materials and property specific. This necessitates multiple calculations using different approaches and settings to describe various properties of a single material.

The paper by Schmitz et al. [4] has created a solid basis on establishing a standard nomenclature and methodology to describe and exchange continuum data of MICRESS microstructures which will be used in the present work. Data exchange is facilitated by using the HDF5 file format [5]. This hierarchical data format is well suited to store and organize data utilizing two types of objects: datasets that can be scalars or multidimensional arrays of a defined type, e.g., float, integer or string, and groups that act as containers holding datasets or other groups. This makes it possible to order data in a hierarchical filesystem-like structure. In addition, each object can be described with attributes allowing for contextual metadata-based description of data. We believe that this flexibility as provided by the HDF5 file format is well suited as a container for atomistic simulation data.

The present work will build upon the proposed notation as suggested in [4]. We expand the definition of descriptors as proposed in [4] by adding another one, which is properties. In this naming scheme, a descriptor is a dataset that describes general information, such as structural data, while a property is a dataset that describes results of calculations. The main difference between both is the amount of metadata used to describe these variables (see further below). The name of a descriptor or property, which can be either an object or a group, starts with a capital letter. In addition, names may be composed of multiple specifiers, e.g., number of atoms is described by the dataset *NumberAtoms*. Another important rule used is that names followed by a number in brackets are vector components. So *Job(1)* is followed by *Job(2)*.

To control the memory footprint of a HDF5 file we suggest to always define the size of a string or character dataset or attribute to exactly match the required length.

Data Structure

As a first step towards bridging the data exchange gap it is necessary to identify a hierarchical data structure most suitable to describe results from atomistic simulations. This hierarchical structure is calculation centric, or following MedeA's naming paradigm, job-centric. This implies that the data of each job is organized within a group. While such a hierarchy appears to be natural for atomistic simulation tools it is not very accessible to others. Consequently, a strategy is required to make the data more generally accessible. We propose to regroup calculated properties from different jobs by the structures, i.e. the specific and characteristic arrangement of atoms. The job-centric data structure is maintained in parallel,

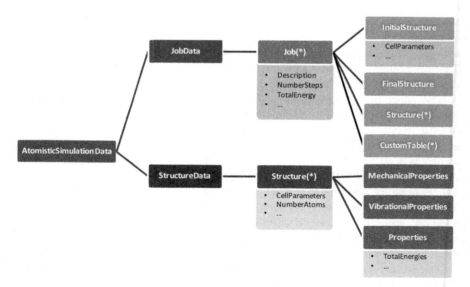

Fig. 2 A possible atomistic simulation data hierarchy of the HDF5 groups is depicted. The group *JobData* contains the primary calculation centric hierarchy. That data can then be mirrored, by using symbolic-links or by copying, into the structure-centric *StructureData* group to allow for an easier access

though. A suitable strategy to deal with this dual concept is to duplicate the job-centric data or to link it into the structure-centric form. Once this is accomplished atomistic simulation data can be easily incorporated into other hierarchies, e.g., the one described by Schmitz et al. [4].

We propose to collect all atomistic simulation data within a group with the name *AtomisticSimulationData*. This will make it easy to identify atomistic results when the HDF5 file is combined with other files that contain data from other sources (see Fig. 2).

The Job-Centric Data Structure

The calculation data are collected in the group *JobData* which is placed in *AtomisticSimulationData* (see Fig. 2). In *JobData* the data from a single calculation procedure, the job, is placed in a group with the name *Job(*)*, with * indicating an index that consecutively increases starting from 1 when additional jobs are added to *JobData*. This index defines the Job ID: *Job(1)* contains all the data from the first calculation procedure while *Job(2)* contains the data from the next one added to the file. The data in a *Job(*)* group can be sorted into general information, properties, structural information, and program specific data. The program specific data may also be a text-file created by the atomistic simulation tool. We will now discuss in detail each of these categories and the associated structure.

General Information

These variables store general information on the calculation setup of a given procedure.

1. *Description*
   The dataset *Description* is a scalar string variable that contains some general description on the calculation procedure. This variable should be user definable.
2. *NumberSteps*
   As a calculation procedure can sometimes contain multiple steps the total number of steps are indicated by this scalar integer.

Properties

Properties are the actual results of the atomistic simulations. The exact makeup of the properties depends on the calculation procedure and on the applied solver, i.e.

the atomistic simulation tool. Focusing on DFT and classical forcefield-based dynamics calculations in MedeA the discussion is narrowed down to calculated properties from the DFT code Vienna Ab initio Simulation Package (VASP) [6, 7] and the molecular dynamics simulation code LAMMPS [8]. However, the discussed definitions can be extended and adapted as required to show results from other solvers as well.

As a procedure can contain multiple steps the properties associated with a given step are indicated by an integer value in brackets at the end of the property name. For example, the dataset *TotalEnergy(1)* contains the total energy as evaluated in the first step while *Pressure(2)* contains the pressure as calculated in the second step. Strictly speaking, this bracketed value is not a vector index as the value in the bracket does not necessarily increase incrementally for any given property. If the pressure is only evaluated in the second calculation step and not in the first one the dataset *Pressure(1)* will not be available.

Since a detailed discussion of all the possible properties would go beyond the scope of the present work an overview on some of the possible properties is given in Table 1.

Structural Information

An accurate description of an atomic structure requires the use of a group of descriptors best organized within a HDF 5 data group in Job(*). As the structure can change during a calculation procedure an initial and, as required, a final structure needs to be described. In addition, the possibility of intermediate structures associated with specific calculation steps should be considered. Therefore, the group *InitialStructure* will contain all descriptors of the initial structure, likewise the group *FinalStructure* those of the final one. Any intermediate structure is defined by descriptors located in *Structure(*)*.

These groups contain a string dataset with the name *Structure.cif* that contains the full structural information in the CIF file format [9]. This allows the user to view the structure with an external program. As with properties a detailed discussion of all descriptors would exceed the scope of this paper. An overview of possible descriptors is given in Table 2. The aim is to use the space group (defined by *SpaceGroupName* and *SpaceGroupID*), the cell parameters (*CellParameters*), the volume (*Volume*), and the Wyckoff positions (*WyckoffPositions*, *WyckoffPositionIDs*, etc.) as structure definition. Note, that some descriptors are only used when required, e.g., ForceFieldAtomType and WyckoffPositionSpins.

Table 1 A list of possible properties sorted by their dimension, scalar or multidimensional array, and the source, general, LAMMPS, VASP or other, is given

	General	LAMMPS	VASP	Other
Scalar	Volume, Density, Pressure, Temperature	TotalEnergy, CoulombEnergy, PotentialEnergy, KineticEnergy, VanDerWaalsEnergy	TotalEnergy	DebyeTemperature, LongitudinalModulus
Multidimensional array	Stress		DOS, BandStructure, FermiSurface,	ElasticConstants, ElasticConstantMatrix, BulkModulus, YoungsModulus, ShearModulus, SoundVelocity, PhononDOS, PhononDispersion

Table 2 Descriptors used to describe structural information are sorted by their dimension, scalar or multidimensional array, and their kind, which can be string, integer or float

	Scalar	Multidimensional array
String	SpaceGroupName, StructureName, EmpiricalFormula, Structure.cif	WyckoffPositionIDs, ChemicalElementNames, ForceFieldAtomType
Integer	SpaceGroupID, NumberChemicalElements, NumberAtoms	
Float	Volume	CellParameters, WyckoffPositions, WyckoffPositionMasses, WyckoffPositionSpins, ForceFieldCharge

Custom Tables

During a calculation procedure custom tables summarizing results may be created. However, HDF5 datasets are homogenous, meaning that they can only contain variables of a single kind, e.g., string, float, etc. Such a homogenous definition does not translate well to a table where different columns might contain data of different kind. To allow for a more flexible table definition we propose to use the group *CustomTable(*)* with the index indicating the calculation step associated with the table contained in a *Job(*)* group. This group then contains the string dataset *Title*, and for each column the datasets *Column(*)* and *ColumnHeader(*)*. With the wildcard * indicating an integer index variable going from 1 to the maximum number of columns defined. *ColumnHeader(*)* will always be string while the kind of the *Column(*)* can vary.

Program Specific Files

A *Job(*)* group may include additional string datasets with the contents of special files used by the atomistic simulation tool. These provide the user with additional information and/or functionality. When using MedeA the *db.backup* string dataset may be included, providing the user with an easy means of reproducing a given calculation protocol on his machine. This *db.backup* dataset just needs to be exported as a file and imported into a MedeA JobServer.

In addition, the dataset *Job.out* provides a text-based summary on the performed calculation and the dataset *Structure.sci*, located in a structure group, contains the full structural information as a MedeA structure file.

Metadata

To track the actual computational setup used to calculate the properties saved in the job centric structure, as described in the previous chapter, we propose to extensively make use of the HDF5 metadata capability. As the quality of calculated properties strongly depends on the actual method and computational setup used, it is important to directly associate each property with the setup. Such a use of metadata increases the transparency as it allows one to reproduce the described property, even if this property has been copied or linked into a different group or file. As with descriptors and properties the list of metadata attributes described in the present work does not necessarily have to be complete and can be extended as required.

General Attributes

Both descriptors and properties use general attributes to track some general job information on the location, job ID and program used.

1. *Program* and *ProgramVersion*
   Both string variables registers which atomistic simulation tool has been used to generate the descriptor or property.
2. *JobID*
   The job ID, that is the integer value in the brackets used for *Job(*)*, is described by this attribute. For example, all datasets and groups located in Job(10) have a job ID of 10.
3. *JobLocation*
   The original location of the dataset is described by this string attribute. If, for example, the dataset or group has been originally written to /AtomisticSimulationData/JobData/Job(10) then this information will be given.

Property Attributes

In addition to general attributes properties always contain the following additional information.

1. *Unit*
   The unit of the property is described by this string variable. It can, for example, have a value of *kJ/mol* or *GPa*.
2. *Solver* and *SolverVersion*
   The solver used to calculate the described property is registered by both string variables.
3. EmpiricalFormula
   The empirical formula of the structure, for which the calculation has been performed, is registered by this string variable. This attribute can for example have the value $CuAu2$.
4. SpaceGroupName
   The space group of the structure is registered with this string using the Herman-Mauguin notation which is known as the international notation [10]. An example for its value is $Fm\bar{3}m$.

Other property attributes are more specific and depend on the calculation setup and used solver. An overview on these can be found in Table 3.

Table 3 Solver dependent metadata attributes for property datasets and groups are sorted by their kind, which can be string, integer or float, and by their source. The source is the program or solver used to calculate a given property and in the present work can be VASP, LAMMPS or other

	VASP	LAMMPS	Other
String	Functional, ExchangeCorrelationFunctional, KMesh, KIntegrationScheme, Precision, Magnetism, Potentials, Projection, CalculationType	ForceField, SimulationTimeUnit, TimeStepUnit, InitialTemperatureUnit, FinalTemperatureUnit, InitialPressureUnit, FinalPressureUnit, Ensemble	Strain
Integer	SmearingFunctionOrder		
Float	KSpacing, SmearingWidth, CutoffEnergy, ElectronicIterationsConvergence, Pressure	SimulationTime, TimeStep, InitialTemperature, FinalTemperature, InitialPressure, FinalPressure, CellConstrains	InteractionRange, AtomDisplacementSize

Structure-Centric Data Structure

The next step is to increase accessibility of the available data by transferring it into a structure-centric form. That is, by placing it in the *StructureData* group located in the *AtomisticSimulationData* group. The data consisting of descriptors and properties of each structure are located in the subgroups *Structure(*)*, with the wildcard * indicating an index that consecutively increases from 1 when additional structures are added to *StructureData*.

The properties themselves can be sorted groups within the *Structure(*)* group. For example, the *ElectronicProperties* group contains electronic properties such as the electronic density of states, the *VibrationalProperties* group contains vibrational properties such as phonon dispersion, the *MechanicalProperties* group contains mechanical properties such as elastic constants, and other properties can be placed in the generic *Properties* group.

Summary and Conclusion

In the present paper we have outlined a strategy on how to bridge the data exchange gap between atomistic and continuum simulation tools. A basis for our discussion has been the work done by Schmitz et al. [4] which describes microstructure data definition within a HDF5 file. However, due to a subtle difference on how these two approaches create and collect data, another strategy is required for atomistic simulation. We start from a calculation centric data definition, that is more suitable for atomistic simulations, and then identify a data structure by which calculation results can be made more accessible to others. Another important ingredient of our approach is to make excessive use of metadata to keep track of the calculation setup used to obtain a property.

The present approach is well suited for atomistic simulations, where a computational procedure is applied to a uniquely defined initial structure, i.e. an arrangement of atoms defined by their element type and coordinates, resulting in a set of computed properties. This requirement is fulfilled for most DFT calculations and forcefield-based molecular dynamics simulations. Other approaches such as Gibbs ensemble Monte Carlo simulations may require an extension of the present concept. In such simulations, the number of particles may change during the course of a simulation and hence the definition of "initial structure" needs to be extended. Another generalization will be needed, if the initial structure is actually an ensemble of structures, for example a set of models of amorphous structures, where the computed properties is a statistical average obtained from the entire ensemble of initial structures. Nevertheless, we believe that the present concept can serve as foundation for building a bridge between discrete models and continuum approaches.

References

1. Committee on Integrated Computational Materials Engineering, National Materials Advisory Board, Division on Engineering and Physical Sciences, National Research Council, *Integrated Computational Materials Engineering: A Transformational Discipline for Improved Competitiveness and National Security* (National Academies Press, 2008), p. 132
2. MedeA, Materials Design, Inc., Angel Fire, NM, USA, 2016
3. M. Christensen, V. Eyert, A. France-Lanord, C. Freeman, B. Leblanc, A. Mavromaras, S. J. Mumby, D. Reith, D. Rigby, X. Rozanska, H. Schweiger, T.-R. Shan, P. Ungerer, R. Windiks, W. Wolf, M. Yiannourakou, E. Wimmer, Software platforms for electronic/atomistic/mesoscopic modeling: status and perspectives. Integr. Mater. Manuf. Innov. **6**, 92–110 (2017)
4. G.J. Schmitz, B. Böttger, M. Apel, J. Eiken, G. Laschet, R. Altenfeld, R. Berger, G. Boussinot, A. Viardin, Towards a metadata scheme for the description of materials—the description of microstructures. Sci. Technol. Adv. Mater. **17**, 411 (2016)
5. The HDF group, http://www.hdfgroup.org. Accessed Feb 2017
6. G. Kresse, J. Furthmüller, Efficient iterative schemes for ab initio total-energy calculations using a plane-wave basis set. Phys. Rev. B **54**, 11169–11186 (1996)
7. G. Kresse, D. Joubert, From ultrasoft pseudopotentials to the projector augmented-wave method. Phys. Rev. B **59**, 1758–1775 (1999)
8. S. Plimpton, Fast parallel algorithms for short-range molecular dynamics. J. Comput. Phys. **117**, 1–19 (1995)
9. S.R. Hall, F.H. Allen, I.D. Brown, The crystallographic information file (CIF): a new standard archive file for crystallography. Acta Crystallogr. **A47**, 655–685 (1991)
10. D.E. Sands, Crystal systems and geometry, in *Introduction to Crystallography* (Dover Publications, Inc., Mineola, New York, 1993), p. 165

A Flowchart Scheme for Information Retrieval in ICME Settings

Georg J. Schmitz

Abstract Retrieving desired information about a specific material may either proceed via querying existing data, such as the general Internet or, if the desired information already exists, specifically dedicated databases. If the desired information is not yet available, it has to be determined by employing one or multiple models to compute the data. If multiple models are required for generating the desired information, interoperability between the models plays a vital role. Interoperability between models implies the need to define a flow of information or a "workflow" and also to specify the timing between the different operations of the different models acting on a state or on subsets of a state description. Ultimately getting such workflows well-defined, operational and also extendable to future decision making processes requires taking some structural considerations into account. An instructive approach is seen in the specification of workflows for decision making in different areas where different types of flowchart tools are needed and used to control and guide the workflow. The proposed flowchart scenario is based on the description of a system state whose evolution is influenced by distinct types of phenomena occurring at different scales. The system state further defines the properties which can be extracted from the system state information. Different types of models/tools operate on this state and a generic classification is proposed which is based on the character/functionality of physical equations, such as (i) evolution equations, (ii) property equations, (iii) equilibrium equations, and (iv) conservation equations.

Keywords Interoperability · Flowcharts · Microstructural state · Model classes · Information retrieval

---

G.J. Schmitz (✉)
Access e.V. at the RWTH Aachen, Intzestr. 5, 52072 Aachen, Germany
e-mail: g.j.schmitz@micress.de; G.J.Schmitz@access.rwth-aachen.de

Introduction

ICME—Integrated Computational Materials Engineering—by its name and its nature draws on the combination of numerous modelling tools and data sources to model the performance and life-cycle of components and products. ICME aims to describe component properties along their production and service life cycles and, for this purpose, draws on models for materials and microstructures or on existing data. Models, which generate data, draw on existing data being generated by other models or, if need be, on other models providing the desired data. The scope of the present paper is to propose a flow chart scenario to structure and organize the information flow in ICME settings.

Information Flow

The first step towards retrieving a specific piece of "desired information" from the Internet, from databases or from simulations is to address several questions about the availability of the desired information such as

- Is a descriptor, i.e. a searchable keyword corresponding to the desired information, available in the namespace?
- Can a descriptor for the desired information be derived from descriptors already existing in the namespace by well-defined operations?
- Is it necessary to amend the namespace?
- Is a value assigned to that particular descriptor in the state description of a model or in any database?

This series of actions is best reflected in a flowchart type of representation, Fig. 1.

Some of the individual steps in this flowchart are briefly explained in the following:

Descriptor available

The "descriptor available"-check searches the actual namespace for the presence of the keyword being sought. In the future, this check may request user confirmation by making suggestions for similar keywords/descriptors, such as in Google search: "You entered *"tet"* … Did you mean *"test"* …?". For a positive reply, a search is then performed for a value which is associated with this particular descriptor. This search may occur in a local dataset, in electronic databases or also other sources of information which are available on the Internet. If a negative reply results, no suitable descriptor is present in the current namespace. Thus measures have to be taken to generate a corresponding descriptor by amending the namespace.

A Flowchart Scheme for Information Retrieval in ICME Settings 59

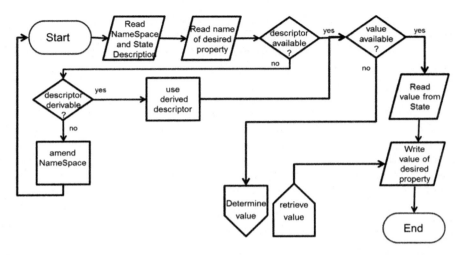

Fig. 1 Flowchart allowing desired information to be sought, which is specified as the value being assigned to the descriptor of the desired property. If the desired descriptor and a corresponding value are available in the data space (e.g. databases or the Internet), the search is immediately and positively terminated and the desired value is returned. If no value for the desired descriptor can be retrieved from existing data, it is necessary to determine this value e.g. by simulations or from experiments in a sub-flowchart

Descriptor derivable

The basic concept of a namespace draws on a minimum number of well-defined descriptors to describe the state or the microstructure of a material. Prior to adding any new descriptor, checks should be performed to determine whether the desired descriptor can be derived based on the existing descriptors. As an example, the "density" of a material, which is commonly understood to be the "MassDensity", can thus easily be derived from the notion of the descriptors "Mass" and "Volume". Some rules to derive descriptors have already been specified, such as normalising by volume, normalising by total value, to name just two. Some further rules are detailed in [1].

AmendNameSpace

If no descriptor is available in the given namespace and also no suitable descriptor can be derived as specified above, amendments become necessary for the given namespace.

Value available

Once a descriptor is available, a search for a value associated with this particular descriptor is performed in the state description, which is stored in a local dataset; such as HDF5 or in other types of electronic databases. The search can be further guided or refined based on the metadata attributes of that particular descriptor. For a positive check, the dataset contains a data object whose name corresponds to the descriptor and comprises the desired value. The desired value, including all its attributes (e.g. units), can then be easily retrieved directly from the file and the overall information search is then positively terminated.

DetermineValue

In the case of a negative reply to the "value available" query, no data object being associated with the descriptor exists in the local or global datasets although the descriptor proves to be valid in the given namespace. As a first step, a corresponding data object then has to be created. In a second step, a value and metadata have to be assigned to this data object. This value may be assigned synthetically, it may be derived from experimental data, it may be calculated from suitable models, etc. The origin of the value should also be stored in metadata attributes related to the descriptor.

Retrieve Value

If a value is available—or has been determined to make it available—it can easily be retrieved by simply reading from the state description. The overall information search is then positively terminated.

Some of the different operations required for determining a value related to a specific descriptor, see Fig. 2, are further detailed in the following:

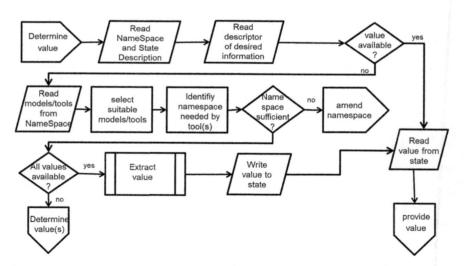

Fig. 2 Flowchart for determining an unknown value. If the values for descriptors are not available they then have to be determined, e.g. by simulation chains. This introduces the need for interoperability between different simulation tools. For explanations, see text

Read models/tools from NameSpace

Each descriptor in the namespace comprises additional metadata information, e.g. on *ModelClasses, ModelIDs, SoftwareToolLabels*. These metadata attributes indicate whether the respective model(s) or simulation tool(s) either provide data or draw on data related to the value of the actual descriptor. Multiple models and tools may thus be related to a single descriptor. For example, the descriptor "position" can be expected to occur in almost any model at any scale. The answer to a query for reading models/tools for a given descriptor will thus return a list of models and/or software tools relevant to the desired information. The search may be refined by, for instance, confining it to specific model classes: e.g. a query for *position* ModelClass *atomistic*, will limit the results to all tools in the namespace dealing with positions of atoms, whereas *position* ModelClass *continuum*, will only return models and tools related to positions of, for instance, grains in a microstructure. Metadata information for the descriptor also allows one to distinguish between *ToolsProviding*, i.e. tools known to provide a value for the data object/value described by the descriptor, and *ToolsUsing*, i.e. tools using the corresponding value.

Select suitable models/tools

From the list of models returned from the previous step, suitable tools for determining the desired property—i.e. "*ToolsProviding*"—have to be selected which can then be used to calculate the desired information. Such a calculation can, however, only be performed if (i) the namespace of the identified tool is available and (ii) values are assigned to the required descriptors in the namespace of the additional tool.

Identify namespace addressed by tool(s)

In future, each software tool and/or model should ideally be provided together with the full namespace required for its operation, or be provided by its execution.

Extract value

Once all the namespaces are available which are required for using (all) the models necessary for determining the desired value, the desired value can be calculated. If one of the models requires one or more additional values associated with other descriptors, the procedure DetermineValue has to be re-performed for these missing values.

Namespaces of tools versus namespaces for models

Determining non-existing values specifically requires drawing on suitable models, tools, methods, equations, relations, databases and other means etc. In turn, these models and tools generally need values of other descriptors/values in order to be solved and thus be capable of providing the desired information. While simulation tools may have complicated and complex namespaces, the namespaces related to materials laws and physics equations are fully determined by the respective mathematics/physics equations or materials relations; which essentially have to be translated into a namespace from a representation in mathematical symbols.

Models—Overview and Classifications

A desired value—if it's not available in an existing dataset—can, in the best case, be determined by applying a single model but, in general, it will require a combination of multiple models. Models, according to the "Review of Materials Modelling V" [2], are defined as physics equations complemented by materials relations together forming the governing equations and are classified into four classes: electronic models, atomistic models, mesoscopic models and continuum models. In view of interoperability, the barriers between such model classes have to be reduced and the current 24 classes of models, as depicted in [2], may perhaps be further reduced and classified, and may be additionally structured in a way which promotes interoperability.

The scenario for such a further classification is based on the description of a system state and its evolution being influenced by all types of phenomena occurring at all scales. The system state further defines the properties and these can be extracted from the system state information using suitable methods/algorithms/models at all scales during a post-processing mode of operation. A description of the system state, which essentially corresponds to the microstructure, has been outlined in [1]. This state description is based on a suitable set of arrays of scalars, vectors, tensors, Fig. 3.

Different types of models/tools can then operate on this state and a further classification scheme becomes possible since it is based on the character/functionality of the model equations:

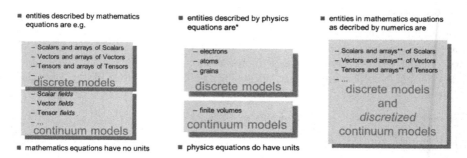

Fig. 3 Scalars, vectors, tensors and arrays of such mathematical entities form the basis of any digital description of a state. **The dimension of the arrays corresponds to the number of objects/features for discrete models, and to the number of numerical cells for discretized continuum fields

- Conservation equations
- Evolution equations
- Equilibrium equations
- Property equations

Most physics models are based on differential equations, which can only be solved if the appropriate initial and boundary conditions are specified. The required initial conditions are already comprehensively defined in the microstructure state description. However, the boundary conditions of the domain have to be explicitly defined and provided separately. Thus, any state description has to provide boundary objects which can be assigned values for the boundary conditions. These values may even be time dependant. This request is already fulfilled by the state description proposed for the microstructures [1].

Conservation equations and other fundamental principles provide strong *constraints* for all types of equations—especially for evolution type equations. Most prominent and important examples are: conservation of energy, momentum, mass/species, angular momentum, charge and positive entropy production.

Equilibrium and balance models describe aspects of an equilibrium state subject to the given conditions. Examples are phase fractions predicted by thermodynamic equilibria, a mechanical equilibrium or the solutions of the stationary Schrödinger equation. Dynamic equilibria result, for example, from models for flux balances. Knowing aspects of the equilibrium state is highly beneficial in regard to formulating an idea of the state to which a given system will relax. Furthermore, information about the equilibrium state may already be sufficient if the kinetics of the evolution process is so rapid that it is not necessary to describe the details of its evolution. Knowledge about equilibrium conditions is most beneficial for controlling and guiding simulation workflows.

Evolution models—hereafter called EVOLVERS—turn the actual state, or at least parts of the actual state, under given conditions into a new state at a later instant of time. Such models accordingly change the state. Evolution models are characterized by any type of time dependency within the physics equation or within the materials relations. Many physics equations are evolution type equations, such as the Schrödinger equation (time dependant), Molecular Dynamics equations, Phase Field equations, Diffusion equations, Navier-Stokes equation, to name but a few.

Models and tools acting as *"post-processors"* on an existing state extract the desired properties from that state. These are called EXTRACTORS in the following. These EXTRACTORS do not alter the state and thus can operate in parallel to EVOLVERS. Examples for EXTRACTORS are mathematical homogenisation models and tools, volume averaging, statistical tools, virtual testing, visualisation tools. In particular, the *extraction of scalar values* from the state description is most important since only *scalar values allow for decisions* based on inequalities or equalities. Such decision type operations are exclusively defined in the space of real numbers and for logical data types. The *extraction of properties* of a given state is the objective of most activities. The properties—and their evolution—are the major feature of interest for any type of application of materials. For instance, extracting the properties of a multiphase, polycrystalline microstructure requires knowledge about the properties of the individual phases and about their spatial arrangement. Knowledge of each of the phase properties can either be readily available in the state description or may be extracted from another available state description of that particular phase which is generated by other models. The *extraction of statistical type data* is always necessary if data is to be transferred to a larger scale, where all the information cannot be digested.

Any EVOLVER requires the *specification of an initial state*. This may be the output state of a preceding tool. However at its very beginning, any simulation chain requires assumptions about an initial state. CREATOR tools can synthesize such a state by assigning values to the set of arrays of scalars, vectors and tensors,

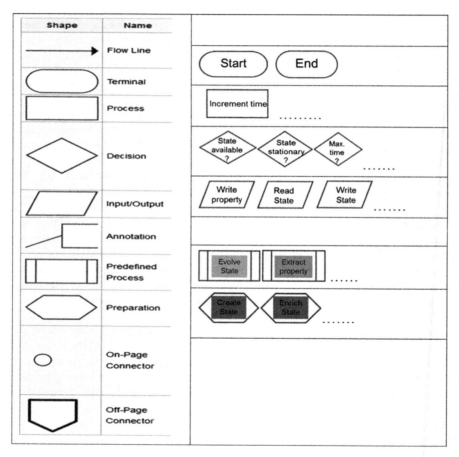

Fig. 4 Typical elements currently used in a flowchart [4] and related models and tools in MoDa *colour* code (see text)

including their units, describing the state. They can also synthesize states at any intermediate step in the simulation chain. Using experimental data to create an initial state can also be considered as a CREATOR type activity.

Simulation Flowcharts

Interoperability between models implies the need to define a flow of information or a "workflow" and also to specify the timing between the different operations of the different models acting on the state or on subsets of the state description. Ultimately getting such workflows well-defined, operational and also extendable to future decision making processes requires further considerations. An instructive and helpful approach is to look at the specification of workflows and decision making in different areas where flowchart tools are specified to control the information flow, Fig. 4.

The colour classes essentially correspond to the colours depicted in the MoDa scheme [3] with "models" (blue), "post-processors" (green) and the "state description" (red). To employ informatics type workflows in these models, states and post-processors have to be supplemented by further tools such as INPUT/OUTPUT and especially by DECISIONS which permit the workflow to be guided. In the overall current scenario, all I/O operations are based on reading/writing to the state description as specified and stored in an HDF5 file. Some examples of simple workflows are depicted in the following Figs. 5 through 8.

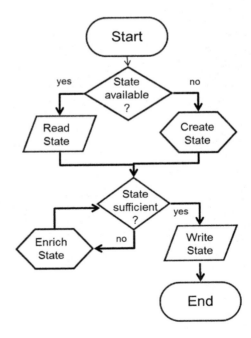

Fig. 5 Simple example of a workflow to create an initial state

Fig. 6 Simple examples of workflows extracting different types of properties

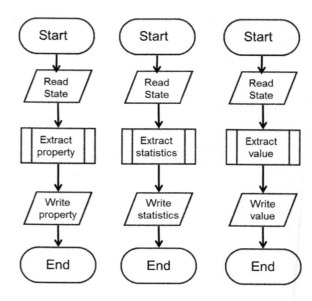

Fig. 7 Simple workflow of an evolution type model including decisions allowing the evolution simulation to be terminated when pre-set criteria are met. A few examples for such criteria could be: Is an equilibrium state reached? Is the state stationary? Is a pre-set simulation time reached?

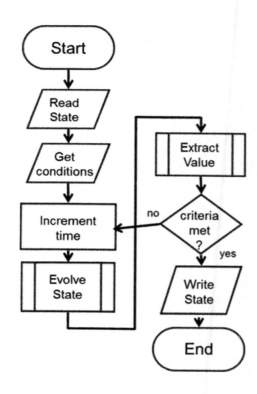

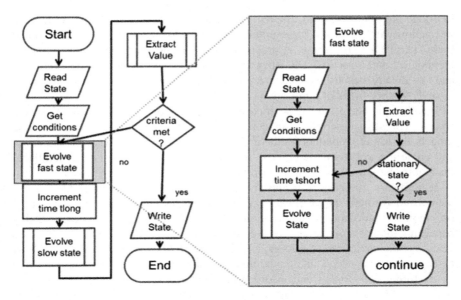

Fig. 8 Example of a workflow to evolve different aspects of a state which relate to phenomena evolving on different timescales. An example may be the evolution of a thermal field (slow) depending on the evolution of an electrical current distribution (fast)

Summary

A scenario for data exchange at the microstructural scale has been outlined. This scenario is based on the definition of a state, which can be evolved and analysed by different types of models and tools. This state essentially corresponds to a comprehensive digital description of the microstructure. It provides the initial conditions for any evolution type model. The output state of such models represents, in turn, a state which can then be further advanced along the simulation chain. Models and tools have been classified into evolution types and extractor types. These allow one to either evolve or process the state, respectively, corresponding to the paradigms: "processing determines microstructure" or "microstructure determines properties", to extract property data from the state.

The digital representation of the state is characterized by a suitable set of arrays of scalars, vectors, and tensors. A nomenclature/namespace for the metadata descriptors for each of the arrays has been proposed [1]. This nomenclature has been elaborated based on continuum models. However, it also considers descriptors for electronic, atomistic, and mesoscopic models.

A major conceptual approach is the hierarchical arrangement of the individual arrays describing the state or the microstructure. This allows the microstructure to be differently represented, such as statistically or spatially, in a coherent way within the same data structure. This inherent hierarchy of local and integral data is essential to bridge the scales between different physical phenomena and enables a seamless interaction with e.g. with, for example, models/tools operating at the component scale. The HDF5 file format is proposed as a specific implementation of such a hierarchical data structure. This open source format is very powerful and versatile, and has already established itself as a standard in computational fluid dynamics and in several other fields of application.

Finally, a concept has been proposed for specifying workflows in simulation chains. The state description, models describing the evolution of the state and models extracting properties are the major building blocks. To allow one to control and guide the workflow, "decisions" have been specially introduced as additional mandatory building blocks of a flowchart.

In summary, the basic concepts for interoperability between simulated, experimental and synthetic microstructures presented here are now available and appear to be viable. These concepts are constructed to also enable/ensure interoperability between models operating on components and processes as well as electronic/atomistic/mesoscopic models. Future activities should deal with further extending these concepts into the community, applying them in practical industrial cases, further developing and incorporating them into workflows and simulation platforms, broadening their scope towards uncertainty propagation and error estimates, generating robust simulation chains, harmonizing namespaces and many other fields.

Acknowledgements The research leading to the results presented in this publication has received funding from the European Union Seventh Framework Programme (FP7/2007–2011) within the ICMEg-CSA (grant agreement no 6067114) and from the Horizon 2020 within the scope of the EMMC-CSA (grant agreement no 723867).

References

1. G.J. Schmitz, B. Böttger, M. Apel, J. Eiken, G. Laschet, R. Altenfeld, R. Berger, G. Boussinot, A. Viardin, Towards a metadata scheme for the description of materials—the description of microstructures. Sci. Technol. Adv. Mater. **17**(1), 410–430 (2016). http://www.tandfonline.com/doi/full/10.1080/14686996.2016.1194166
2. DeBaas/Rosso (eds.), Review of Materials Modelling IV. http://ec.europa.eu/research/industrial_technologies/modelling-materials_en.html
3. MoDa: elements in materials modelling. http://ec.europa.eu/research/industrial_technologies/modelling-materials_en.html
4. https://en.wikipedia.org/wiki/Flowchart

An Ontological Framework for Integrated Computational Materials Engineering

Sreedhar Reddy, B.P. Gautham, Prasenjit Das,
Raghavendra Reddy Yeddula, Sushant Vale and Chetan Malhotra

Abstract ICME is expected to significantly reduce the dependence on trial and error based experimentation cycles for materials development and deployment in products. However, modeling and simulation is a knowledge intensive activity. In an integrated design, choosing right models for different phenomena, at right scales, with right parameters, and ensuring integration across these models is a non-trivial task. The gaps in modeling and simulation need to be filled with tacit knowledge and co-engineered with product knowledge. Therefore, an IT platform having capabilities such as, (a) a repository of building-block models, templates and workflows with an intelligent means to choose and compose right workflows for a given problem, (b) a knowledge engineering framework for knowledge management, (c) a simulation services framework for simulation tool integration and simulation execution, (d) tools for decision support, optimization, robust design etc., is essential for scaling up ICME for industrial applicability. This requires a unifying semantic foundation. Ontologies can provide the common substrate for integration of different models, the common language for information exchange, and the means for capturing and organizing knowledge. However, ontology engineering is a challenge when we consider the diversity of the material systems, products, processes and mechanisms involved in ICME. This calls for a flexible ontological framework that provides a means for modeling the generic structure of a subject area (e.g. materials) and a means for instantiating subject specific ontologies from this generic structure. We describe a model driven framework and how it has been used for developing an enabling platform for ICME.

Keywords Domain specific search engine · Information retrieval · Information extraction · Materials engineering

---

S. Reddy (✉) · B.P. Gautham · P. Das · R.R. Yeddula · S. Vale · C. Malhotra
TCS Research, 54B, Hadapsar Industrial Estate, Pune, India
e-mail: sreedhar.reddy@tcs.com

© The Minerals, Metals & Materials Society 2017
P. Mason et al. (eds.), *Proceedings of the 4th World Congress on Integrated Computational Materials Engineering (ICME 2017)*,
The Minerals, Metals & Materials Series, DOI 10.1007/978-3-319-57864-4_7

Introduction

Integrated computational materials engineering (ICME) is a new paradigm for integrated design of materials, products and manufacturing processes using modeling and simulation, knowledge-guided decision making and data-driven reasoning. ICME is widely recognized as a paradigm changer that is expected to significantly reduce the dependence on trial and error based experimentation cycles. This is expected to result in (a) faster development of new materials, and (b) significant improvement in quality and time-to-market of products by integrating material design with product design. However, industrialization of this approach has many roadblocks to overcome [1]. Modeling and simulation is a highly knowledge intensive activity. Models exist at multiple dimensions and scales representing diverse physical phenomena. In an integrated design, one has to worry about a multitude of phenomena and the corresponding models. Choosing right models for these phenomena, at right scales, with right parameters, and ensuring integration across these models is a non-trivial task. Without strong automation, scaling up ICME is going to be a difficult problem.

With this motivation, we are developing an IT platform called TCS PREMAP [2, 3] at Tata Consultancy Services. Our goal is to use this platform to industrialize the benefits of the ICME approach, with a special focus on integrated design of products and materials. In view of the vast diversity of material systems and component/product application categories, the platform consists of a set of domain dependent and domain-independent components as shown in Fig. 1.

On the right side of the figure are the components that are domain dependent and those on the left are domain independent. A domain may refer to a material category with associated manufacturing processes and/or a product category. Domain specific components include models of various kinds, design templates, design rules, design cases, etc. Domain independent infrastructure includes, among other things, (a) knowledge engineering framework for knowledge management,

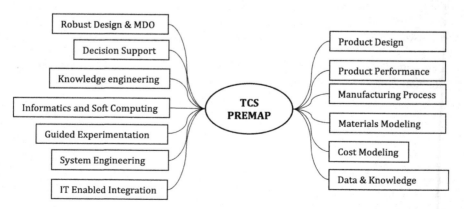

Fig. 1 Domain independent (*left*) and domain dependent (*right*) components of the platform

(b) simulation services framework for simulation execution and simulation tool integration, (c) tools for robust design and multidisciplinary optimization techniques (MDO), (d) decision support systems (e.g., the compromise decision support problem construct), and (e) design of experiments and combinatorial experimentation tools to drive both simulation and experimental studies.

Building all these capabilities into the platform in an integrated manner requires a unifying semantic foundation. Domain ontology provides such a foundation. It serves as the common substrate for integrating different models. It serves as a means for capturing and organizing knowledge. However, ontology varies from subject to subject, and, being a generic platform, TCS PREMAP has to cater to a wide range of subjects. For instance, ontology of steel is different from ontology of a fibre reinforced composite material. This calls for a flexible ontology engineering framework that enables us to create and evolve subject specific ontologies without hard coding them into the platform. We use model driven techniques to engineer such a framework. In this paper we present the modeling framework underlying the TCS PREMAP architecture and give a brief overview of some of the aspects automated using model driven techniques.

TCS PREMAP Modeling Framework

TCS PREMAP uses a reflexive modeling framework [4] to bootstrap its modeling infrastructure.

Reflexive Modeling Framework

An information system can be seen as a collection of parts and their relationships. A model of an information system is a description of these parts and relationships in a language like Unified Modeling Language (UML) [5], Web Ontology Language (OWL) [6], etc. The modeling language itself can be described as a model in

Fig. 2 Modeling layers

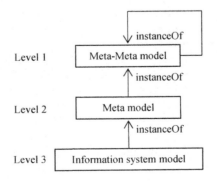

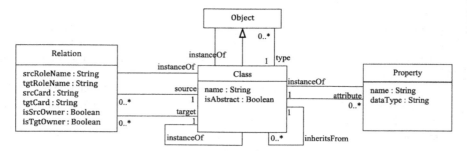

Fig. 3 A reflexive meta-meta model

another language. The latter language is the meta model for the former as shown in Fig. 2. A model at each level is an instance of the model at the previous level.

The meta-meta model is the base model in the hierarchy. It is reflexive, i.e. it is capable of describing itself, and is the basis for bootstrapping models at all levels. Figure 3 shows this model which is compatible with Meta Object Facility (MOF) [7]. Everything in a model is an object. An object is described by its class. A class is specified in terms of a set of properties and relations. An object is an instance of a class that has property values and links to other objects as specified by its class. Since everything is an object, a class is also an object. A class is specified by another class called metaclass. In Fig. 3, the class *'class'* is a metaclass which is an instance of itself. Any class that inherits from the class *'class'* is also a metaclass. A meta model specification consists of a set of metaclasses and their relations, and a set of constraints and rules to specify consistency and completeness checks on its instance models. Due to the reflexive nature of the meta-meta model, there is no inherent limit on the number of modeling layers that can be supported.

Ontology Modeling Framework

Domain ontology provides the semantic foundation for capturing knowledge about a domain. It is essential to model the domain at the right level of abstraction (1) to maximize the utility of knowledge and (2) to ensure information exchange and integration happen at the right level of abstraction. However, ontology varies from subject to subject, and, being a generic platform, TCS PREMAP has to cater to a wide range of subjects. To address this, we have conceptualized domain models at two ontological levels—a meta level and a subject level as shown in Fig. 4.

Models relevant to ICME can be broadly categorized into three subject areas—materials, products and processes. Corresponding to these subject areas we have three related meta models—material meta model, product meta model and process meta model. Subject specific ontologies are created as instances of these meta models. We illustrate this with an example.

Fig. 4 Domain ontology levels

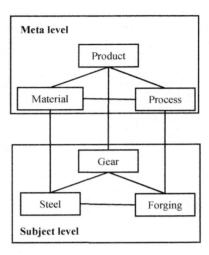

Figure 5 shows a part of the component meta model, which is a part of the product meta model. The meta model has meta classes *Component, Material, GeometricFeature, FunctionalFeature, Parameter,* etc. A component has a geometry and a set of functional and geometric features. These features may be described in terms of a set of parameters. A component may be made from one or more materials; similarly different geometric features of the component may be made from different materials.

Figure 6 shows Gear ontology as an instance of this meta model. The figure shows classes in <class>:<meta-class> format and objects in <object>: <class> format. A gear is a component whose geometry has features such as hub, web, rim and teeth. Its function is to transmit motion in the same or a different direction and a change in rotational speed. The geometric feature 'hub' has diameter and width as parameters (parameters of other features are omitted from the diagram). The figure also shows a specific gear (NanoCarGear) with its dimensions, as an instance of the gear ontology.

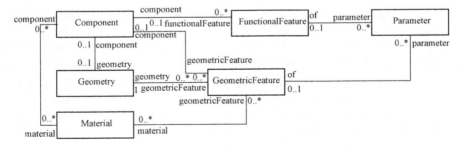

Fig. 5 Component meta model

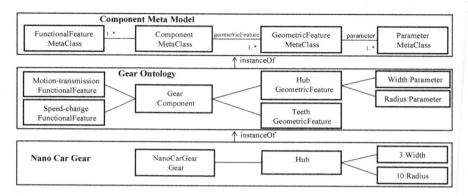

Fig. 6 Component modeling layers

This layered modeling architecture provides several benefits:

- **Extensibility**. It lends extensibility to the platform by enabling new subjects to be created as instances of meta models and integrate them with existing subjects. For instance, to extend the platform to support the design of composite materials, we create composites ontology as an instance of the materials meta model. Similarly to support the design of an engine block, we create engine block ontology as an instance of the products meta model. Subject specific ontologies thus become first class entities in the platform. Since all subjects of a subject area are described using the same meta model, it provides the semantic basis for capturing and discovering generalization relations among these subjects using ontological reasoning.
- **Knowledge management**. It provides a means to organize domain knowledge systematically. Knowledge that is applicable to all subjects of a subject area is captured at the meta model level, knowledge applicable to a design subject is captured at the subject model level and knowledge specific to a design instance is captured at the instance level. For example, knowledge that is common to all materials is kept at the material meta model level, knowledge pertaining to steels is kept at the steel ontology level, and knowledge specific to a steel grade is kept at the instance level.
- **Integration**. It provides a systematic means for integration across different domains at different levels. Meta model integration enables design subject level integration and subject model integration enables design instance level integration. Given below is a sample of a few simple integration rules:
  - Product—Material: Materials used in a component of an assembly should be a subset of the materials identified for the assembly (meta level rule).
  - ManufacturingProcess—Material: Carburizing temperature in the carburization process of steels should be above the austenization temperature of the steel under consideration (subject level rule).

– Product—ManufacturingProcess: Open-die forging process should not be used where near-net-shape products are desired (subject level rule).
- **Automation**. Meta models specify the semantics of subject models and subject level models specify the semantics of instance models. Since all these models are machine interpretable they lend themselves to a great deal of automation. This has been exploited within TCS PREMAP to automate the generation of a large part of the platform implementation such as persistence management, user interfaces, simulation services, knowledge services, etc.

Model Driven Engineering in TCS PREMAP—A Few Examples

As mentioned above, TCS PREMAP uses model driven generation techniques to automate several features of the platform. We discuss a couple of these features in this section.

Simulation Tool Integration

A design workflow consists of design of several process steps such as forging, machining, carburization, quenching, tempering, etc. Each of these processes has its own simulation model. In integrated design simulation, these models have to be simulated in an integrated manner, with right information flowing from one model to the other. This is done by mapping the inputs and outputs of each simulation tool to the domain ontology, as shown in Fig. 7. It is then possible to validate a process chain for information integrity by checking that right information is flowing to the right process step. From these mappings it is also possible to generate input/output adapters for plug-and-play integration of simulation tools.

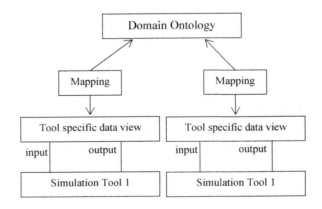

Fig. 7 Simulation tool integration

Fig. 8 Data integration

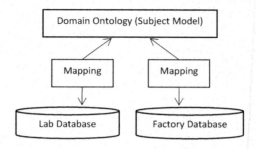

Data Integration

In TCS PREMAP we want to be able to utilize data (on materials, processes, etc.) from a number of sources. The data sources may include laboratory databases, factory floor databases, or third party proprietary data. To integrate these data sources we map them to subject model ontologies using Global-as-view (GAV) [8, 9] or Local-as-view (LAV) [10] schemes (Fig. 8). Subject model ontology then serves as the unified interface to access all data sources uniformly. A query on the subject model is translated into an equivalent extract-transform-load (ETL) data flow graph which is responsible for extracting data from individual sources and suitably combining them to produce the query result.

Conclusion and Future Work

We have given an overview of the computational platform we are developing to support ICME and briefly discussed the model-driven engineering design principles underlying its architecture. We have identified the domain modeling challenge and presented an ontology modeling framework that has been developed to address this challenge. We have also touched upon how model driven techniques have been used to automate the features of the platform. The core platform is ready for proof of concept implementations. We are currently working on advanced knowledge engineering features such as context sensitive knowledge retrieval, context sensitive question answering, knowledge guided machine learning, etc., which will be integrated into the platform in due course.

References

1. A.W.A. Konter, H. Farivar, J. Post, U. Prahl, Industrial needs for ICME. JOM **68**(1): 59–69 (2015)

2. M. Bhat, S. Shah, P. Das, P. Kumar, N. Kulkarni, S.S. Ghaisas, S.S. Reddy, in *PREMAP: Knowledge Driven Design of Materials and Engineering Process*, eds. A. Chakrabarti, R.V. Prakash. ICoRD'13, Lecture Notes in Mechanical Engineering (Springer India, 2013), pp. 1315–1329
3. B.P. Gautham, A.K. Singh, S.S. Ghaisas, S.S. Reddy, F. Mistree, *PREMAP: A Platform for the Realization of Engineered Materials and Products*, eds. A. Chakrabarti, R.V. Prakash. ICoRD'13, Lecture Notes in Mechanical Engineering (Springer India, 2013), pp. 1301–1313
4. V. Kulkarni, S. Reddy, A. Rajbhoj, Scaling up model driven engineering—experience and lessons learnt, in *MoDELS*, vol. 2 (2010), pp. 331–345
5. Unified Modeling Language, http://www.omg.org/spec/UML/2.5
6. Web Ontology Language, https://www.w3.org/OWL/
7. Model Object Facility, http://www.omg.org/spec/MOF/2.0
8. J.D. Ullman, Information integration using logical views, in *Database Theory—ICDT'97* (Springer, Berlin, 1997), pp. 19–40
9. M. Lenzerini, Data integration: a theoretical perspective, in *Proceedings of the Twenty-first ACM SIGMOD-SIGACT-SIGART Symposium on Principles of Database Systems* (ACM, 2002)
10. A.Y. Halevy, Answering queries using views: a survey. VLDB J. **10**(4), 270–294 (2001)

European Materials Modelling Council

Nadja Adamovic, Pietro Asinari, Gerhard Goldbeck,
Adham Hashibon, Kersti Hermansson, Denka Hristova-Bogaerds,
Rudolf Koopmans, Tom Verbrugge and Erich Wimmer

Abstract The aim of the European Materials Modelling Council (EMMC) is to establish current and forward looking complementary activities necessary to bring the field of materials modelling closer to the demands of manufacturers (both small and large enterprises) in Europe. The ultimate goal is that materials modelling and simulation will become an integral part of product life cycle management in European industry, thereby making a strong contribution to enhance innovation and competitiveness on a global level. Based on intensive efforts in the past two years within the EMMC, which included numerous consultation and networking actions with representatives of all stakeholders including Modellers, Software Owners, Translators and Manufacturers in Europe, the EMMC identified and proposed a set of underpinning and enabling actions to increase the industrial exploitation of

---

N. Adamovic (✉)
TU Wien, Institute of Sensor and Actuator Systems (ISAS), Vienna, Austria
e-mail: nadja.adamovic@tuwien.ac.at

P. Asinari
Department of Energy, Politecnico di Torino, Torino, Italy
e-mail: pietro.asinari@polito.it

G. Goldbeck
Goldbeck Consulting Ltd, St Johns Innovation Centre, Cambridge, UK
e-mail: gerhard@goldbeck-consulting.com

A. Hashibon
Fraunhofer Institute for Mechanics of Materials IWM, Freiburg, Germany
e-mail: adham.hashibon@iwm.fraunhofer.de

K. Hermansson
The Ångström Laboratory, Chemistry Department, Uppsala University,
Uppsala, Sweden
e-mail: kersti@kemi.uu.se

materials modelling in Europe. EMMC will pursue the following overarching objectives in order to bridge the gap between academic innovation and industrial application:

- enhance the interaction and collaboration between all stakeholders engaged in different types of materials modelling, including modellers, software owners, translators and manufacturers,
- facilitate integrated materials modelling in Europe building on strong and coherent foundations,
- coordinate and support actors and mechanisms that enable rapid transfer of materials modelling from academic innovation to the end users and potential beneficiaries in industry,
- achieve greater awareness and uptake of materials modelling in industry, in particular SMEs,
- elaborate Roadmaps that (i) identify major obstacles to widening the use of materials modelling and (ii) elaborate strategies to overcome them.

Keywords Materials modelling · Models · Coupling/Linking · Interoperability · Metadata schema · e-CUDS · Open simulation platform · Marketplace · Validated software · Data repositories · Translation · Economic impact

Introduction

The development of new and improved materials is a significant innovation driver for the global competitiveness and sustainability of European industry and society in general. It has been demonstrated in many individual cases that materials modelling is a key enabler of R&D efficiency and innovation. Companies reported [1] that computational modelling benefits include reduced R&D time and cost, more efficient and targeted experimentation, more strategic approach to R&D,

---

D. Hristova-Bogaerds
Dutch Polymer Institute (DPI), Eindhoven, The Netherlands
e-mail: d.hristova-bogaerds@polymers.nl

R. Koopmans
Koopmans Consulting GmbH, Zurich, Switzerland
e-mail: rudy.koopmans@gmail.com

T. Verbrugge
Dow Benelux B.V, Hoek, The Netherlands
e-mail: tverbrugge@dow.com

E. Wimmer
Materials Design, Le Mans, France
e-mail: ewimmer@materialsdesign.com

Fig. 1 The different enabling and underpinning activities of the EMMC

a route to performance optimisation, wider patent protection, improved supply chain control and early understanding of application performance aiding faster and more assured market introduction. Nevertheless modelling today is still not always an essential part of or critical tool in creative materials design or business decision making on product innovation. Thus, the overarching goal of the European Materials Modelling Council (EMMC) is to stimulate and enhance the use of materials modelling in industry, in particular SME's. It implies fostering ongoing efforts, addressing observable hurdles, enhance the visibility, inspire improved models and step up the contribution of materials modelling to strengthening the global competitiveness of the European manufacturing industries and complementing the drive to the fourth industrial revolution, known as Industry 4.0.

The aim of the EMMC is to establish current and forward looking complementary activities necessary to bring the field of materials modelling closer to the demands of manufacturers (both small and large enterprises) in Europe. The ultimate goal is that materials modelling and simulation will become an integral part of product life cycle management in European industry, thereby making a strong contribution to enhance innovation and competitiveness on a global level. Based on intensive efforts in the past two years within the European Materials Modelling Council (EMMC) which included numerous consultation and networking actions with representatives of all stakeholders including Modellers, Software Owners, Translators and Manufacturers in Europe, the EMMC identified and proposed a set of underpinning and enabling actions to increase the industrial exploitation of materials modelling in Europe (Fig. 1).

The EMMC will build on these findings by supporting activities such as open[1] simulation platforms, information infrastructure, materials marketplace and

---

[1]Open refers to the ability to combine different open source and proprietary elements, not to open source necessarily.

business decision support systems, databases, multiscale coupling science, education and translation activities and validation. Our vision is that only when considering all these elements together sufficient synergies can be created to address the most challenging questions and overcome the hurdles for further industrial application of materials modelling. In particular this will enable manufacturers to utilise modelling in their business decisions, much as this is done today with experiments. This will help close the innovation gap ("valley of death") between materials modelling and industrial application and enhance a knowledge driven product development process.

EMMC Objectives

EMMC will pursue the following overarching objectives in order to establish and strengthen the underpinning foundations of materials modelling in Europe and bridge the gap between academic innovation and industrial application:

1. **Enhance the interaction and collaboration between all stakeholders engaged in different types of materials modelling, including modellers, software owners, translators and manufacturers;**
   addresses the need to bring all stakeholders together so that complex industrial problems can be tackled successfully by drawing on multiple expertise. This will also enable identifying gaps in modelling and subsequently exerting collaborative efforts to cure them more efficiently than is possible today.
2. **Facilitate integrated materials modelling in Europe building on strong and coherent foundations;**
   address a need to build, utilising this new collaboration, stronger foundations of modelling in Europe. This includes integrating material models with data repositories, data driven approaches (big data) and machine learning, integrated open simulation platforms and interoperability. Openness and interoperability will allow integrating open source and closed source commercial software and data equally which will lead to a boost in materials modelling due to more comprehensive exploitation of all available knowledge.
3. **Coordinate and support actors and mechanisms that enable rapid transfer of materials modelling from academic innovation to the end users and potential beneficiaries in industry;**
   address a need to advance the translation and transfer process of innovation from academy to manufacturers. Improved transfer mechanisms will be stimulated to empower translators in academy, research organisations and R&D departments to harvest materials modelling foundations more diligently for the benefit of industry. This includes infrastructure such as Marketplace, which is a central hub for materials modelling in Europe, model validation repositories, enhanced training, and most importantly defining the role of Translators as specific

stakeholders. There often a lack of awareness to the precise capabilities of modelling particularly in SMEs.
4. **Achieve greater awareness and uptake of materials modelling in industry, in particular SMEs;**
support and coordination actions in terms of workshops, case studies will be made to raise awareness, and consequently utilisation of materials modelling in all industry in Europe.
5. **Elaborate Roadmaps that (i) identify major obstacles to widening the use of materials modelling in European industry and (ii) elaborate strategies to overcome them;**
addresses future research needs that will stem directly from the actions of the EMMC. These will lead to more targeted research to close gaps in materials modelling and bring it up to the next level.

Taking all of the foundation and transfer elements together, the EMMC will be in a position to create impact in industry. The EMMC main objectives will be achieved via actions that make use of a range of accompanying measures, such as workshop organisation which bring together multiple stakeholders and target specific underpinning and enabling actions for wider adoption of materials modelling in industry. The actions will also include targeted surveys as a means to reach out for different stakeholders and gather rich information on a plethora of key topics. Interest and expert groups for managing and maintaining metadata standards document, training workshops, and summer/winter schools (for young researchers).

EMMC Stakeholder Groups

The underpinning vision of the EMMC is that only when bringing all stakeholders together, sufficient synergies can be created to address the most challenging questions in materials modelling today and in the future. The key concept is to increase the interaction between all stakeholders to stimulate novel exploitation avenues of materials modelling based on collating expertise, models, knowledge and data. The materials modelling community consists of several stakeholders that are all represented in the EMMC:

Manufacturers: This group represents the interests of (current and future) end-users of materials modelling in small and large European manufacturing industry. It gathers key company representatives across industrial sectors, from consumer goods to industrial chemicals, from polymers to alloys etc. The objective of the manufactures in participating in the present project is to (i) clearly articulate commercial end-users needs, i.e. articulate what barriers need to be overcome to introduce materials modelling into their business cycle or enable an enhanced and more efficient/higher quality use of it, as well as what future developments they foresee to be necessary, (ii) provide case studies for the testing, validation and assessment of models, workflows, platforms and software. Actions will be taken in order to identify key areas of company interest for materials modelling solutions and how those can be achieved and act as a sounding board and participate in European consultation initiatives.

Translators: The successful application of materials modelling in industry depends heavily on translating industrial problems back into modelling questions, i.e. performing a process in the opposite direction to the value chain. Today, this role is performed by different actors, including R&D staff in large enterprises, application scientists in software companies, scientists in research institutions as well as individual consultants. The aim of the EMMC is to promote activities that optimize the roles of the translators by creating a set of open and transparent conditions and best practices as well as a resource of neutral competences available to all industry more widely, in particular to SMEs.

Software Owners: Software owners are defined as those stakeholders (academic or commercial) who actively make their software available to third parties by a wide range of licensing schemes. This stakeholder group includes academic software owners who offer their software freely as open source code, and proprietary software owners who sell their software to industry. A key objective of software owners is the transfer of materials modelling technology to end users, in particular manufacturing industry. One of the objectives is to identify where the current policies and programmes are supporting the industrial exploitation of academic and proprietary software, where there are gaps and where there are obstacles. The planned activities aim to provide guidelines on quality assurance in software development which aims to support the process of transferring academic software to the manufacturing industry.

Modellers: This stakeholder group consists of the developers of materials electronic, atomistic, mesoscopic and continuum models and respective solvers, as well as developers of coupling and linking schemes. The scope of this stakeholder group encompasses two main tracks of efforts: on the one hand, the stimulation of improved and wider exploitation of existing models, and on the other, the analysis of the state of the art and establishment of a Roadmap for further research necessary for the development of new or improved, more accurate, reliable (yet computationally feasible) models of industrial relevance.

Overall EMMC Approach

The main overarching objective and goal of the EMMC is to allow European Industry to reap the benefits of materials modelling more effectively and vigorously. To this end, the EMMC has identified a number of core activities that are vital to increase the industrial exploitation of materials modelling in Europe. These

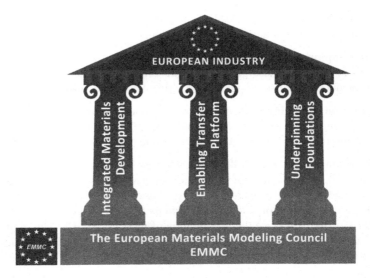

Fig. 2 The three underpinning and enabling actions constituting the three pillars on which advancement of European Industry through advanced integrated materials modelling rests

activities can be structured into three main pillars on which the advancement of European Industry rests on as shown schematically in Fig. 2.

1. **Underpinning Foundations**: Stronger, more robust, better validated and more versatile materials modelling foundations. The existing Foundations in terms of discrete and continuum models, open simulation platforms, interoperability based on metadata schema will be further strengthened, and roadmaps established for future actions.
2. **Enabling Transfer Platform**: Designing the systems and mechanisms to transfer materials modelling to a range of industries. A Transfer Platform will address key technological, organisational and human capital gaps. The underlying main concept stems from the recognition that bringing materials modelling benefits to manufacturers requires a new collaborative and integrative approach that reaches out through the limits of each modelling or manufacturing community. These actions include a European Marketplace hub to ease the access of industry to materials modelling and data repositories, development of the Translators role and function, Training and validation of software.
3. **Integrated Materials Development**: Facilitating a more holistic materials development targeted at achieving tangible impact in different industrial applications. The ultimate goal is that materials modelling and simulation will become an integral part of product life cycle management in European industry, thereby making a strong contribution to enhance innovation and competitiveness on a global level.

The main approach of the EMMC is to establish and enforce these actions by undertaking coordination and support actions for:

Model development and validation—aims at identifying model gaps, establishing Roadmaps to help cure these gaps and validate materials models as a joint activity between modellers, industrial stakeholders, translators and software owners.

Interoperability and integration—aims at providing operational infrastructures for interoperability of different codes and data of materials facilitating the integration of multiple models and tools to build integrated modelling workflows.

Modelling marketplace and data repositories—aims at promoting platforms providing ICT infrastructure for integration of materials models, development of workflows and access to concerted data repositories and collaboration platforms. It will engage in coordination and support actions to raise awareness for curation of materials data and will drive the proliferation of repositories. The Marketplace will also provide a central hub for materials modelling activities in Europe, including Training and Translation services for Manufacturers.

Translation and training from companies—aims at promoting and defining the role of Translators. EMMC will develop methodologies for Translators and translation processes adapted to various industrial sectors. Translators should be well knowledgeable with the capabilities of a wide range of materials modelling paradigms.

Industrial software deployment—aims at stimulating academic software exploitation, conformity to professional standards, user-friendliness and industrial exploitation. The goal is to enhance the level of software development, in particular in academy so as to enable more rapid transfer to industrial applications. Workshops on basic software writing will be recommended for new modellers as well as workshops on version control and testing suites. The EMMC will organize a white paper on software quality and software engineering advice. Furthermore, the activities will enable all stakeholders to optimise the process and the infrastructure which are necessary for the successful and sustainable industrial deployment of materials modelling software.

Industrial integration and economic impact—aims at assessing the economic impact of materials modelling on industrial prosperity, defining and documenting case studies (i.e. documenting successful as well as unsuccessful modelling cases in industry). The goal is to raise and measure the materials modelling impact with a specific emphasis on coordinating the development of Business Decision Support Systems (BDSS). These platforms should allow an end-user to select, connect and through multi-objective optimisation drive for options that make business decisions more data driven and acquire a higher level of efficiency for developing business cases as fitting into the industrial value chain.

From the current materials modelling state of the art the hurdles will be identified through industry consultation ensuring critical input to EMMC. It will ensure a feedback loop of continuous improved materials models and access facilities for advancing the ease of materials modelling implementation in industry. EMMC will

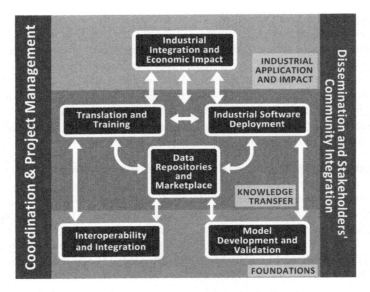

Fig. 3 The interconnectivity and relations between different EMMC actions

bring together Modellers (discrete/continuum), Translators, and Manufacturers to tackle each activity in a collaborative manner, actively contributing to increase the interaction between all stakeholders. The interconnectivity and relations between these activities is shown schematically in Fig. 3.

Industrial Relevance of EMMC Activities

The EMMC will enable industrial end-users to increase the quality and rate of innovation of their products by:

- Improved access to advanced materials modelling tools and databases.
- Enhance interaction and collaboration between all stakeholders.
- Better access to the large pool of knowledge of materials modelling which exists in Europe.
- Higher efficiency through integrated solutions of discrete and continuum models and materials databases.
- Clearly defined metrics by key performance indicators (KPI's).
- Clearer evidence of best practice in materials modelling.

The EMMC will achieve these as a result of:

1. Addressing (and providing roadmaps for) key issues in the underpinning foundations required to build a more integrated, open and accessible landscape, including the materials models themselves as well as their linking and coupling

to support complex workflows, the interoperability and discoverability of models and data.
2. Facilitating an Enabling Transfer Platform consisting of human capital actions (Translators and Training), coordinated and more accessible knowledge sources and improved end-user software development and exploitation.
3. Supporting a more compelling business case for the application of materials modelling in industry via clearly specified, sector specific case studies, quality attributes, cost-benefit analysis and coordination and support for the integration of modelling into industrial decision support pathways.

a. **Materials modelling and related databases will become much more accessible and usable.**

Integration of communities, improved access and usability for a wider range of stakeholders entails agreeing on common nomenclature and metadata schema so that information can easily flow around and is categorised and searched in a unified manner. EMMC will pursue this goal by building on previous and ongoing actions such as the Review of Materials Modelling (RoMM) [2] and the Organisation of Modelling Data (MODA) and Metadata [3]. EMMC will lead to greatly enhanced access and use of materials modelling as a result of: (a) supporting the adoption of coherent terminology, which means that scientists that are not experts in modelling codes will find it much easier to understand and hence access what materials modelling does, (b) metadata schema that achieve a semantic level of interoperability, which means faster integration and easier access for people developing solutions, (c) de facto standards for data structures and (d) modelling platform developments, which increase efficiency; EMMC will coordinate a reference design for Open Simulation Platforms supporting operation across different models, databases and the sub-disciplines of materials modelling. It means that codes can quickly be integrated, hence become more easily accessible and usable, (e) data repository design based on the above metadata schema which will facilitate translation and easier access to information across different fields, and (d) establishment of an ICT platform to manage the materials modelling knowledge and information in Europe, including new sustainable collaboration channels that break through current community barriers. By means of the above actions, EMMC addresses the obstacle that "the interoperability of software and data is a major barrier resulting in limited use of (integrated) simulation software by non-simulation experts" [4]. It will lead to reducing the efforts required for designing advanced integrated workflows impacting therefore the utilisation of materials modelling in industry. In addition, the increased use of databases will allow novel routes to develop new materials with specific properties harvesting existing and future materials data.

b. **Better access to the large pool of knowledge of materials modelling which exists in Europe.**

Industry requires a one-stop-shop, a gateway to all materials modelling relevant information. This includes in the first instance materials data linked to Models and

Software Solutions used to generate the data, validation information on the models and verification of software solutions. Such information allows stakeholders a complete view of the relevance and quality of a particular data element and its applicability in a specific workflow. Added to that will be human resources by linking of data and models to respective Modellers (both creators and general experts) and relevant Translators, allowing Manufacturers direct, more targeted interaction with Modellers and Translators. Finally, this information is augmented with additional documentation resources such as show-cases, education and training, case-studies and linked publication repositories. EMMC will coordinate and network with relevant stakeholders, technology platforms, excellence centres and projects to work towards the design and establishment of a European Materials Marketplace as an open Gateway connecting for information and knowledge management in materials modelling. This will support federated resources of materials data and information repositories, curation of data and information, active collaboration between stakeholders, computational modelling services and education and training. This will impact material modelling in Europe on all levels, in particular enhancing the Human Capital factor by empowering all stakeholders and bringing them together regardless of geographical or community barriers. It will impact the proliferation of knowledge source and material data repositories, speed of publication and dissemination as well as increased novel collaboration channels.

c. **Higher efficiency through integrated solutions of discrete and continuum models and materials databases for industrial use.**

As mentioned in a TMS study [5], successful execution of a project in industry depends upon a broad range of tools and methodologies that must be reliable, cost effective, verified and validated to ensure the accuracy of the results. EMMC will coordinate the definition and acceptance of standards for metadata and simulation platforms supporting the efficient integration of modelling components for different length- and time-scales into comprehensive modelling systems. It will build on foundations laid by the EMMC and the European Multiscale Materials Modelling cluster and ensure that an interoperability level is agreed on, which would form the basis for implementations across all levels of materials modelling as well as materials data. The Roadmap of Materials Models will identify opportunities and gaps for coupling and linking of models. It will be an essential compendium of methods and spawn more targeted research. Within the EMMC, interactions between industrial end-users and the EMMC partners will identify and focus on areas where urgent and highly relevant industrial problems can be addressed with materials modelling of the most mature technology readiness level. In other words, the EMMC will at first emphasise application areas, which have the highest and most immediate industrial and societal impact.

d. **Increased efficiency and industrial effectiveness of materials models in industry and research.**

Many industrially urgent and unsolved problems can be tackled with today's materials modelling capabilities provided that these problems can be translated into modelling strategies leading to tangible result. The EMMC will support the development of the role Translators, who play a critical role in achieving a higher and faster impact of materials modelling. EMMC will support the development of methodologies for Translators, translation processes specific to different industry sectors, and implementation of methodologies for using model workflows, including identification of software components connecting models operating on different scales. A strengthened, more widely available and trustworthy Translator function will enable industry to decide on where and how best to use modelling to greatest effect. Training courses with a focus on how to use the progress in materials modelling and simulation in order to support industrial innovation will contribute to increased efficiency and industrial effectiveness. Training will also increase the awareness in industry about modelling software and data resources available. Furthermore, modelling experts and translators will become more efficient and effective as a result of EMMC interoperability actions as they will be able to choose and utilise models much more efficiently. Further actions of the EMMC that impact efficiency and industrial effectiveness include improved quality and end-user ready (commercial or open) software, easier access to information and knowledge via metadata and marketplace, best practice cases, know-how about what works and what doesn't and hence focus on relevant key performance indicators.

e. **Establishment of technical and business-related quality attributes (Key Performance Indicators) that inspire trust in materials modelling.**

The use of modelling in industry today is hindered by a relatively low 'level of acceptance' which is often due to the difficulty to appreciate what modelling actually can and cannot bring and how exactly it will affect the commercial offering of differentiated products. Consequently, the uncertainty and risk for a major engagement (i.e. investment) in materials modelling is often unclear. This is compounded as no single model or software exists to address all the critical (often conflicting) performance criteria and various options that define a differentiated product offering. EMMC will counter this and create impact by compiling a catalogue of case studies to provide tangible facts demonstrating the impact of materials modelling in particular applications. The EMMC will consider the business case for modelling by spawning a cost-benefit analysis and demonstrate the value of materials modelling by using suitable technical and business related KPIs. This work will build on the survey and report [1] on the economic impact of materials modelling which identified already a range of KPI categories. EMMC will also work closely with the forthcoming BDSS projects and EMMC supported Translators in the definition of technical KPI models as functional of the materials modelling state variables, and coordinate the way in which the technical aspects of an industrial problem are mapped to a multiscale modelling workflow, and represented as a KPI model. The Translator will support the end-user in deciding business oriented quality attributes, which may depend on the modelling state variables as well as for instance costs, and processing time, which usually depend on the material variant currently simulated in the modelling workflow. The role of EMMC in inspiring trust will be to support, coordinate and disseminate demonstration examples of implementation of KPI driven workflows in industrial applications. Also, a training guide for industrial operators using these methodologies will be developed. Further EMMC actions that will contribute to a greater level of trust include laying out clearly the current strengths and current gaps in the Roadmap, a best-practice guide for standardizing the error estimates of materials models, and model validation guidelines.

f. **Industrial best practice (methodologies) for end-users increasing speed of development in industries**

The EMMC will support best practice by a range of actions. Methodology development on Use Cases defined together with manufacturing industry organisations will be carried out by Translators who will define best practice in specific industry applications. The actions on Training support workforce development regarding skills required to launch effective materials modelling supported programmes and will increase the confidence, speed and success of implementation and hence development in industry. The EMMC will work with stakeholders to define the expectations of a materials modelling function in an organisation more clearly (the What, When and How) and define measurable impact levels that this function can be appraised by. Experience from Computer Aided Drug Discovery shows the benefits of this process in terms of trust as well as performance [6]. Working Groups will be formed and Workshops organised to discuss and endorse recommendations and guidelines for best practice. Practitioners will be supported to include a careful economic analysis of the cost and savings associated with implementing materials modelling. EMMC will gather requirements and coordinate the planned Business Decision Support projects to provide both best practice guidelines and a coordinated platform for an economic analysis.

Conclusion

EMMC will undertake the actions in order to close the gap between materials modelling and manufacturing. It will aspire to eliminate or at least significantly contribute to substantially reduce the innovation valley of death by strengthening and incorporating the underpinning and enabling actions into the modelling landscape in Europe. The definition of the role of Translators, the development of data repositories based on standards and interoperability, increased collaboration between all Stakeholders and validation actions are all only but one part of the whole picture.

References

1. G. Goldbeck, C. Court, *The Economic Impact of Materials Modelling, Indicators, Metrics, and Industry Survey* (2016). doi:http://dx.doi.org/10.5281/zenodo.44780
2. https://bookshop.europa.eu/en/what-makes-a-material-function–pbKI0616197/?CatalogCategoryID=lR4KABst5vQAAAEjxZAY4e5L
3. https://emmc.info/moda-workflow-templates/
4. S. Glotzer et al., *International Assessment of Research and Development in Simulation-Based Engineering and Science* (World Technology Evaluation Center, Baltimore, 2009)
5. Integrated Computational Materials Engineering (ICME): Implementing ICME in the aerospace, automotive, and maritime industries. Warrendale, PA: The Minerals, Metals & Materials Society (2013)
6. D. Loughney, B.L. Claus, S.R. Johnson, To measure is to know: an approach to CADD performance metrics. Drug Discov. Today **16**(13/14), 548 (2011)

Facilitating ICME Through Platformization

B.P. Gautham, Sreedhar Reddy, Prasenjit Das and Chetan Malhotra

Abstract Integrated Computational Materials Engineering (ICME) is poised to be integral to the engineering design processes in the future where product engineering will be carried out in close association with materials and manufacturing engineering. This is already being manifested in newer technologies such as additive manufacturing and composite materials where the boundaries between product and process and material are sufficiently blurred. In order to successfully leverage ICME, we need enabling platforms that should allow for seamless integration of product design with process design and materials design and should allow for all three to be investigated, analyzed and optimized simultaneously to be able to obtain the right material for the right product to be manufactured in the right way. It should also provide for a unified and flexible language for expressing the problem domain and allow for the integration of modeling and simulation tools, product and materials databases as well as machine learning, data-analysis and optimization algorithms into the design process. Most importantly, such a platform should be context aware and knowledge enriched. It should provide a strong semantic basis for expressing and capturing knowledge related to the problem domain and a means to reason with this knowledge in a context-sensitive manner to provide context-appropriate guidance to the designer during the design process. The current paper proposes a basic structure for such a platform and how it is being realized as TCS-PREMAP.

Keywords ICME · Model-driven engineering · IT platforms · Ontology · Design workflows · Knowledge engineering

---

B.P. Gautham · S. Reddy · P. Das · C. Malhotra (✉)
TCS Research, Tata Research Development and Design Centre,
54 Hadapsar Industrial Estate, Pune 411015, India
e-mail: chetan.malhotra@tcs.com

Introduction

Motivation

A key challenge for the manufacturing industry is to reduce design time and bring out more differentiated products over shorter cycles. In order to do this, the industry needs to bring in a structured design process, leveraging state of the art modeling and simulation tools and digital technologies while bringing all aspects of design into a single window spanning different departments. Also, the design engineer needs to be empowered with knowledge from a variety of sources such as in-house knowledge, data from past designs, standard procedures, etc. Besides this, the knowledge available with in-house experts on product development as well as modeling and simulation processes also needs to be captured systematically and made available to engineers during the design process.

Integrated Computational Materials Engineering (ICME) [1] is a new paradigm which seeks to bring the cutting edge in modeling and simulation supported by systems engineering and knowledge engineering tools into the product design process. It seeks to integrate design horizontally across the product lifecycle as well as vertically across length scales from a materials perspective. It also seeks to enable greater collaboration between designers and the removal of silos between design departments.

Challenges in Implementing ICME

While significant effort in bringing ICME to the industry would be directed at developing more accurate and sophisticated materials and process models across different length scales, successful adoption of ICME would also require the development of enabling platforms which can successfully integrate and leverage these models to solve industrial problems. The key challenges in implementing an ICME-driven approach to design exist in four broad areas: (a) Integration of simulation tools, (b) Integration of teams, (c) Information handling, and, (d) Standardization and reuse of information at industrial scale.

Integration of Simulation Tools Horizontal integration across the product lifecycle requires tying together different process steps, from sourcing to processing, to quality assurance, to in-use performance and finally end-of-life processing. Design decisions at any process stage could depend on decisions taken at other process stages. Similarly, at any given stage of material design or process simulation, a number of tools at various length/time scales need to be leveraged. For example, in a steel plant, simulation of continuous casting phenomena would involve a fluid-flow and solidification module integrated with a thermo-mechanical module to analyze the solidified section at the macroscopic scale as well as a microstructure evolution module at the microscopic scale supported by modules for thermodynamics and precipitation. It would also need to connect to simulation modules at the previous step of ladle refining and the latter steps of rolling [2]. Hence, information needs to be able to seamlessly travel from one scale to the other and one step to another over the simulation value chain without having to resort to writing tool-specific wrappers every time a tool needs to be accessed.

Integration of Teams Product design and development today is done in silos. The functional design of a product is usually done by the product team, the design of the material to be used within the product is done by another team while the design of the process to make the product is done by a separate team. Design silos with rigid handshakes can prove to be costly resulting in sub-optimal design or overdesign. For example, it is important to know the limitations of the plant and equipment as well as the constraints on cost during the alloy development phase in a steel industry. Similarly, process optimization needs to go hand-in-hand with the material development phase for better realization of products. Thus, there is a need for greater integration between product design, materials design and process design teams and there is a need for a system that is able to constantly shift and merge different design contexts and domains and leverage different ICME tools across the value chain.

Information Handling An engineer at any decision point uses a number of inputs in the design process. In order to structure the design and decision process, we need to establish a system that allows for seamless integration of knowledge sources, simulation tools and decision-making algorithms into the design process. There is a need for a consistent way of handling knowledge, a priori knowledge as well as knowledge that gets generated continuously during the design process such as results from experiments and simulations performed as part of the design process. For example, a number of simulation control parameters constitute an important part of the knowledge base of an industry's design analysis department which are often not persisted and get regenerated every time a new simulation is to be performed. Persisting enterprise knowledge requires appropriate semantic bases for handling information and data and methods for harnessing the generated knowledge by making it available in a context-aware manner.

Standardization and Reuse Product and process engineers and researchers should be able to retrieve and reuse any information captured or generated during a design process that has already been undertaken in the past. There should be a standard way of representing design processes as well as the knowledge and data that goes into them. Ideally, this information needs to be stored hierarchically so that information that is applicable at a higher level should naturally flow to more specific levels in the hierarchy. The process for storing such information has to be inherently flexible and scalable so as to be able to span diverse application domains. This requires standardization of the terminology used, which is largely missing in the materials science engineering community.

How to Address These Challenges?

The above requirements are varied but the underlying common requirement is a highly flexible and scalable IT platform within which information, knowledge and design processes from multiple domains can be represented and integrated in a standardized manner [3]. The platform should provide for interfaces through which multiple users simulation tools and data sources can interact and exchange information. It should allow for creation of complex design workflows which can retrieve information from different knowledge sources and call appropriate simulation tools at appropriate points within the design process. It should provide a means to extract and integrate knowledge from different sources and deliver it to designers in a context appropriate manner to help them make right decisions. This functionality should be available as a reusable framework upon which different domain-specific, design-goal-specific applications can be built.

The needs of the above platform can be summarized as follows:

- A common, flexible and scalable language to represent, span and integrate diverse domains
- Ability to capture design process workflows at an abstract level and use them for solving concrete problems
- Ability to integrate various simulation tools, databases and decision-support systems
- Ability to capture, represent and leverage knowledge during the design process
- Ability to easily create design applications
- Ability to learn from iterations of the design process

Proposed Platform for Implementing ICME

Figure 1 gives the conceptual structure of a proposed software platform for implementing ICME. In the following paragraphs we briefly explain each aspect of the platform and how it has been implemented in TCS PREMAP.

Domain Representation

The foundational layer in the architecture is the Domain Representation layer which provides a common language that can be used to represent all entities within the domain and the relationships between them. This language has to be highly flexible and scalable. The same language should also be able to represent knowledge related to the domain so that it can be seamlessly leveraged while answering questions related to the domain.

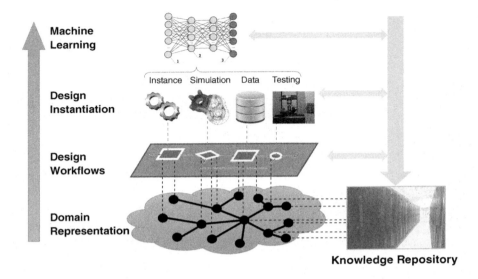

Fig. 1 Conceptual structure of the proposed IT Platform for implementing ICME

Figure 2 provides a high-level view of the domain representation structure of TCS PREMAP. It comprises three hierarchical ontological levels—the meta-level, the subject-level and the instance-level (ontology is a computer-understandable language for representing information). The models at the meta-level are connected and form an integrated set of meta-models. This hierarchical structure of models is important since from a given meta-model, we can create a whole range of subject-level models and from a single subject-level model, we can derive numerous instances. The meta-model definitions form the fundamental language of the platform. The platform then allows users to define their own domain by defining models at the subject-level. Since the model structure allows for object-orientation, any entity at the subject-level could itself be derived from a more generic entity at the subject-level. Hence, the subject-level allows for creating a further internal hierarchy based on specialization. At the instance level in the model hierarchy, individual instances of the components, processes and materials are created and represented using specific values. Hence, for every combination of subject level models, numerous instances can be created for exploring the design space.

Representation of Design Workflows

The process of asking questions related to the domain is via Design or Engineering Workflows. Using Design Workflows, generic questions are asked and the generic process for answering them is represented. For example, we can ask, what is the

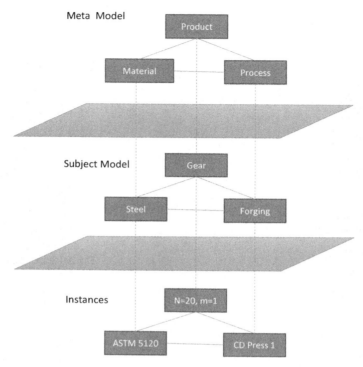

Fig. 2 Hierarchy of ontological levels for domain representation

process followed in a given plant to troubleshoot a particular problem in product quality. What are the steps involved and how can the decisions be taken to systematically eliminate possible sources of deviations in the quality. The process that is followed in executing the above steps, taking decisions and systematically coming up with the solution is represented as a Design Workflow. While assembling Design Workflows, one needs to access and refer to different entities in the domain. Hence, the Design Workflows need to be tightly integrated with the Domain Representation layer at every step of creating the workflows.

In TCS PREMAP, design workflows are represented using a workflow meta-model consisting of constructs to represent steps, connectors and gateways. Workflow steps indicate the processing steps in the workflow. They could represent actions such as user interaction, simulation, database lookup, invocation of a rule, a decision-support algorithm, etc. Gateways provide a means to model decision flows. A diverging gateway represents a decision point where alternate paths can be taken based on decision variables. A converging gateway merges decision paths. All data elements in the workflow are mapped to domain ontology entities thereby providing semantics to the information exchanged between processing steps in the workflow.

Design Instantiation

Once abstract workflows for asking and answering design questions are set up, one can create instances of the workflows. While creating instances of design workflows, the corresponding entities in the domain also get instantiated. For example, if we are currently restricting our interest to the manufacturing of a gear, then through the workflows we have a way of asking questions related to the manufacture of the gear, but before we do that, we need to actually instantiate a specific gear. At this stage, we also need to instantiate the sources of information and decision enablers that would be flowing into the design workflows. For example, to be able to take a certain decision during the design of the gear, a simulation needs to be performed to predict the stresses at a certain point within the gear [4]. At the abstract workflow level, we only need to specify that there exists a step where a simulation will be performed which will predict the given stress for which a location has been created during the domain representation. During the instantiation step, the actual connection to the simulation tool gets established, the data that is expected by the simulation tool is passed in the format that it expects it in, the results of the simulation such as internal stresses, etc., are fetched and are stored at appropriate locations within the instance of the gear that was created as part of the instantiation step and the decision step based on these stress value is taken. Simultaneously, as part of the design procedure, one may need to access data sources which may be passive data stored earlier or could also be live data that is being gathered from the plant or an online data repository.

Within TCS PREMAP, an instantiated design workflow executes in the context of the design problem which is created by instantiating the corresponding subject-model entities. The life cycle of these instantiated entities is managed by a set of classes (e.g. Java classes) generated from the subject model. These entities carry information pertaining to the design problem on hand. The generated classes manage the storage, retrieval, persistency and consistency of this information. User screens are also automatically generated by mapping screen controls to subject model entities. The generated screens store and fetch user inputs and outputs from the corresponding entity instances. Similarly simulation service wrappers are also automatically generated by mapping a simulation tool's inputs and outputs to subject model entities.

Learning from Past Design Runs

Over time, a designer would have performed a large number of executions of the instantiated design workflows. A particular workflow would be run with different input design parameter values and resulted into different sets of results. The next step is to establish a model between the inputs and the results using machine learning algorithms which can be stored in the Knowledge Repository so that it can

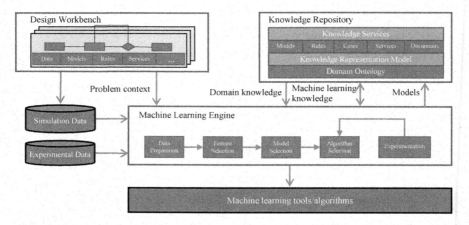

Fig. 3 Overview of the machine learning framework

be included as a check in future design workflows for pre-eliminating designs that are highly likely to be sub-optimal. Learning from past design problems and using these learnings for future design problems is an essential ingredient of a next-generation engineering-design platform..

Figure 3 gives an overview of the machine learning framework proposed in TCS PREMAP. As problems are solved in a design process, the data generated in the process along with the associated design context are captured in a database. This data is then mined to learn models using machine learning techniques. The machine learning process itself has several stages—data preparation, feature selection, model selection, experimentation, etc. All these are domain-knowledge dependent, and this knowledge is context dependent. The machine learning framework uses context-based knowledge retrieval to retrieve the relevant knowledge to make these decisions. After models are learnt, they are stored back into the knowledge repository under the same context. These models can then retrieved and reused in the future when working on similar design problems.

Knowledge Capture, Representation, Retrieval and Inclusion into the Design Process

Just as a machine-learnt model can be stored in the Knowledge Repository, the design workflows themselves constitute knowledge and can also be stored in the Knowledge Repository and so do workflow instantiations which can also be promoted to the Knowledge Repository as instances of successful/unsuccessful designs. Hence, the designs, the way they are arrived at, and the learnings during the design process, all become knowledge available for leveraging during future design runs and can be stored in the Knowledge Repository.

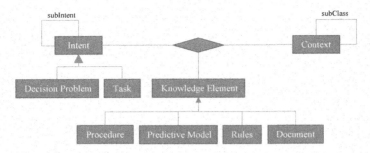

Fig. 4 Knowledge model—Overview

The goal of the knowledge engineering framework in TCS PREMAP is to enable knowledge-guided decision making within design process by delivering context-appropriate knowledge to designers to help them make right design decisions. We use a knowledge modeling approach to achieve this. Figure 4 gives a high-level representation of the knowledge model [5]. Knowledge serves an intent in a context. The intent could be to solve a decision problem or to perform a task. Knowledge could be in the form of a procedure, a predictive model, a set of rules, etc., that need to be executed to serve the intent. Knowledge could also be represented by a document that provides necessary information to the designer. The domain ontology provides the vocabulary to express knowledge.

The knowledge retrieval process starts by identifying the designer's intent. It then collects information relevant to the intent from the design process. This information is used for identifying the matching context and retrieving the knowledge relevant for the <intent, context> combination. The retrieved knowledge is then suitably adapted to the context of the design process and processed (if it is a document, it is simply shown to the designer).

Conclusions

The manufacturing industry is poised to bring in the next wave of digitization in its design processes. ICME represents a new paradigm where silos between departments are removed and collaborative design is effected using seamless inputs from simulation tools, databases, knowledge engines and decision-support algorithms. However, in order to effectively leverage ICME, there is a requirement for a platform which provides for a common language to capture design information as well domain knowledge, a way to capture the design steps and decisions in reusable workflows, easy connection to different simulation tools and data sources, and the integration of past as well as machine-learnt knowledge into the design process.

In the current paper we have elaborated each of these components in terms of their structure, technology and function and have also shown how they can be combined to effect the next generation of product, process and materials design using the TCS PREMAP platform.

References

1. Integrated Computational Materials Engineering: A Transformational Discipline for Improved Competitiveness and National Security, *NRC Report* (2008) (The National Academies Press, National Research Council, Washington, DC, 2008)
2. J. Allen, F. Mistree, J. Panchal, B.P. Gautham, A.K. Singh, S.S. Reddy, N. Kulkarni, P. Kumar, Integrated realization of engineered materials and products: a foundational problem, in *Proceedings of the 2nd World Congress on Integrated Computational Materials Engineering* (ICME 2013), Warrendale, PA: The Minerals, Metals & Materials Society, pp. 277–284 (2013)
3. B.P. Gautham, A.K. Singh, S.S. Ghaisas, S.S. Reddy, F. Mistree, *PREMAP: A Platform for the Realization of Engineered Materials and Products* eds. by A. Chakrabarti, R.V. Prakash. ICoRD'13, Lecture Notes in Mechanical Engineering (Springer India, 2013), pp. 1301–1313
4. B.P. Gautham, N. Kulkarni, D. Khan, P. Zagade, R. Uppaluri, S. Reddy, Knowledge Assisted Integrated Design of a Component and Its Manufacturing Processes, in *Proceedings of the 2nd World Congress on Integrated Computational Materials Engineering* (ICME 2013), Warrendale, PA: The Minerals, Metals & Materials Society, pp. 291–296 (2013)
5. R.R. Yeddula, S. Vale, S. Reddy, C.P. Malhotra, B.P. Gautham, P. Zagade, A Knowledge Modeling Framework for Computational Materials Engineering, in *Proceedings of The 28th International Conference on Software Engineering & Knowledge Engineering, SEKE* (2016)

Bridging the Gap Between Bulk Properties and Confined Behavior Using Finite Element Analysis

David Linder, John Ågren and Annika Borgenstam

Abstract Theoretically and empirically based models of materials properties are crucial tools in development of new materials; however, these models are often restricted to certain systems due to assumptions or fitting parameters. When expanding a model into alternative systems it is therefore necessary to have sufficient experimental data. When working with composite or highly confined materials, such as layered structures or coatings, this can be problematic as most available data is on bulk materials. The present work displays the potential of using Finite Element Method (FEM) simulations as a tool to understand experimental observations and expand existing models to new systems using only bulk properties of the constituent phases. The present work focuses on the effect of geometrical constraints on the indentation behavior of elasto-plastic materials as an example on how FEM may be used to better understand experimental observations in composite or layered materials. The results may also be integrated into phenomenological models, expanding their application range.

Keywords Indentation behavior · Confined hardness · Finite element analysis

Introduction

The Finite Element Method (FEM) is a commonly used computational method, apart from simulations such as properties of components and their behavior during application [1], the method may also be used in computational reproduction and

---

D. Linder (✉) · J. Ågren · A. Borgenstam
Department of Materials Science and Engineering, KTH Royal Institute of Technology, Stockholm, Sweden
e-mail: davlind@kth.se

J. Ågren
e-mail: john@kth.se

A. Borgenstam
e-mail: annbor@kth.se

simulation of materials testing, such as hardness measurements [2] and tensile testing [3]. Single phase and composite material models are implemented in commercial software such as Abaqus. The implemented composite material models are practical when dealing with materials where the macroscopic properties are known and the overall material response to e.g. mechanical load is of interest. In order to achieve a better understanding of the influence of relative phase fraction and properties of the constituent phases on the macroscopic behavior of a composite material, actual composite microstructures with explicitly defined phases, with corresponding properties, may be used in simulations.

In many composite materials, the size (or thickness) of the constituent phases can vary down to the range of a few hundred nm or even less. At these length scales, the deformation behavior of a phase will differ from that in a bulk material. Models that are based on elastic mismatch or differences in hardness at larger scales may thus not be valid in these materials. As materials development progresses towards more complex structures, it is not sufficient to study only thin films or wires, one also need to consider the effect of 3D-constraints and confinement.

The aim of the current study is to take the first steps towards expanding existing methods to simulate indentation hardness to 3D-confined materials by FEM using different axisymmetric models with geometries ranging from bulk samples to large-sheet thin films and finally thin films confined by surrounding material.

Theoretical Background and Practical Considerations

Material properties are closely related to the microstructure of a material, which in turn is a result of chemical composition and processing parameters such as cooling rate. There are many ways to include these effects into hardness modeling, for example by the use of empirical relations for specific systems or more general models such as models on solution strengthening [4] and precipitation hardening [5]. Empirical and semi-empirical models linking bulk properties and geometrical factors, such as film thickness, to measured hardness [6, 7] are very practical when considering bulk properties and macro hardness or the effective hardness of a specific geometry. Furthermore, the well-known size effect of small depth indentation, i.e. indentation depths in the range of micrometers and below, has been modeled by Nix and Gao [8] by linking measured hardness values to a characteristic length-scale. In order to incorporate the effect of nano-scale indentation (in the range of 100 nm and below) Huang et al. [9] adjusted the model by including a second characteristic length-scale. They also incorporated the size-effect into FEM-modeling by limiting the flow stress through the application of geometrically necessary dislocation (GND) density. The modified model shows a good fit to experimental data even at indentation-depths below 100 nm. Despite the existence of models to describe the size-effect during very shallow indentation, this is rarely incorporated in the models used for e.g. thin film hardness but Beegan et al. [10] have shown how to do this for the Korsunsky [6] and Puchi-Cabrera [7] models.

In order to correctly model indentation hardness in composite materials all the different effects such as composition, constituent phases, area function of the indenter, material pile-up/sink-in around the indenter and geometrical constraints (in 3 dimensions), as well as the indentation size effect, need to be incorporated. In this work, several simplifications have been made to limit the different contributions to the hardness in order to study the effects of geometrical constraints on the indentation behavior of a soft phase embedded in a hard phase. In order to verify the model and the material parameters, experimental bulk and thin film indentation data have been used as comparison.

Computational Method

As the focus of this work is on the influence of geometrical constraints on indentation in soft materials, the hard phase is for simplicity assumed to act as rigid and unmoving and may thereby be replaced in the FEM model by fixed boundaries of the soft phase. Further, the bulk properties of the soft material is averaged over the entire volume and thereby assumed to be homogeneously distributed, isotropic and size-independent. As the bulk properties are taken from experimental data this may not be an entirely correct description as e.g. precipitates may influence the local properties, however, as the aim is to qualitatively model the indentation behavior this should not be of great concern.

The simulations in this study have been performed in Abaqus/CAE 6.14, using CAX4 elements (axisymmetric stress) with full integration. A mesh of quad-dominated elements was used in the analysis. A fine mesh is applied just beneath the indenter and, in order to reduce the computational time, a coarser mesh is applied further from the indenter, as shown in Fig. 1. An axisymmetric model

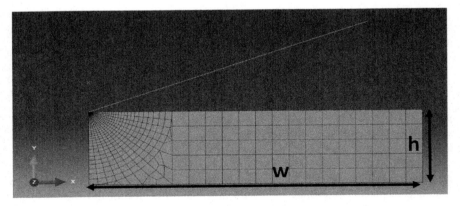

Fig. 1 Example of the "3D-confined" geometry used in the FEM-analysis, w and h mark the variable width and thickness of the sample material

was built to resemble a nanoindentation setup with a conical indenter where indentation is performed in the center of a sample. The conical indenter has a half-angle of 70.2° to approximate nanoindentation with a Berkovich tip [11]. For simplicity the indenter has been modeled as rigid, this is a fair assumption for indentation for soft materials but needs to be used with caution for harder materials. The bottom surface is fixed in all directions and the side is fixed in the indentation direction. The indentation is displacement controlled and the analysis is divided into a loading and a unloading step.

The indentation hardness (H) is calculated by

$$H = \frac{F}{A} \quad (1)$$

where F is the load from the indenter and A is the contact area between the indenter and the sample which often is described by a indenter-specific area function that needs to be corrected according to the quality of the tip and the sink-in or pile-up of material around the indent. The contact area (A) is directly calculated by the model, thereby avoiding the problems related to sink-in and pile-up, and the reaction force on the indenter is used as the indentation load (F). F and A are saved for each time-step and then used to calculate the hardness (according to Eq. (1)) as a function of indentation depth.

Material Properties in the Model

As we are interested in elasto-plastic materials it is necessary to define the plasticity of the material in addition to the Young's modulus and Poisson ratio. In these simulations the Hollomon relation [12],

$$\sigma = K * \varepsilon^n \quad (2)$$

has been used to describe the plastic response. σ is the true stress, ε the true strain and n the strain hardening exponent. K is the strength coefficient calculated as

$$K = \frac{\sigma_Y}{\left(\frac{\sigma_Y}{E}\right)^n} \quad (3)$$

where E is the Young's modulus and σ_Y is the Yield stress of the material. The strain hardening exponent has been fitted to experimental hardness values or taken directly from references (as in the case of the Al listed in Table 2).

In order to find, and verify, the strain hardening exponents (n) for different materials, bulk indentation has been performed to compare experimental data to the calculated values. These n are then used in the investigation of the effect of geometric constraints on the indentation behavior of the materials. Three types of confined geometries have been used in this study; "bulk" geometry for verification

Table 1 Geometrical parameters used for the different geometries

Geometry	Abbreviation	w (nm)	h (nm)
Bulk	B	10,000	10,000
Thin film	TF250	10,000	250
Thin film	TF500	10,000	500
3D-confined	3D	1,000	250

and reference hardness values, "thin film" geometry where hardness is evaluated with respect to material thickness and "3D-constrained" geometry to investigate the combined effect of width (w) and thickness (h) on indentation behavior. Figure 1 shows an example of a geometry and mesh used in the analysis. The geometric variables w and h are marked by black arrows. The geometrical parameters used in this study are listed in Table 1.

Results and Discussion

Indentation of Bulk Materials

The calculated bulk hardness values of different materials are listed in Table 2 along with the corresponding material parameters. The hardness values calculated for Ni (soft) and Ni (hard) both correspond well to the lower and upper limit of experimental Ni hardness according to [13]. The bulk hardness of Al also fits the values reported in [2] reasonably well. The stress fields below the indenter reached as far as approximately 10 times the indentation depth, i.e. roughly a quarter of the total depth of the bulk geometry, thus confirming that the bulk indentation should be unaffected by the boundaries.

Indentation of Thin Films with Rigid Constraints

Figure 2 shows the hardness increase (relative to bulk hardness) with increasing indentation depth in Ni and Al thin films (250 nm thick) when applying rigid boundary conditions. There is a clear difference in the hardening behavior of the

Table 2 Materials and parameters used for the FEM analysis as well as the resulting bulk hardness value using conical indenter with a half-angle of 70.2° and 225 nm indentation depth

Material	Ref.	Young's modulus (GPa)	Yield stress (MPa)	Poisson ratio	Strain hardening exponent	Simulated hardness (GPa)	Experimental hardness (GPa)
Ni (soft)	[13]	200 [13]	70 [13]	0.3 [13]	0.25 (fit)	0.75	0.78 [13]
Ni (hard)	[13]	200 [13]	115 [13]	0.3 [13]	0.25 (fit)	1.08	1.08 [13]
Al	[2]	70 [2]	168 [2]	0.33 [2]	0.05 [2]	0.45	0.52 [2]

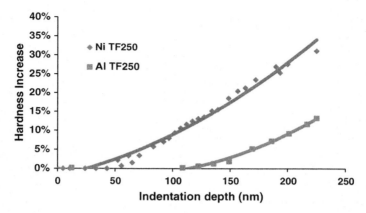

Fig. 2 Hardness increase compared to bulk hardness as a function of indentation depth for 250 nm thick thin films of Ni (*orange diamonds*) and Al (*blue circles*) using fully rigid constrains

two materials with the Ni film showing an increase of almost 30% at an indentation depth equal to 90% of the film thickness whereas the Al film only shows an increase of around 15%. The relative hardening is thereby related to the bulk properties of the material. Given that this is the hardening when using rigid constraints, this should be the upper limit of hardening for these materials using very stiff substrates. Comparing to the experimental data from Saha et al. [2], where they measured the hardness of an Al-film on a glass substrate, a 15% increase in hardness seems like a reasonable value at this relative indentation depth. As it is possible to reproduce these results rather well alternative geometries may also yield reasonable results.

Indentation of 3D-Confined Materials

The effect of 3D-constraints (radius of 1 μm) is shown in Fig. 3. The comparison to a thin film of the same thickness (250 nm) shows that the influence of 3D constraints could have a rather significant influence on the hardness of the material. At small indentation depth the influence is negligible; however, as the normalized indentation depth reaches around 75% of the film thickness, the effect of the 3D constraints becomes noticeable. The hardening at 90% normalized indentation depth is increased from around 15% to around 20%. This effect may be explained by the confinement of the stress-fields as the material is deformed. Figure 4 shows the Mises-stress of the thin film and the 3D-confined geometries at small (a, b) and large (c, d) indentation depths (d) compared to the film thickness (h). At small indentation depth, the stress fields are the same for the two geometries. At larger indentation depth, the stress fields reach the fixed edge of the 3D-geometry and are thereby unable to propagate further, this leads to a stress concentration in the already deformed material.

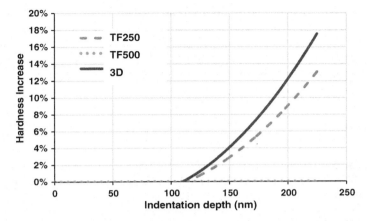

Fig. 3 The effect of 3D constraints (*orange*) on the indentation hardness of Al with respect to indentation depth compared to a thin films with the thickness 500 nm (*green*) and 250 nm (*blue*), the radius of the 3D-constrained material is 1 μm

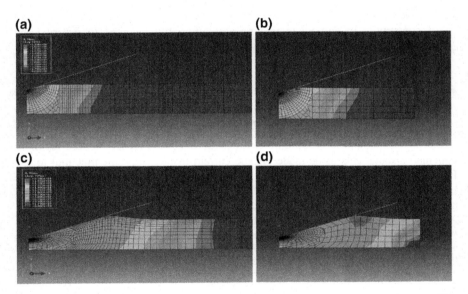

Fig. 4 Mises stress fields in TF250 (**a**, **c**) and the 3D (**b**, **d**) geometries of Aluminum at 20% (**a**, **c**) and 80% (**c**, **d**) normalized indentation depth. The stress states are color-coded with *blue* corresponding to 0 MPa and *red* corresponding to 210 MPa

The qualitative (if not yet quantitative) results of thin film hardness is captured by the model, by further extending this to "3D-constraints" the experimentally observed indentation behavior in composites (such as cemented carbides where the measured binder hardness [14] is significantly higher than corresponding bulk hardness [15]) may be better understood.

The effect of confinement on the nano-indentation (with indentation depths in the range of a couple of hundred nm) hardness seems to be limited unless the confinement is in length-scales of 0.5 µm and below. However, this is a range which is found in many composite materials and the effect is therefore important to consider when modeling the mechanical properties of such composites.

Summary

The indentation behavior of confined materials have been investigated by FEM-analysis. Different geometries have been used to investigate the influence of geometrical confinement and the results show the potential of using FEM to achieve a better understanding of the mechanical response of composite materials as well as the constituent phases. The results must, however, be interpreted with some caution due to the simplifications made in the model. The effects of rigid constraints on the indentation behavior for different materials can be summarized as follows:

- The use of rigid constraints should be able to provide an upper limit of the effect of constraints in composite materials and possibly be used to describe the indentation behavior in composites with very large difference in stiffness of the constituent phases. Comparison to experimental data suggests that this may be the case.
- The influence of 3D-constraints on the indentation behavior becomes more significant the larger the indentation depth as the deformation and the stress fields interact with the confining boundaries.

Further investigation of the influence of different parameters is needed for more conclusive statements to be made but the current results show the potential to use FEM to better understand and predict the mechanical response of constituent phases in composite materials.

Acknowledgements This work was carried out in the NoCo project, within the strategic innovation programme Metallic Materials, co-financed by the Swedish Innovation Agency Vinnova, and in collaboration with the VINN Excellence Center Hero-m. The authors would like to express their gratitude towards Prof. Jonas Faleskog (KTH Royal Institute of Technology) for his expertise and helpfulness.

References

1. The Minerals, Metals & Materials Society (TMS), Modeling Across Scales: A Roadmapping Study for Connecting Materials Models and Simulations Across Length and Time Scales (TMS, Warrendale, PA, 2015). Electronic copies available at http://www.tms.org/multiscalestudy
2. Saha et al., J. Mech. Phys. Phys. Solids **49**, 1997–2014 (2001)

3. Joun et al., Comp. Mat. Sci. **41**, 63–69 (2007)
4. Fleischer, Acta Metal. **11**, 203–209 (1963)
5. Mott et al. Proc. Phys. Soc. **52**, 86 (1940)
6. Puchi-Cabrera, Surf. Coat. Technol. **160**, 177–186 (2002)
7. Nix et al., J. Mech. Phys. Solids **46**, 411–425 (1998)
8. Huang et al., J. Mech. Phys. Solids **54**, 1668–1686 (2006)
9. Korsunsky et al., Surf. Coat. Technol. **99**, 171–183 (1998)
10. Beegan et al., Thin Solid Films **516**, 3813–3817 (2008)
11. Sakharova et al., Int. J. Solids Struct. **46**, 1095–1104 (2009)
12. Trans Hollomon, AIME **162**, 268 (1945)
13. CES Edupack 2015 (Granta Design Limited, 2015)
14. Engqvist et al., Tribol. Lett. **8**, 147 (2000)
15. Roebuck et al., Mat. Sci. Eng. **66**, 179–194 (1984)

Ontology Dedicated to Knowledge-Driven Optimization for ICME Approach

Piotr Macioł, Andrzej Macioł and Łukasz Rauch

Abstract Development of new materials, products and technologies with the ICME approach requires challenging computations, controlled by optimization algorithms. A computational time might be decreased with a "knowledge-driven optimization"—an optimization process is controlled not only by a numerical algorithm, but also by a Knowledge Based System. That requires development of a common language, able to cover communication between numerical models without sophisticated translators. There are several formalisms of knowledge representation, but the most common ones are based on First Order Logic (FOL) and Description Logic (DL). None of them meets all the requirements of knowledge management in ICME processes. We present an approach to development of an environment for knowledge management, combining DL and FOL. An exemplary multiscale problem is described, as well as an OWL2 based ontology and rules controlling an optimization process.

Keywords Ontology · Multiscale modelling · Integrated models · Knowledge based system · Knowledge-driven optimization

Introduction

Integrated Computational Materials Engineering (ICME) is an approach to design products, technologies and materials within an integrated environment. The goal is to find the best available combination of a material and a technology to manufacture

---

P. Macioł (✉) · Ł. Rauch
Faculty of Metals Engineering and Industrial Computer Science, AGH University of Science and Technology, al. A. Mickiewicza 30, 30-059 Kraków, Poland
e-mail: pmaciol@agh.edu.pl

Ł. Rauch
e-mail: lrauch@agh.edu.pl

A. Macioł
Faculty of Management, AGH University of Science and Technology,
al. A. Mickiewicza 30, 30-059 Kraków, Poland
e-mail: amaciol@zarz.agh.edu.pl

a desirable product. One can describe the ICME process as an optimization, with a multiscale model applied to evaluate an objective function. A successful design process requires resolving of two fundamental problems—the high computational demands (arising from high computational requirements of multiscale models, multiplied by running in an optimization loop) and the integrity of knowledge involved in the process, arising from complexity and heterogeneity of ICME processes. A technological process is a system of several sub-processes. All of the mutual dependencies, the constraints and the sources of ambiguities should be taken into account. Hence, management of knowledge, taking into consideration all the aspects of materials, processes and modeling issues is a necessary component of ICME.

ICME requires also coherent software tools for designing and modeling, meeting two contradictory requirements: the flexibility (to be able to support diverse technological challenges) and the internal consistency (to ensure a seamless communication between various numerical models). Furthermore, both decreasing of a number of runs necessary to achieve an optimization goal and decreasing of computational costs of a single run are most welcome. Initial iterations of optimization do not require as high reliability as later ones. Moreover, in multi-step processes, not all stages need to be modeled with high accuracy. Hence, overall computational costs may be decreased by adaptation of the numerical model to the actual requirements [10, 12]. Typically, interfaces between sub-models are analyzed by researchers and hard-coded within integrating procedures. This approach fails when the number of sub-models grows. Alternatively, the modeling environment should be able to automatically agree the communication interfaces between sub-models. This task is reachable only when a common language is used for all the involved interfaces.

Run-time adaptation of the multistep numerical model itself is a non-trivial task. First, the applicability range of the sub-model in question must cover the expected process conditions. Secondly, the sub-model reliability and the accuracy must harmonize with a distance of a particular optimization step from the quasi-optimum of the problem. If more than one sub-model fulfill both conditions, the less computationally demanding one is chosen. In our approach, a structure of an integrated model is controlled by a Knowledge Based System (KBS).

Summarizing, the core of the proposed approach is knowledge management. There are three perspectives: (i) the management of knowledge of a process, a material and modeling techniques, (ii) the information passing between sub-models and (iii) the rules for an optimal choice of components of integrated model of a process.

In this paper an example based on an optimization of a manufacturing process for auto body parts made of the Advanced High Strength Steels (AHSS) is discussed. The detailed description of the process can be found in [16]. The discussed process consists of a multi-pass hot rolling (the first step) and a laminar, controlled cooling (the second step). The hot rolling step is aimed at reducing thickness of a sheet, as well as preparing a requested microstructure and thermal conditions for the laminar cooling step. The second step is aimed at obtaining a required microstructure. In the original work, for the hot rolling, a simplified model considering a number of passes and cooling conditions between passes had been applied. Empirical equations for phase transformations (the Avrami approach) had been used for modeling of microstructure evolution during the laminar cooling.

Knowledge Management

All the introduced issues are the Knowledge Management problems [4]. Knowledge management can be considered as a methodology of management of information and knowledge in institutions or as a system of IT tools, supporting or partially replacing human beings, executing tasks requiring intelligent behavior. Methods of knowledge representation are divided into two groups, the *ontological* and the *rule-based* ones. Ontology is a kind of conceptualization (called specification of conceptualization) [2, 3]. In the presented research this form of a knowledge representation is applied for management of knowledge of process, material and modeling techniques and for information passing between autonomous sub-models.

Ontologies, especially those based on Description Logic (DL) are strongly connected to the concept of *Semantic Web*. They are also frequently applied in other fields. Unfortunately one of the important limitations of DL is the lack of effective and reliable reasoning mechanisms. Hence it is necessary to integrate an ontology language with a rule-based one. Simultaneously, expressing of the rules with more than one variable is difficult if possible. This limitation is not present when First Order Logic (FOL) and Horn clauses are applied (e.g. [5, 8, 9]). Unfortunately, FOL has its own limitations, connected mainly to difficult describing of relations between predicates—what can be easily and naturally done with *concept hierarchies and roles* in DL. The need of integration of rule-based and ontological languages had been identified by designers of the Semantic Web. The response to this issue was initially development of the *RuleML* language and later the *Semantic Web Rule Language* (SWRL), a "bridge" between DL and FOL. Unfortunately expressiveness of the SWRL is large enough to make this language undecidable. In effect, the hope of eliminating of the FOL limitations with DL had failed. Recently several hybrid solutions has been proposed (e.g. [1]). They are focused on automatic conversion of DL knowledge to the Horn clauses form. Such approach is not suitable for the given problem due to the need of defining relations between concepts (classes), what is difficult or even impossible with DL. Hence, in the proposed approach, the original knowledge base grounded on the ontology is extended with a model describing rules arising from these relations. This intermediate ontology module is applied for developing of the knowledge base compatible with the REBIT System, based on Horn clauses and developed by AGH-UST [12].

Ontology

The first part of the presented research has been focused on formulation of the ontology, classifying a description of a problem domain with DL. Among the available DL languages and the adequate IT tools, OWL2 language was chosen. In DL, a domain is described by *concepts, hierarchies of concepts* and *roles* (specific relations between concepts and individuals of concepts). OWL2 is a realization of these assumption,

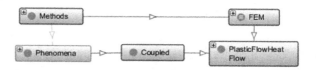

Fig. 1 The object properties defined for the different abstraction levels

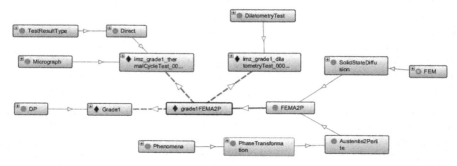

Fig. 2 The exemplary individuals and instantiated object properties

with slightly different nomenclature (a *concept* in DL is a *class* in OWL2, a *role* in DL is an *object property* in OWL2), simultaneously extending capabilities of domain describing (i.a. via introducing of *data properties*). OWL2 do not require strict distinguishing between the so-called *TBox* (a terminological box, a structure of knowledge) and *ABox* (an assertional box, particular cases). A knowledge modeling expert decides, is the particular real object a class or an individual.

In this paper only the subset of the whole ontology is presented. The whole ontology is available online [11]. There are five main "categories" of classes: Material, Experiment, Model, Technological process and Simulation (of technological process). Each of these "superclasses" includes hierarchies of subclasses, connected with *is-a* relations. The second component of the ontology is a set of *object properties*, binding particular classes. These properties can be defined on different abstraction levels. For example, the object property *solvesProblemOf* binds the abstract classes *Method* and *Phenomena*. The subproperty *solvesCoupledPlasticFlowHeatTransfer* binds the *FEM* (subclass of *Method*) and *PlasticFlowHeatFlow* (subclass of *Phenomena*) classes (Fig. 1). The class hierarchies and the relations between them (object properties) defines permissible items, which can be used by knowledge developers to precisely and unambiguously define particular cases. This task is done by adding *individuals* to particular classes. Besides classes, also object properties are instantiated, binding individuals belonging to particular classes (Fig. 2).

Numerical Models

The integrated model of the process consists of two modules, the hot rolling and the laminar cooling. Detailed description of the integrated model can be found in [14]. The hot rolling module takes into consideration heat flow and plastic deformation in macroscale and dynamic recrystallization in mesoscale. Depending on expected accuracy, 1D, 2D or 3D FEM-based models may be used in macroscale and an Avrami-based Internal Variable model, a Cellular Automata (CA) model or an Artificial Neural Network (ANN) based metamodels in mesoscale (description of the metamodeling technique can be found e.g. in [6]). The laminar cooling module may use 1D/2D/3D FEM models for macroscopic heat transfer, Avrami/CA/ANN models and a 2D FEM model of carbon diffusion for phase changes modeling in mesoscale as well as an empirical equation and a thermodynamic model for precipitation kinetic in mesoscale. Each of the sub-models is described as an individual within the ontology. The description includes applicability ranges, accuracies, predictive capabilities and computational requirements. Detailed example of this approach (thermo-mechanical treatment of aluminum alloy) can be found in [13].

Generation of Rules with an Ontology

As mentioned above, three phases may be distinguished in knowledge management. While the DL solution is sufficient for two first tasks, the remaining one, configuring of the multiscale/multistep model during the optimization process requires a different approach. The structure of the numerical model is modified by the KBS. Hence, the KBS must be able to unambiguously answer the question "which sub-model should be used". Since DL languages are not able to meet this requirement, FOL reasoning engines must be used. Unfortunately, automatic converting of DL knowledge into FOL rules is very difficult if possible. In the presented approach, an additional layer, transcribing DL-based knowledge into rules is developed.

The first step of translation is converting of the generic ontology into the functional one, compatible with FOL rules. The original classes, the object properties and the data properties are converted to "templates" of rules. Each generic object property is transformed into the set of functional object properties. In our research we applied REBIT System, developed by AGH-University of Science and Technology. Its leading idea is representing of facts with variables. The vocabulary of REBIT System consist of variables defined by standard or predefined types by different multiplicity and ordering, user-defined functions, constants, terms representing instances of variables, constant and function in *atomic formulas* (*an atomic formula*—elementary logical expression consisting of a left hand term, a right hand term and the relational operator) and finally rules consisting of a precondition (conjunction of atomic formulas) and a conclusion assigning a new value to the selected variable. The excerpt of OWL2 model of REBIT vocabulary is shown in Figs. 3, 4 and 5.

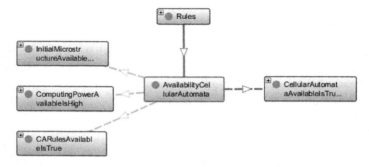

Fig. 3 The *AvailabilityCellularAutomata* rule

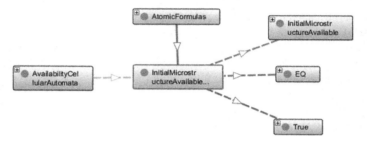

Fig. 4 The exemplary precondition's atomic formula

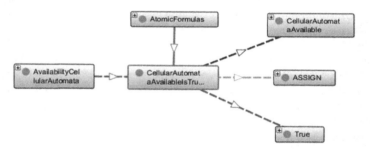

Fig. 5 The exemplary conclusions atomic formula

The generic ontology is a base for the functional ontology. Each item in the generic one is converted to a set of preconditions and a conclusion, but without particular values, which must be entered by a domain expert, acquired from an existing databases or evaluated by knowledge acquisition methods. The generic ontology provides consistency of the functional ontology and subsequently the rule-based knowledge base. Each variable in the knowledge base has its counterpart in the generic ontology ("leaves" of class hierarchies in the generic ontology) or, in some cases, the proper individual. Similarly each rule (relation between precondition and conclusion) has its counterpart, an adequate object property in the generic ontology.

```
RULE AvailabilityCellularAutomata
IF InitialMicrostructureAvailable = True AND
ComputPowerAvailable = High AND CARulesAvailable = True
THEN CellularAutomataAvailable = True

RULE AvailabilityInitialMicrostructure_01
IF InitialMicrostructureFromMicrograph = True
THEN InitialMicrostructureAvailable = True

RULE MicrographAvailable
IF GradeMicrograph = GradeComputation AND
GrainSizeCompare(MeanGrSi, MeanGrSiComp, Deviation) = True AND
PhaseComposCompare(PhaseCompos, PhaseComposComp) = CloseEnough
THEN InitialMicrostructureFromMicrograph = True

RULE CARules
IF EstimatedProcessTemperature IN CAModelTemperatureRange AND
EstimatedCoolingRate IN CAModelCoolingRateRange AND
MaterialGradeRules = MaterialGradeComputation
THEN CARulesAvailable = True
```

Fig. 6 The exemplary rules for the functional ontology

In the Fig. 6 the exemplary rules generated from the functional ontology are presented. The REBIT rules model provides reliable reasoning mechanism, based on deductive reasoning principles. Our experiences show that knowledge base model presented above allows also reasoning when variables must be represented with linguistic or ranged values, on the ground of fuzzy reasoning [7, 10].

The presented problem could be, with some restrictions, treated as a Multi-Criteria Decision Making (MCDM) problem. A comparison of rule based reasoning methods with classical MCDM methods was presented i.a. in [15]. MCDM methods requires providing of a complete set of criteria for each reasoning process. The REBIT's ability to optimize knowledge inference is a significant advantage, allowing to avoid unnecessary investigations and decreasing computational requirements.

A reasoning process may be followed with an example. The rules presented in Fig. 6 are applied for cooling of a modern bainitic steel [14]. The availability of a CA model is investigated. Let assume, that estimated process temperature is 850 °C and the applicability range of the temperature variable is between 600 °C and 800 °C (temperature range of the austenite-ferrite transformation). Hence, *CARules* rule cannot be fired. There are no other rules able to determine the variable *CARulesAvailable*. Consequently, the *AvailabilityCellularAutomata* cannot be fired. The REBIT system is able to determine that any further variables, like mean grain size, material phase composition, etc. do not have to be evaluated. This ability distinguish the REBITs approach from classical optimization methods and might significantly decrease necessary computational time.

Conclusions

Presented research is aimed at development of the knowledge management system, integrating DL and FOL knowledge representations. This system will be able to manage knowledge associated with an ICME process. Currently existing systems are usually limited to DL or FOL approach, while few existing systems combining both representations lose benefits of at least one of them. Developing of a consistent, broad and useful knowledge base (regardless of a chosen representation) is a very difficult challenge. Hence, if usefulness of developed knowledge is limited by its representation, the necessary efforts are believed to be unreasonable. The presented approach is expected to change this perception, mainly by decreasing the time necessary for optimization process, simultaneously increasing reliability of numerical computations. Not less important advantage is simplifying of development of multiscale/multistep models (due to the common communication language) and better knowledge reuse (due to its human-friendliness, high expressiveness and an reliable, FOL-based reasoning mechanism). However, this goal is not very close yet. The presented approach has been verified only with very limited knowledge range and in an "laboratory environment". Knowledge acquiring, managing and transforming is still inconvenient. More thorough tests are necessary, as well as development of user-friendly IT tools, supporting all operations on knowledge.

Acknowledgements This work is supported by the NCN, project No. 2014/15/B/ST8/00187.

References

1. G. Antoniou, C.V. Damasio, B. Grosof, I. Horrocks, Combining rules and ontologies. A survey. Technical Report (2005). http://rewerse.net/deliverables/m12/i3-d3.pdf
2. F. Baader, D. Calvanese, D.L. McGuinness, D. Nardi, P.F. Patel-Schneider, *The Description Logic Handbook: Theory, Implementation, and Applications* (2003)
3. T.R. Gruber, A translation approach to portable ontology specifications. Knowl. Acquis. **5**(2), 199–220 (1993)
4. C. Holsapple (ed.), *Handbook on Knowledge Management 1*. Springer (2013)
5. A. Iqbal, H.C. Zhang, L.L. Kong, G. Hussain, A rule-based system for trade-off among energy consumption, tool life, and productivity in machining process. J. Intell. Manuf. **26**(6), 1217–1232 (2015)
6. J. Kusiak, M. Sztangret, M. Pietrzyk, Adv. Eng. Soft. **89**, 90–97 (2015)
7. A. Macioł, Knowledge-based methods for cost estimation of metal casts. Int. J. Adv. Manuf. Technol. 1–16 (2016)
8. A. Macioł, P. Macioł, Intelligent hybrid system for forging process design support. Steel Res. Int. **79**, 521–528 (2008)
9. A. Macioł, P. Macioł, S. Jedrusik, J. Lelito, The new hybrid rule-based tool to evaluate processes in manufacturing. Int. J. Adv. Manuf. Technol. (2015)
10. A. Macioł, M. Wilkus, J. Duda, P. Macioł, U. Rauch, Rule-based method for choosing of wear model in hot forging simulation. Arch. Metall. Mater. p. press (2017)
11. P. Macioł, *Laminar Cooling Ontology* (2017). http://home.agh.edu.pl/~pmaciol/wordpress/wp-content/uploads/lcooling.zip

12. P. Macioł, R. Bureau, C. Poletti, C. Sommitsch, P. Warczok, E. Kozeschnik, Agile multiscale modelling of the thermo-mechanical processing of an aluminium alloy. In: *Esaform 2015*, vol. 651–653, pp. 1319–1324 (2015)
13. P. Macioł, K. Regulski, Development of semantic description for multiscale models of thermo-mechanical treatment of metal alloys. JOM **68**(8), 2082–2088 (2016)
14. Ł. Rauch, K. Bzowski, R. Kuziak, J. Kitowski, M. Pietrzyk, The off-line computer system for design of the hot rolling and laminar cooling technology for steel strips. J. Mach. Eng. **16**(2), 27–43 (2016)
15. B. Rebiasz, A. Macioł, Comparison of classical multi-criteria decision making methods with fuzzy rule-based methods on the example of investment projects evaluation bt, in *Intelligent Decision Technologies SE - 47*, ed. by R. Neves-Silva, L.C. Jain, R.J. Howlett Smart Innovation, Systems and Technologies, vol. 39 (Springer International Publishing, 2015), pp. 549–561
16. D. Szeliga, U. Sztangret, J. Kusiak, M. Pietrzyk, Optimization as a support for design of hot rolling technology of dual phase steel strips, in *NUMIFORM 2013*, vol. 1532 (AIP Publishing, 2013), pp. 718–724

Integration of Experiments and Simulations to Build Material Big-Data

Gun Jin Yun

Abstract In this paper, a method for extracting stress-strain databases from material test measurements is introduced as one of the potential Integrated Computational Materials Engineering (ICME) tools. Measuring spatially heterogeneous stress and strain evolutionary data during material tests is a challenging and costly task. The proposed method can extract a large volume of spatially heterogeneous stress and strain evolutionary data from experimental boundary measurements such as tractions and displacements. For the purpose, nonlinear finite element models are intrusively implemented with artificial neural network (ANN)-based material constitutive models. Then a specialized algorithm that can auto-progressively train ANN material models guided by experimental measurements is executed. Any complex constitutive law is not presumed. From the algorithm, ANN gradually learns complex material constitutive behavior. The training databases are gradually accumulated with self-corrected stress and strain data predicted by the ANN. Finally, material databases are obtained. For an example, visco-elastoplastic material databases are obtained by the proposed method.

Keywords Material Big-Data · Self-learning simulation · Artificial intelligence neural network · Nonlinear finite element analysis

Introduction

Recent research in Integrated Computational Materials Engineering (ICME) area has recognized pressing needs for construction of material Big-Data to accelerate development of emerging state-of-the art materials [1–3]. Material design requires advanced mechanics theories [4, 5] and multiscale models [6], and high-fidelity computational tools [7]. Not only do validation and verification of these theories

---

G.J. Yun (✉)
Department of Mechanical & Aerospace Engineering, Seoul National University,
Building 301 Room 1308, Gwanak-ro 1, Gwanak-gu, Seoul 08826, South Korea
e-mail: gunjin.yun@snu.ac.kr

© The Minerals, Metals & Materials Society 2017
P. Mason et al. (eds.), *Proceedings of the 4th World Congress on Integrated Computational Materials Engineering (ICME 2017)*,
The Minerals, Metals & Materials Series, DOI 10.1007/978-3-319-57864-4_12

and computational models require material Big-Data but also Big-Data can facilitate creations of new knowledge and mathematical models through informatics, statistics and artificial intelligence, etc. [8–12]. Although measurement technologies have been greatly advanced, it is still costly to obtain material Big-Data from experimental tests. Applications of artificial intelligence to material Big-Data are expected to enable fast development of emerging materials.

In this paper, a new computational approach that can generate material Big-Data is introduced. In the method, artificial neural network (ANN) [13] substitutes for material constitutive models in general nonlinear finite element analyses. Since ANN material models need a batch of stress and strain history data, self-learning simulation (SelfSim) methodology was used to create material Big-Data of a large volume of stress and strain history data [14]. Boundary measurements from experimental structural or material tests are inputs to the SelfSim. During SelfSim training, ANN models are gradually evolved learning true material response by the reference experimental boundary measurements. Predictability of the SelfSim trained ANN material constitutive models can be improved continuously whenever new experimental test data are available. Once ANN model is sufficiently trained with comprehensive experimental data, it can be readily used as a material constitutive model in nonlinear finite element models subjected to different loading histories. However, the trained ANN material models is applied to generate history stress-strain data. It has not been attempted to apply the SelfSim method to generate material Big-Data. For a demonstration, a Neo-Hookean hyper-elasticity model with rate-dependent hysteresis was utilized in this paper.

ANN Based Material Model and SelfSim Computational Algorithm

In this section, a new ANN material model is highlighted, and the primary components of the SelfSim analysis and integration idea with experiments are described.

ANN-Based Material Model

The ANN-based material model is defined as

$$\sigma_n = \sigma_{NN}([\varepsilon_n, \varepsilon_{n-1}, \sigma_{n-1}, \varsigma_n]: [NN\ Architecture]) \quad (1)$$

In order to achieve the single-valuedness between inputs and outputs of the hysteretic ANN-based constitutive model, one internal variable $\varsigma_n = \sigma_{n-1}\varepsilon_{n-1} + \sigma_{n-1}\Delta\varepsilon_n$ is included [15]. To deal with finite strain problems, the ANN material model should directly update Cauchy stresses with respect to a material reference framework and integrate the objective Jaumann stress rate with respect to time in order to obtain co-rotational stress tensors at each increment. The updated Cauchy stresses can be expressed as:

$$^n\sigma_{ij} = \sigma_{NN}\left([\Delta R_{ip}{}^n\varepsilon_{pq}\Delta R_{qj}^T, \Delta R_{ip}{}^{n-1}\varepsilon_{pq}\Delta R_{qj}^T, \Delta R_{ip}{}^{n-1}\sigma_{pq}\Delta R_{qj}^T, SV_{ij}]: [\text{NN Architecture}]\right) \quad (2)$$

where ΔR is the incremental rotational matrix used to account for the rigid body rotation; the superscript T indicates transpose of a matrix. ABAQUS automatically conducts this simple rotational transformation of stress and strain tensors before UMAT is called. Jacobian material stiffness matrix (**C**) is updated using an explicit expression which is a function of inputs and outputs as well as the ANN parameters, such as weight factors, scale factors and activation function values from each neuron. In finite strain problems, a large volume change due to geometric nonlinearity may slow down the convergence rate. Thus, the Jaumann stress rate of the Kirchhoff stress $(J\sigma)_{ij}[e_i \otimes e_j]$ in the co-rotational configuration is used for the stress update in this paper. Therefore, the Jacobian matrix is formulated as [16]

$$\mathbf{C} = \frac{1}{J}\frac{\partial \Delta(J\sigma)}{\partial \Delta\varepsilon} = \frac{1}{J}\frac{\partial(J^{n+1}\sigma_{ij} - J^n\sigma_{ij})}{\partial \Delta^{n+1}\varepsilon_{kl}} = \frac{1}{J}\frac{\partial J^{n+1}\sigma_{ij}}{\partial \Delta^{n+1}\varepsilon_{kl}}$$
$$= \frac{1}{J}\frac{\partial J^{n+1}\sigma_{ij}}{\partial^{n+1}\sigma_{pq}^{NN}}\frac{\partial J^{n+1}\sigma_{pq}^{NN}}{\partial^{n+1}\varepsilon_{rs}^{NN}}\frac{\partial^{n+1}\varepsilon_{rs}^{NN}}{\partial \Delta^{n+1}\varepsilon_{kl}} = \frac{1}{J}\frac{S^\sigma}{S^\varepsilon}\frac{\partial J^{n+1}\sigma_{pq}^{NN}}{\partial^{n+1}\varepsilon_{rs}^{NN}} \quad (3)$$

$$\text{where } S^\sigma = \frac{\partial J^{n+1}\sigma_{ij}}{\partial^{n+1}\sigma_{pq}^{NN}} \text{ and } \frac{1}{S^\varepsilon} = \frac{\partial^{n+1}\varepsilon_{rs}^{NN}}{\partial \Delta^{n+1}\varepsilon_{kl}}$$

where J indicates the determinant of the deformation gradient tensor **F** and S^σ and S^ε indicate the scale factors corresponding to stresses and strains, respectively.

ANN material models are intrusively implemented into nonlinear finite element analysis procedures. The incremental iterative equilibrium equation is expressed as

$$\mathbf{K}_t \Delta \mathbf{d}_n^{(k)} = \mathbf{F}_n^{ext} - \mathbf{I}_n^{(k-1)} = \mathbf{F}_n^{ext} - \sum \int \mathbf{B}^T \sigma_{NN} dV \quad (4)$$

where $\mathbf{K}_t$ is the tangent stiffness matrix; d is the displacement vector; $\mathbf{F}_n^{ext}$ is the external force vector at load step n and $\mathbf{I}_n^{(k-1)}$ is the internal resisting force vector at the load step n and the iteration step $(k-1)$ and **B** is the strain-displacement matrix.

SelfSim Computational Algorithms

Experimental boundary measurements i.e. reaction forces and displacements are measured from a structural or material test. Two ANN-based nonlinear FE models

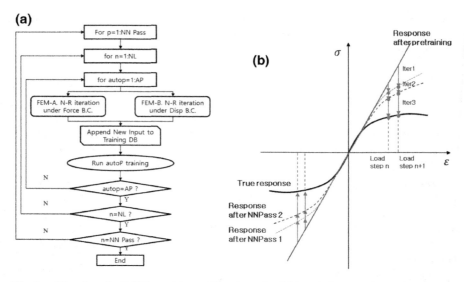

Fig. 1 **a** Flowchart for SelfSim training and **b** schematic ANN model's auto-progressive learning of true material response

are prepared. One ANN-based FE model is subjected to the measured boundary reaction forces and called FEM-A. The other ANN-based FE model is subjected to the measured boundary deformations and called FEM-B. In addition to the two loops in the conventional nonlinear finite element analyses by Newton-Raphson (NR) algorithm, SelfSim requires two additional iteration loops, which are NN Pass and auto-progressive (autoP) training. A schematic computational flowchart that includes the additional loops for SelfSim is depicted in Fig. 1a.

In each of the load incremental steps, NR iteration is performed until reaching equilibrium states. This NR iterations for convergence to equilibrium are performed multiple times. Such multiple NR iterations is called the autoP training, which continues until the predetermined number is reached or a convergence criterion is satisfied [14]. Sweeping all load steps is called one NN Pass. Multiple NN Passes are necessary since the ANN-based material model may not be trained with one NN Pass. In Fig. 1b, iter1, iter2 and iter3 indicate gradual training of the ANN model toward true stress-strain response through autoP training. During the SelfSim simulation, stress-strain history data at all material points are appended into the training database. After SelfSim training, the ANN model can be used in forward nonlinear FE analyses. This generates material Big-Data in terms of stress-strain history data.

Example of Building Material Big-Data

For a simulated material testing (Fig. 2a), a laminated rubber bearing subjected to pure cyclic shearing was selected in which an ABAQUS built-in hysteretic Neo-Hooke model [17] with rate-dependency was used. An FE model for the reference simulation was also constructed using 150 8-node 3D solid elements (C3D8H) for rubber materials and 75 8-node solid elements (C3D8H) for shim plates assuming a plane strain condition. A maximum 167% shear deformation (e.g. ratio of horizontal shear displacement to the height of specimen) was applied. On the contrary, an FE model for the SelfSim training was constructed using 150 4-node plane strain elements (CPE4H) with an ANN-based UMAT subroutine for rubbery material and 75 4-node plane strain elements (CPE4H) with a linear elastic material model for steel shim plates (Fig. 2b). The rubber bearing is subjected to pure cyclic shearing displacements on the top surface without eccentric vertical loads at two different frequencies (1 and 10 Hz). The reference data from two cycles (the first cycle with amplitude 40 mm and the second cycle with amplitude 100 mm) at 1 Hz frequency were used in the SelfSim training. The reference boundary information for the FEM-A and FEM-B includes force and displacement histories at all nodes on the top surface and at four points (at the center of each mid rubber layer) on each side. In all simulations, the geometric nonlinearity is considered. Table 1 summarizes parameters used for the ANN material model.

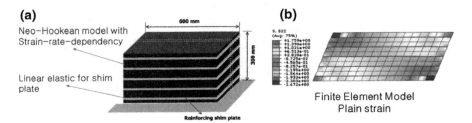

Fig. 2 a Laminated rubber bearing and **b** plain strain finite element model used for SelfSim simulation

Table 1 Parameters used in SelfSim analysis for Numerical Test II

	Name of parameters	Values
ANN model and SelfSim parameters	Architecture	16-25-25-4
	Number of auto-progressive cycles	3
	Number of epochs for pre-training	1000
	Scale factors for strains (ε_{xx}, ε_{yy}, ε_{zz} and ε_{xy})	(2,2,0.1,10)
	Scale factors for stresses (σ_{xx}, σ_{yy}, σ_{zz} and σ_{xy})	(500,500,500,500)
	Scale factors for internal state variables	(400,400,400,400)
	Max strains used in pretraining data (e_{radius})	0.01
	Tolerance for auto-progressive cycle	0.05

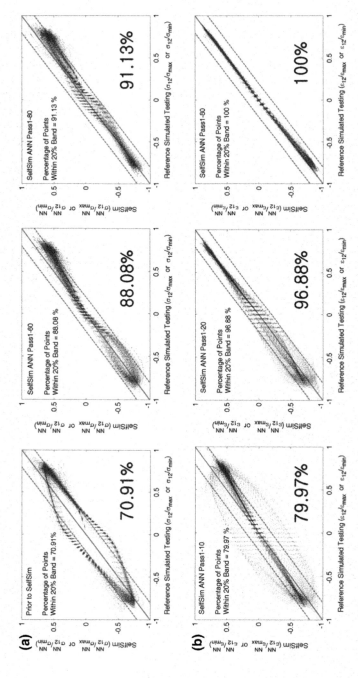

Fig. 3 Collections of **a** shear stress and **b** shear strain data at different load steps of SelfSim training

For the SelfSim training, only one ANN Pass was performed since the SelfSim-trained ANN model after one ANN Pass was capable of predicting the target material constitutive behavior with sufficient accuracy. Stresses and strains history data were extracted from the forward analysis along all of the load steps at every integration point, and they were normalized with max or min values. The ANN material model at different loading steps was used in a forward analysis. Positive stress components were normalized using the maximum stress, and negative stress components are normalized using the minimum stress. As depicted in Fig. 2, the stress components become more accurate as the ANN model is more trained by the SelfSim. The closer the points are distributed to the solid diagonal line, the more accurate the predictions by the ANN model are. For example, in the case of shear stress component, percentage of the number of points that fall within the 20% error band has increased from 70.91 to 91.13%.

A gradual learning of the logarithmic strain (true strain) value from the forward analysis was also observed in Fig. 3b. It indicates that the SelfSim analysis can effectively generate realistic strain distributions within the entire domain of the boundary value problem throughout the load steps. For instance, the shear strain component was almost perfectly predicted with high accuracy, as shown in Fig. 3b.

Conclusions

In this paper, a new approach for building material Big-Data from limited experimental test measurements was introduced. In the method, an ANN model substituted for the material constitutive model in the general nonlinear finite element model. An algorithmic technique called SelfSim auto-progressively trains the ANN material model referring to experimental boundary measurements. From the SelfSim training, the ANN based material model could be trained and subsequently used in forward nonlinear FE analyses. From the forward analysis, ANN generates a Big-Data consisting of stress-strain history data. From an example with a laminated rubber bearing, the SelfSim training methodology was proven to be an efficient way to generate material Big-Data and gradual accuracy improvements of the SelfSim trained ANN model could be demonstrated.

References

1. M.M. Rapporteur, *Big Data in Materials Research and Development: Summary of a Workshop* (National Research Council of The National Academy, Washington, D.C., 2014)
2. K. Rajan, Materials informatics: the materials "gene" and Big Data. Annu. Rev. Mater. Res. **45**, 153–169 (2015)
3. K. Rajan, *Informatics for Materials Science and Engineering: Data-driven Discovery for Accelerated Experimentation and Application* (Butterworth-Heinemann, Amsterdam, 2013)

4. X.M. Wang, B.X. Xu, Z.F. Yue, Micromechanical modelling of the effect of plastic deformation on the mechanical behaviour in pseudoelastic shape memory alloys. Int. J. Plast. **24**, 1307–1332 (2008)
5. M.F. Horstemeyer, A.M. Gokhale, A void-crack nucleation model for ductile metals. Int. J. Solids Struct. **36**, 5029–5055 (1999)
6. K. Matous, M.G.D. Geers, V.G. Kouznetsove, A. Gillman, A review of predictive nonlinear theories for multiscale modeling of heterogeneous materials. J. Comput. Phys. **330**, 192–220 (2017)
7. S.M. Arnold, B.A. Bednarcyk, A. Hussain, V. Katiyar, *Micromechanics-Based Structural Analysis (FEAMAC) and Multiscale Visualization within Abaqus/CAE Environment* (NASA Glenn Research Center, Cleveland, OH, United States, 2010)
8. D. Farrusseng, F. Clerc, C. Mirodatos, R. Rakotomalala, Virtual screening of materials using neuro-genetic approach: concepts and implementation. Comput. Mater. Sci. **45**, 52–59 (2009)
9. J.M. Schooling, M. Brown, P.A.S. Reed, An example of the use of neural computing techniques in materials science—the modelling of fatigue thresholds in Ni-base superalloys. Mater. Sci. Eng. Struct. Mat. Prop. Microstruct. Process. **260**, 222–239 (1999)
10. X.H. Yang, W. Deng, L. Zou, H.M. Zhao, J.J. Liu, Fatigue behaviors prediction method of welded joints based on soft computing methods. Mater. Sci. Eng. Struct. Mater. Prop. Microstruct. Process. **559**, 574–582 (2013)
11. C. Kamath, O.A. Hurricane, Robust extraction of statistics from images of material fragmentation. Int. J. Image Graph. **11**, 377–401 (2011)
12. C.A.C. Coello, R.L. Becerra, Evolutionary multiobjective optimization in materials science and engineering. Mater. Manuf. Process. **24**, 119–129 (2009)
13. J. Ghaboussi, J. Garrett, X. Wu, Knowledge-based modeling of material behavior with neural networks. J. Eng. Mech. ASCE **117**, 132–153 (1991)
14. G.J. Yun, J. Ghaboussi, A.S. Elnashai, Self-learning simulation method for inverse non-linear modeling of cyclic behavior of connections. Comput. Methods Appl. Mech. Eng. **197**, 2836–2857 (2008)
15. G.J. Yun, J. Ghaboussi, A.S. Elnashai, A new neural network-based model for hysteretic behavior of materials. Int. J. Numer. Methods Eng. **73**, 447–469 (2008)
16. G.J. Yun, A.F. Saleeb, S. Shang, W. Binienda, C. Menzemer, Improved SelfSim for inverse extraction of non-uniform, nonlinear and inelastic constitutive behavior under cyclic loadings. J. Aerosp. Eng. **25**, 256–272 (2012)
17. ABAQUS/Standard H, *A General Purpose Finite Element Code* (Karlsson & Sorense, Inc., Hibbitt, 2004)

Part II
ICME Design Tools and Application

ICME-Based Process and Alloy Design for Vacuum Carburized Steel Components with High Potential of Reduced Distortion

H. Farivar, G. Rothenbucher, U. Prahl and R. Bernhardt

Abstract Carburized steel components are usually quenched from a hardening temperature, which lies in a complete austenitic phase, to room temperature. This leads to a microstructure comprised of mostly martensite plus bainite giving rise to unwanted heat-treatment-induced distortion. However, having a soft phase of ferrite dispersed throughout the microstructure can be quite effective in this regard. This is attributed to the capability of ferrite in accommodating the plasticity resulted from austenite-to-martensite transformation expansion. In the context of this work, it is demonstrated that how a proper selection of chemical compositions and a hardening temperature can greatly suppress the associated distortion. Hence, in order to systematically design a new steel alloy which fits to the above mentioned conditions, an ICME-based methodology has been employed. Thus, a series of calculations have been carried out by means of the well-known thermodynamic-based software Thermo-Calc® and the scripting language of Python. The austenite to ferrite phase transformation kinetics is also captured by the software DICTRA® generating a virtual TTT (Time-Temperature-Transformation) diagram which is subsequently utilized for further finite element simulations in the software Simufact.forming®. The carburizing process, the following phase transformations and the effect of the developed microstructure on the final distortion are simulated in macro-scale through Simufact.forming. The finite-element-based results of the Simufact.forming have in turn been enhanced by the results of the above-mentioned thermodynamic-based computational tools. At a later stage the simulation outcomes are experimentally validated by employing Navy C-Ring specimens.

Keywords ICME · Process and steel alloy design · Carburizing · Distortion · Simulation · Navy C-ring

---

H. Farivar (✉) · U. Prahl
Integrated Computational Materials Engineering Department, Steel Institute,
RWTH Aachen University, Intzestraße 1, 52072 Aachen, Germany
e-mail: Hamidreza.Farivar@IEHK.RWTH-Aachen.de

G. Rothenbucher · R. Bernhardt
Simufact Engineering GmbH, Tempowerkring 19, 21079 Hamburg, Germany

Introduction

Because machinability is important simultaneously with strength in gears used in the transmissions and suspensions of automobiles and construction equipment, the majority of gears used in these applications are manufactured by a process of forging and machining followed by surface hardening by carburization, using case hardening steels. In recent years, higher strength in these gears has been strongly desired from the viewpoints of weight reduction and downsizing. In automotive applications, heat-treatment-induced distortion of gears has also become a problem due to demand for reduced noise [1].

Carburized steel components are usually quenched from a hardening temperature, which lies in a complete austenitic phase, to room temperature. This leads to a microstructure comprised of mostly martensite plus bainite giving rise to unwanted heat-treatment-induced distortion. However, having a soft phase of ferrite dispersed throughout the microstructure can be quite effective in this regard. This is attributed to the capability of ferrite in accommodating the plasticity resulted from austenite-to-martensite transformation expansion. Therefore, in order to properly design a steel alloy with such a described concept, numerical thermodynamic-based tools such as ThermoCalc® and DICTRA® have been utilized. The design approach concentrates on critical transformation temperatures namely Ae3 (equilibrium temperature at which transformation from austenite into ferrite begins) and Aem (equilibrium temperature at which precipitation of cementite in austenite begins for hypereutectoid steels) and kinetic of phase transformation as function of chemical composition. The aim of numerical alloy design is to produce a microstructure in the core of case-carburized steel consisting of ferrite by properly adjusting its chemical composition in order to evaluate the final heat-treatment-induced distortion compared to those grades with martensitic-bainitic microstructure. During conventional heat treatments both carburized surface and core of the specimen remain in a single austenitic phase region. However, basically, by adjusting the alloying elements the critical transformation temperatures can be controlled. For instance, by adding ferrite stabilizer elements (Si, Mo, Al and…) the Ae3 temperature is elevated and makes it possible to produce a dual-phase microstructure of austenite and ferrite in the core of carburized specimen at the hardening temperature, while the carburized layer is still in a single austenitic phase. Having a soft ferrite phase in the core of carburized components lowers the usual dimensional changes associated with such typical case hardening processes, Fig. 1.

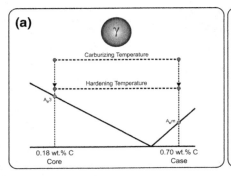 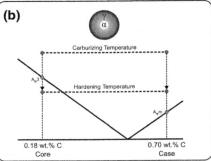

Fig. 1 Schematic representation of **a** typical carburizing heat-treatment, **b** modified carburizing heat-treatment leading to a ferrite containing phase in core by means of adjusting the chemical compositions and hardening temperature

Numerical Alloy Design

In order to design a desired alloy with an optimum chemical composition which meets the goals described earlier, the software Thermo-Calc® has been used. An optimum chemical composition here is interpreted as a set of alloying elements which fulfills the following criteria:

- Maximum possible difference between the critical transformation temperatures (ΔT = Ae3 − Aem). Hence, a certain volume fraction of ferrite in the core of component can be reached quicker at a given hardening temperature;
- Lowest possible hardening temperature (Hardening temperature = Aem + 20 °C). The lower hardening temperature, the less distortion potential;
- Sufficient hardenability;
- Reasonable time for reaching a specified volume fraction of ferrite in the core;
- Reasonable time for carburizing the case to a specified amount.

To come up with such a satisfactory chemical composition, the alloying contents of a set of widely-used case hardening steels have been selected as reference. Accordingly, an interval for each element is defined based on which the respective element content is systematically altered. Furthermore, the corresponding critical temperatures are calculated and extracted. Obviously, it is not possible to calculate and extract the critical transformation temperatures (Ae3 and Aem) for each chemical set separately and a set of codes essentially need to be written. By means of such codes which have been developed in a scripting tool, Python®, different macro files are created according to each individual chemical composition so that they can be imported to the Thermo-Calc® software. The macro files are made executable and further submitted to Thermo-Calc® through the Linux shell, Bash,

Fig. 2 Overview of the numerical platform created for the purpose of optimum alloy design

Fig. 2. Furthermore, all calculated critical temperatures are collected and post processed.

With respect to the selected increment for changing the elemental values, the total number of chemical sets are exceeding than 6000. Obviously not all of them are acceptable according to our previously defined criteria. Having applied the first 3 criteria, the total number of chemical sets reduced to 25 sets and the rests are excluded. The remaining sets of alloying elements can provide enough difference in critical temperatures which satisfy the first three criteria as explained above ($\Delta T = 50$ °C, 85 °C if Carbon concentration in the case is selected as 0.80 and 0.70, respectively).

In order to estimate the hardenability of the above chemical compositions the so-called Ideal Diameter parameter (DI) has been calculated. Quantitative measure of a steel's hardenability is expressed by its DI. DI values are an excellent means of comparing the relative hardenability of two materials [2]. Hence, the ASTM Specification A255-02 (Standard Test Method for Determining Hardenability of Steel) is selected as our database. Ideal diameter (DI) is calculated using the Eq. (1) with individual values. It is assumed that all of chemical compositions sets have an identical grain size, ASTM grain size number 7. It should be noted that this formula is an empirical relation and it has been used for the purpose of comparison of the effect of various alloying elements on the hardenability behaviour of material.

$$DI = Base\ Hardenability \times f_{Si} \times f_{Mn} \times f_{Cr} \times f_{Mo} \times f_{Ni} \qquad (1)$$

As a final chemical composition the following set has been selected, Table 1. It is worth mentioning that the amount and mass ratio of Al and N have been designed (Al/N = 3) in a way so that they can effectively contribute to the austenite grain size control by means of their pinning effect, Zener force.

Table 1 Chemical composition of the designed steel (in wt%)

C	Si	Mn	P	S	Al	Cr	Mo	Ni	N	Fe
0.18	0.40	0.75	0.010	0.020	0.035	0.35	0.50	0.10	0.010	Balanced

Results and Discussion

Case-Hardening Process

Based on the selected chemical composition, Table 1, a material has been casted in vacuum inductive furnace at the Steel Institute (IEHK) of the RWTH Aachen University. The material was initially homogenized sufficiently and then followed by hot forging in multiple steps and final air cooling to room temperature. The hot-forged block was subsequently normalized and directly annealed to get a coarse ferritic-pearlitic microstructure with improved machinability. As it was initially explained in this paper, in order to investigate the effect of microstructure on the final distortion, the so-called Navy C-Ring specimens, Fig. 3b, have been manufactured out of the hot-forged and annealed material. The specimens were subsequently subjected to the heat-treatment cycle as it is depicted in Fig. 3a in a dual-chamber vacuum carburizing furnace (ALD Vacuum Technologies GmbH®) located at the Steel Institute of RWTH Aachen University. Navy C-Ring specimens have frequently been addressed in the literature and employed by many researchers to study the variations in components dimensions associated with typical case hardening heat-treatments in which the amount of distortion of a quenched sample is studied by the change in the gap width [1, 3–7].

The investigated Navy C-Rings are carburized under vacuum at 950 °C by means of the so called pulse carburization method in which the carburizing process has been partitioned to multiple consecutive boost/diffusion cycles enhancing the

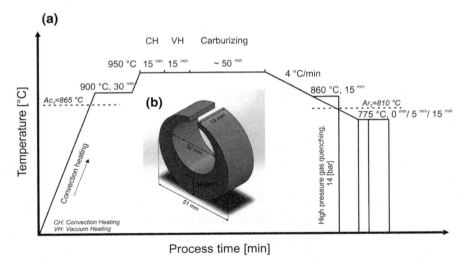

Fig. 3 Schematic representation of **a** the applied heat-treatment cycle, **b** employed Navy C-Ring

process effectiveness [8–10]. The corresponding carburizing parameters have been set so that the case hardness depth equals to 0.5 mm (CHD = 0.5 mm). As it is illustrated in Fig. 3a, when the carburizing part is ended, the first set of specimens are cooled down to reach a hardening temperature of 860 °C and soaked at this temperature for 15 min prior to quenching to room temperature. Since this hardening temperature is well above the Ar3 temperature (temperature at which ferrite starts to transform out of parent austenite during the cooling), therefore, no ferrite is stable and hence, the corresponding microstructures both in the core and carburized surface are in a single austenitic phase. On the contrary, however, the second, third and fourth set of the samples have been cooled down more to reach a lower hardening temperature of 775 °C and for three different holding times over which austenite decomposes to ferrite leading to different volume fractions, respectively. Whereas, the high carbon-containing layer of these carburized samples are still in a single stable austenitic phase. It has to be mentioned that in order to perform the above mentioned experiments, four individual separated tests have been carried out accordingly. Moreover, for the sake of having sufficient statistics, in each single experiment three Navy C-Rings are employed and further characterized.

Microstructural Investigation

The microstructures of the carburized and quenched test pieces have been characterized by means of light optical microscope (LOM) to study the developed microstructures. Therefore, the samples have been embedded and etched in a 3% Nital etchant to reveal the constituent phases. Figure 4 represents the microstructure in the core of the carburized and quenched Navy C-Rings in their thickest part (Body). As it can be seen in Fig. 4a, the sample which has been quenched from the hardening temperature of 860 °C contains bainite and martensite. Whereas, the samples which have been quenched from the hardening temperature of 775 °C show the existence of ferrite phase whose fraction increases proportionally with the holding time.

Fig. 4 Typical LOM pictures in the thickest part (Body) of the carburized and quenched samples from the hardening temperature of **a** 860 °C soaked for 15 min, **b** 775 °C no soaking time, **c** 775 °C soaked for 5 min and **d** 775 °C soaked for 15 min

Distortion Measurement

As it was earlier mentioned in this work, the amount of distortion in the case-hardened Navy C-Rings is inspected by the change in their gap distance, Fig. 5a. The specific geometry of this test piece makes it possible to investigate the effect of microstructure in the thickest part (labeled as Body) on the dimensional change of the quenched sample. In other words, from the thickest part of a Navy C-Ring upwards the sample becomes thinner in its two sides. Hence, these two thinner parts can act like two levers through which the influence of microstructural characteristics is magnified and reflected in its upper part (labeled as Head). In order to precisely measure the distortion of the heat-treated samples, just prior to the thermal cycle and just after the quenching process they have been inspected by mans of a coordinate-measurement machine (CMM, UPMC 850 Carat of Fa. Zeiss) with 0.8 μm accuracy, Fig. 5b, and thus, the net effect of the case hardening process is quantified.

In Fig. 6 the correlation between the volume fraction of ferrite formed in the thickest part of the Navy C-Rings (Body) and the measured absolute gap change has been presented. It clearly shows that the maximum measured gap change belongs to the case in which no ferrite formed (860 °C soaked for 15 min prior to quenching), whereas for the samples quenched from 775 °C a smaller amount of distortion is measured with increasing the volume fraction of ferrite. Furthermore, according to Fig. 6, no tangible influence on suppressing the gap change for the samples quenched from 775 °C with no soaking time have been measured. However, by increasing the ferrite fraction, the effect of this soft phase on suppressing the gap change becomes more obvious. The smaller gap change measured in the case of the samples which were quenched from 775 °C and for longer soaking times can be attributed to a higher fraction of the soft ferrite phase which accommodates the volume expansion of its adjacent bainitic and martensitic constituents during the

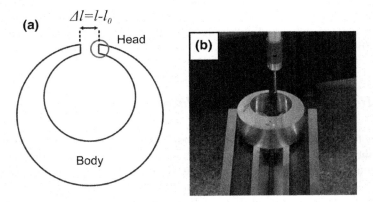

Fig. 5 a Illustration of the employed Navy C-Ring and its gap change, **b** CMM utilized for distortion measurement

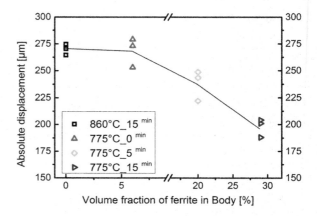

Fig. 6 Measured absolute gap change as function of ferrite volume fraction for the carburized and quenched Navy C-Rings

quenching. A lower hardening temperature also plays a role to suppress the quenched-induced distortion but to a lower degree. This can be inferred when the samples with an identical quenching temperature of 775 °C but two different holding times (i.e.: two different ferrite fraction) are compared.

Austenite to Ferrite ($\gamma \rightarrow \alpha$) Phase Transformation

Based on the applied carburization temperature and the process duration time an average prior austenite grain size has been experimentally obtained and was further used in DICTRA® in order to set an initial austenite cell size. A series of isothermal simulations have subsequently been carried out in different temperatures employing relevant thermodynamic (TCFE7) and kinetic (MOBFE2) databases to numerically predict the critical temperatures in which ferrite starts to transform out of austenite (Ar3). The simulation results basically generate a virtual Time-Temperature-Transformation (TTT) diagram predicting the $\gamma \rightarrow \alpha$ phase transformation kinetics which is quite essential and will be further called during the finite element simulations (Fig. 7).

Case Hardening Simulation in Macro-scale

In order to understand the effect of microstructure on the final distortion induced due to the applied case hardening process a set of finite element simulations have been set up in Simufact.forming® software. According to the chemical composition listed in Table 1, the corresponding carbon diffusivity coefficients as function of temperature and carbon content have been calculated in DICTRA® and further

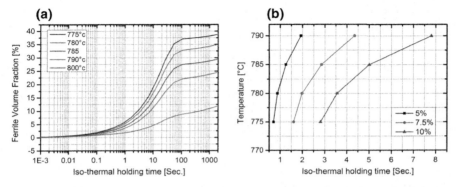

Fig. 7 **a** Evolution of ferrite out of austenite over time for different temperatures, **b** virtual TTT diagram for 5%, 7.5% and 10% ferrite fraction

implemented in Simufact.forming® to simulate carbon diffusion during the carburizing process, Fig. 8.

Moreover, in order to numerically estimate the critical temperature in which ferrite starts to transform out of austenite (Ar3) the generated virtual Time-Temperature-Transformation (TTT) diagram as it was explained in section "Austenite to Ferrite ($\gamma \rightarrow \alpha$) Phase Transformation" has been utilized to support the FE simulations. Additionally, an appropriate heat transfer coefficient (HTC) has been used for the finite element simulations with respect to the boundary conditions (quenching medium, Nitrogen pressure during the quenching) of the vacuum furnace employed for conducting the real hardening process. Figure 9a indicates the ferrite phase fraction developed in the core of the sample after 15 min holding time at 775 °C. It is worth mentioning that the outer surface of the sample due to its higher carbon content even after decreasing the temperature remains austenitic.

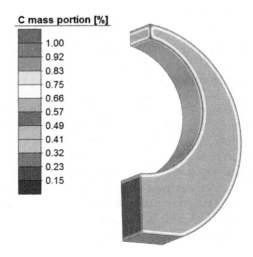

Fig. 8 Typical simulation of carbon diffusion on the carburized Navy C-Ring

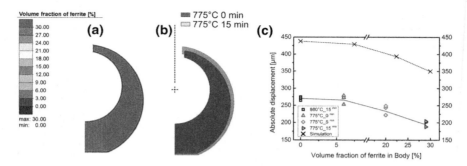

Fig. 9 **a** Ferrite fraction developed in the core of the sample after 15 min holding time at 775 °C, **b** deviation of the sample's head which are soaked for 0 and 15 min at 775 °C and **c** comparison of the measured and simulated variations in gap distance

In Fig. 9b the predicted deviation of the sample's head has been demonstrated for the samples soaked for two different holding times (0 and 15 min) and quenched from a single temperature of 775 °C. It is readily observed that the final reduced distortion is mainly due to the ferrite formation and the reduced quenching temperature plays a minor role. In Fig. 9c the simulated gap changes have been compared with those which were obtained experimentally. It is seen that the simulation results are in a perfect qualitative agreement with the measured distortion. These microstructural-dependent simulations confirm that ferrite can decrease the macroscopic dimensional change of the Navy C-Rings. It has to be mentioned that although the decreasing trend of the distortion is perfectly captured by the simulations, however, the absolute amount of the gap changes are over predicted and require further development.

Conclusion

In the context of this work an ICME-based methodology has been employed in order to design an optimum chemical composition of a case hardening steel with high potential of reduced quenching-induced distortion. In order to meet this goal, a set of calculations on the basis of a numerical platform have been performed in Thermo-Calc® and DICTRA® which are coupled with the relevant thermodynamic and kinetic databases. By means of the developed platform a set of chemical compositions have been selected in a way so that ferrite transformation out of parent austenite becomes possible while temperature is dropped from carburizing segment down to the hardening temperature. Besides, the hardening temperature has been designed lower than that which is typically selected. Lowering hardening temperature increases the risk of network-like carbides on the carburized layer due to its higher carbon content which subsequently deteriorates the final mechanical properties. However, this issue has also been foreseen and according to the employed alloy design methodology the formation of detrimental carbides on the carburized layer is suppressed. The performance of the designed material has further been examined through distortion investigation of the Navy C-Ring specimens manufactured out of the developed steel. The existence of a soft ferrite phase, to a major degree, in addition to a lowered hardening temperature, to a minor degree, give rise to a final reduced quenching-induced distortion. This was also analyzed by means of the finite element-based software Simufact.forming® which in turn was supported by the thermodynamic and kinetic simulations of DICTRA® and Thermo-Calc®. The finite-element simulations are in a good qualitative agreement of the experimentally obtained data. It is interestingly predicted that with increasing the fraction of ferrite the final change in the sample's head is decreased. In the presented work, it has successfully been demonstrated that by coupling computational softwares integrated on a numerical platform and according to a desired application an appropriate material and a heat-treatment process can properly be engineered.

Acknowledgements The authors gratefully acknowledge the financial support of this work provided by the German Federal Ministry of Education and Research (BMBF) under the context of the Indo-German Science and Technology Center (IGSTC).

References

1. K. Fukuoka, K. Tomita, T. Shiraga, Examination of surface hardening process for dual phase steel and improvement of gear properties (2010)
2. H.K.D.H. Bhadeshia, R.W.K. Honeycombe, *Steels: Microstructure and Properties*, 3rd edn. (Elsevier, Butterworth-Heinemann, Amsterdam, 2006)
3. G. Totten, M. Howes, T. Inoue, *Handbook of Residual Stress and Deformation of Steel*. ASM International, Materials Park, Ohio 44073–0002 (2002)
4. C.M. Amey, H. Huang, P. Rivera-Díaz-del-Castillo, Distortion in 100Cr6 and nanostructured bainite. Mater. Des. **35**, 66–71 (2012). doi:10.1016/j.matdes.2011.10.008
5. A.D. da Silva, T.A. Pedrosa, J.T. Gonzalez-Mendez et al., Distortion in quenching an AISI 4140 C-ring—predictions and experiments. Mater. Des. **42**, 55–61 (2012). doi:10.1016/j.matdes.2012.05.031
6. A. Clark, R.J. Bowers, D.O. Northwood, Heat treatment effects on distortion, residual stress, and retained austenite in carburized 4320 steel. MSF **783–786**, 692–697 (2014). doi:10.4028/www.scientific.net/MSF.783-786.692
7. E. Boyle, R. Bowers, D.O. Northwood, The use of Navy C-Ring specimens to investigate the effects of initial microstructure and heat treatment on the residual stress, retained austenite, and distortion of carburized automotive steels, ed. by E. Boyle, R. Bowers, D.O. Northwood. SAE World Congress & Exhibition. SAE International, 400 Commonwealth Drive, Warrendale, PA, United States (2007)
8. J. Rudnizki, B. Zeislmair, U. Prahl et al., Thermodynamical simulation of carbon profiles and precipitation evolution during high temperature case hardening. Steel Res. Int. **81**(6), 472–476 (2010). doi:10.1002/srin.201000048

9. M. Zajusz, K. Tkacz-Śmiech, M. Danielewski, Modeling of vacuum pulse carburizing of steel. Surf. Coat. Technol. **258**, 646–651 (2014). doi:10.1016/j.surfcoat.2014.08.023
10. P. Kula, K. Dybowski, E. Wolowiec et al., Boost-diffusion vacuum carburising—process optimisation. Vacuum **99**, 175–179 (2014). doi:10.1016/j.vacuum.2013.05.021

Study of Transient Behavior of Slag Layer in Bottom Purged Ladle: A CFD Approach

Vishnu Teja Mantripragada and Sabita Sarkar

Abstract Purging of argon gas in the molten metal bath is a process that is regularly involved in secondary steel making operations. The injected gas imparts momentum to the liquid metal, which induces high turbulence in the molten metal and helps in homogenization of the bath composition and temperature, and facilitates the slag metal interactions. In this study, a computational fluid dynamics (CFD) based numerical investigation is carried out on an argon gas stirred ladle to study the flow and interface behavior in a secondary steel making ladle. A transient, three phase coupled level-set volume of fluid (CLSVOF) model is employed to track the slag-metal, gas-metal and slag-gas interfaces. The transient behavior of slag layer deformation and open eye formation is studied for different slag layer to metal bath height ratios at various argon gas flow rates.

Keywords Slag eye · Steel making ladle · Multiphase flows

Introduction

Argon gas is purged from the bottom of the ladle during secondary steel making operations. The argon gas jet dissociates into many bubbles after entering the steel bath, and a gas-liquid plume is formed. If this plume has enough momentum, it breaks open the slag layer and exposes the molten metal inside to the atmosphere. This exposed metal region on the slag surface is known as open-eye. Open-eye

---

V.T. Mantripragada (✉) · S. Sarkar
Department of Metallurgical and Materials Engineering,
Indian Institute of Technology Madras, Chennai 600036, India
e-mail: mvishnutejaa@gmail.com

formation has both advantages and disadvantages, depending on the ladle refining process used. For example, large slag eye opening enhances the dissolution of oxygen into the steel, resulting in reoxidation. Here, open-eye formation is not desirable. However, during processes like desulphurization, the open eye formation facilitates more slag-metal contact, facilitating faster slag-metal reactions. In this case, open-eye formation is advantageous.

Significant amount of research work on the dependence of slag eye area on the gas flow rate, density and viscosity of slag, location of the gas injector etc. were reported in the literature [1–5]. However, only the slag eye opening size was reported in these studies and the transient behavior of the slag layer was not studied. Also, all these studies have been done by performing experiments on scaled down water models. Liquids such as petroleum ether, benzene, paraffin oil etc. were used in these cold model experiments as upper phase liquids, to mimic the behavior of the slag layer. The ratio of densities of molten steel and argon as well as molten steel and slag were not exactly replicated in all these cold model experiments, using water, air and the above upper phase liquids. Due to this reason, accurate prediction of slag eye area in industrial scale ladles has been unsuccessful [3].

CFD based numerical studies have been performed on industrial scale ladles to understand the open-eye formation behavior [6, 7]. These were performed for obtaining the average open eye area at various gas flow rates and other parameters such as slag thickness, molten metal bath depth, position of the gas injector etc. The transient nature of the turbulent flow during single plug bottom gas injection has been studied numerically, by tracking argon bubbles in Lagrangian framework [7]. The volume average velocity and turbulent kinetic energy of the molten steel were considered as the parameters for transient analysis, and the time taken by the system to reach a steady state was reported. However, the transient nature of slag layer deformation was not considered in this study.

Numerical studies on industrial ladle as well as experimental studies on scaled water model were performed to study the transient nature of the flow and to predict the slag layer opening size, using combined bottom gas bubbling and electromagnetic stirring [8]. The time taken by the system to reach a steady state, during electromagnetic stirring was identified. However, transient analysis of the flow during bottom plug gas injection has not been reported in this work.

In the present work, a coupled level-set volume of fluid (CLSVOF) method is used to model three dimensional, transient, three phase turbulent flow and to study the slag layer deformation during argon gas purging in a 160 Tonne capacity steel making ladle [9, 10]. SST k-ω turbulence model is used to describe the turbulent flow in the ladle. The transient nature of the open-eye is studied and the time averaged slag eye area is obtained at various values of argon gas flow rate and slag thickness.

Mathematical Modeling

Governing Equations

Volume of fluid (VOF) method: In volume of fluid method, the tracking of the interface is done by solving volume fraction continuity equation for the slag and argon phases, which is given by

$$\frac{\partial}{\partial t}(\alpha_i \rho_i) + \nabla \cdot (\alpha_i \rho_i \vec{v}_i) = 0 \qquad (1)$$

where, α, ρ are the volume fraction of a phase in a control volume and density respectively and $\vec{v}$ represents the flow velocity. The index "i" represents argon and slag phases, indicating that the above equation is solved only for these two phases. Once this equation is solved for obtaining the volume fractions of both these phases, the volume fraction of steel in a cell can be obtained as follows:

$$\alpha_{Steel} = 1 - (\alpha_{Slag} + \alpha_{Argon}) \qquad (2)$$

Level-set method: Level set function is defined as a signed distance to the interface. For a two-phase flow system, it is expressed as

$$\varphi(x) = \begin{cases} +d & \text{if } x \in \text{Primary Phase} \\ 0 & \text{if } x \in \text{Interface} \\ -d & \text{if } x \in \text{Secondary Phase} \end{cases} \qquad (3)$$

where, d is the distance from the interface. Thus, on the interface, the value of the level-set function φ is equal to zero. The level-set equation can be written as

$$\frac{\partial \varphi}{\partial t} + \nabla \cdot (\vec{v}\varphi) = 0 \qquad (4)$$

Both level set method and volume of fluid method are popular interface tracking methods, used for modeling flows involving immiscible fluids and complex interfaces. The level-set function is smooth and continuous. Hence the interface curvature, surface tension effects etc. can be accurately estimated. However, the level-set method does not follow volume conservation. This limitation is overcome by using VOF method, which conserves volume fraction efficiently. VOF model, however is not spatially continuous and a smooth interface is not obtained for complex topologies [11]. To overcome the deficiencies of both these methods, a coupled level-set volume of fluid (CLSVOF) method is considered for the current work.

Momentum Equation: The surface tension force is added as a source term to the momentum equation. The momentum equation is given as follows

$$\frac{\partial}{\partial t}(\rho \vec{v}) + \nabla \cdot (\rho \vec{v} \vec{v}) = -\nabla P + \nabla \cdot [\mu \nabla(\vec{v})] + \rho \vec{g} - \sigma \kappa \delta(\varphi) \vec{n} \qquad (5)$$

where, P is the pressure, $\vec{g}$ is the acceleration due to gravity, σ is the surface tension between two phases, κ is the curvature of the interface and $\vec{n}$ is the local normal to the interface. $\delta(\varphi)$ is a function, which depends on the level-set function as well as the interface thickness. Here, ρ, μ are the volume fraction averaged density and dynamic viscosity of the phases. That is, a single momentum equation is solved all over the computational domain, considering the volume averaged properties of a control volume.

Standard SST k-ω model is used to predict the turbulent nature of the flow [12]. It effectively uses the k-ω model in the near wall region and k-ε model in the free stream regions, by using blending functions. Similar to the momentum equation, a single set of turbulence equations is solved throughout the computational domain, by considering the volume averaged properties.

Geometry and Boundary Conditions

A three-dimensional computational domain was considered, which is shown in Fig. 1. The ladle was filled with molten steel, with slag layer floating on the surface of the bath. Argon gas was injected into the molten steel bath from the inlet, located at the bottom of the ladle. Computational Fluid Dynamic (CFD) techniques were used to solve the governing equations.

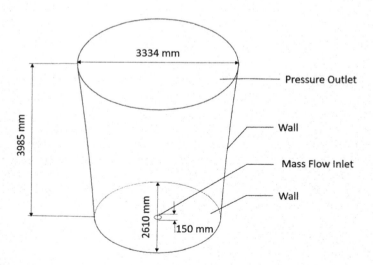

Fig. 1 Schematic of computational domain along with boundary conditions

Table 1 Fluid properties and operating conditions

Property	Value
Ladle capacity	160 T
Steel density	7020 kg/m^3
Steel viscosity	0.006 kg/m-s
Slag density	3500 kg/m^3
Slag viscosity	0.03 kg/m-s
Argon viscosity	8.9e–5 kg/m-s
Surface tension slag-steel	1.15 N/m
Surface tension steel-argon	1.82 N/m
Surface tension slag-argon	0.58 N/m
Slag thickness to metal bath depth ratio $\left(\frac{\Delta L}{L}\right)$	0.03–0.06
Reynolds number (Re)	6158–10,263
Operating temperature	1873 k
Total flow time	300 s

The computational domain was divided into a total of 657,882 cells. The surface of the geometry is divided into two dimensional tetrahedral elements and the rest of the geometry is divided into hexahedral elements. The different boundaries of the computational domain are marked and the corresponding boundary conditions are also indicated in Fig. 1.

If 'L' is the depth of molten steel bath in the ladle and 'A_{in}' is the area of the inlet, then the argon gas mass flow rate 'Q' at the inlet is represented in terms of Reynolds number as

$$\text{Re} = \frac{\rho_{Ar} Q L}{\mu_{Ar} A_{in}} \qquad (6)$$

The physical properties and operating conditions of argon gas purging employed in this study are shown in Table 1.

Results and Discussion

Numerical simulations were carried out for a total flow time of 300 s, at various operating conditions as discussed in Table 1. An attempt was made to study the transient behavior of the slag-eye deformation during argon gas injection. The cross-sectional view of the ladle in the form of phase distribution contours at different times during the flow are shown in Fig. 2. The instantaneous stream lines of the flow are superposed on these phase distribution contours.

It can be observed from these contours that around the gas-liquid plume and below the slag layer opening, there is a recirculating current. In Fig. 2, even though

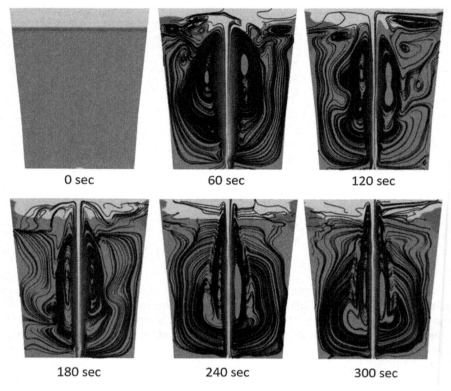

Fig. 2 Stream lines and phase distribution contours at Re = 10263 and $\frac{\Delta L}{L}=0.0293$.

it looks like the slag phase is pushed near to the walls as the flow progresses, the slag opening in similar other cross-sections narrows down, leading to an ellipse shaped open eye formation. This elliptical shape of the slag opening is observed at higher Reynolds numbers.

The slag opening area is averaged over the entire flow time to obtain the time averaged slag eye area. For the range of operating conditions considered in the current study, the average slag eye opening area was computed and its maximum value was found to be about thirty percent of that of the total slag layer area. The contours of average slag layer opening at various Re and slag thickness to bath depth ratios are shown in Fig. 3. The open eye cross-section changes from circular to elliptical shape as the Reynolds number increases, which can also be observed from this figure. This change in the shape of the cross section, which is often accompanied by an increase in the slag eye area occurs when the inertial forces on the slag layer dominates the surface tension and viscous forces. For processes like desulphurization, more slag-metal contact area is crucial. From Fig. 3, it can be observed that high slag-metal contact can be obtained at high Reynolds numbers

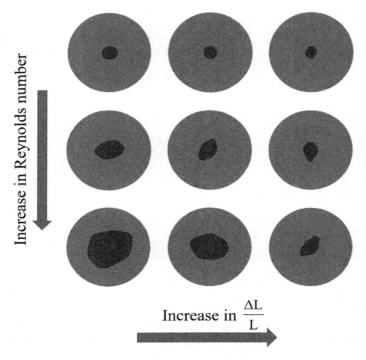

Fig. 3 Contours of average slag-eye opening at various Re and $\frac{\Delta L}{L}$ values

and low ratios of slag thickness to the metal bath depth. However, if preventing reoxidation of the steel is of importance, then high slag thickness to bath depth ratios and operating the ladle at low Reynolds numbers is recommended.

The variation of slag-eye area with time at different Reynolds numbers is obtained for two different slag thickness to metal bath ratios, and is plotted in Fig. 4. It can be observed from this figure that the slag opening increases as the flow begins and it becomes stationary as the process reaches steady state. It can also be observed that the slag eye area rises initially and as the flow progresses, it falls back and becomes stable. Slag layer is highly viscous in nature. The momentum of the argon plume at lower Reynolds numbers is easily dissipated in the slag layer, leading to a small open eye area. However, at higher Reynolds numbers and particularly at low ratios of slag thickness to metal bath depth, the viscous dissipation is dominated by the buoyancy and inertial forces of the argon plume, leading to larger open eye area.

A graph showing variation of the time averaged slag layer opening with increase in the Reynolds number is plotted at various values of the ratio of slag thickness to metal bath depth, which is shown in Fig. 5. The time averaged slag eye opening

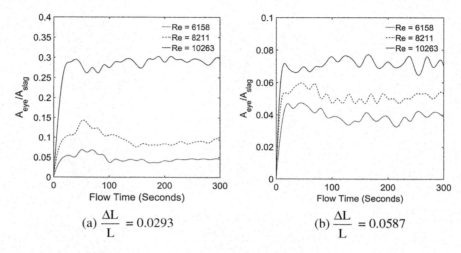

Fig. 4 Transient behavior of slag layer during argon gas purging

area is normalized with the initial area of the slag layer, i.e. the area of the slag layer before purging argon gas. It can be observed that with an increase in the Reynolds number, the slag eye opening area increases. Also, when the ratio of slag thickness to the bath depth is low, the average slag opening area is high. It is also observed that with an increase in the Reynolds number, the rate of change of average slag eye area is more at lower values of slag thickness to bath depth ratio. As discussed earlier, at high Reynolds numbers and low slag thickness to metal bath ratios, the viscous effects of the slag layer are dominated by the inertial effects of the argon bubble plume. When this happens, we can observe a sudden increase in the slag opening size, which can be observed in Fig. 5.

Fig. 5 Variation of slag eye area with Reynolds number

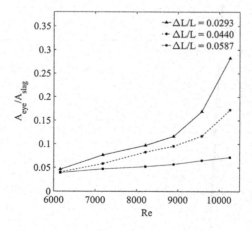

Conclusions

Multiphase coupled level-set volume of fluid (CLSVOF) simulations were performed to investigate the transient behavior of the slag layer during bottom purging of argon gas in steel making ladles. The results show that the slag eye size increases initially and reaches a steady state, after which the slag eye size remains constant. It is seen that the slag eye area increases as the Reynolds number increases or as the ratio of slag thickness to the bath depth decreases. To facilitate the slag-metal interactions, it is recommended to operate the ladle at high Reynolds numbers. If preventing oxygen contamination of molten steel is crucial, then low Reynolds numbers and high slag thickness to bath depth ratios are recommended.

References

1. K. Yonezawa, K. Schwerdtfeger, Spout eyes formed by an emerging gas plume at the surface of a slag-covered metal melt. Metall. Mater. Trans. B **30**(3), 411–418 (1999)
2. D. Mazumdar, J.W. Evans, A model for estimating exposed plume area in steel refining ladles covered with thin slag. Metall. Mater. Trans. B **35**(2), 400–404 (2004)
3. K. Krishnapisharody, G.A. Irons, Modeling of slag eye formation over a metal bath due to gas bubbling. Metall. Mater. Trans. B **37**(5), 763–772 (2006)
4. K. Krishnapisharody, G.A. Irons, A study of spouts on bath surfaces from gas bubbling: part 1. Experimental investigation. Metall. Mater. Trans. B **38**(3), 367–375 (2007)
5. M. Perananddhanthan, D. Mazumdar, Modeling of slag eye area in argon stirred ladles. ISIJ Int. **50**(11), 1622–1631 (2010)
6. C.A. Llanos, S.G. Hernandez, J.A.R. Banderas, J.D.J. Barreto, Modeling of the fluid dynamics of bottom argon bubbling during ladle operations. ISIJ Int. **50**(3), 396–402 (2009)
7. H. Liu, Z. Qi, M. Xu, Numerical simulation of fluid flow and interfacial behavior in three-phase argon-stirred ladles with one plug and dual plugs. Steel Res. Int. **82**(4), 440–458 (2011)
8. U. Sand, H. Yang, J. Eriksson, R.B. Fdhila, Numerical and experimental study on fluid dynamic features of combined gas and electromagnetic stirring in ladle furnace. Steel Res. Int. **80**(6), 441–449 (2009)
9. M. Sussman, E.G. Puckett, A coupled level set and volume-of-fluid method for computing 3D axisymmetric incompressible two-phase flows. J. Comput. Phys. **162**(2), 301–337 (2010)
10. T. Menard, S. Tanguy, A. Berlemont, Coupling level set/VOF/ghost fluid methods: validation and application to 3D simulation of the primary breakup of a liquid jet. Int. J. Multiph. Flow **33**(5), 510–524 (2007)
11. ANSYS Fluent User's Manual, Release 16.0, ANSYS Inc (2016)
12. F.R. Menter, Two equation eddy-viscosity turbulence models for engineering applications. Amer. Inst. Aeronaut. Astronaut. **32**(8), 1598–1605 (1994)

Developing Cemented Carbides Through ICME

Yong Du, Yingbiao Peng, Peng Zhou, Yafei Pan, Weibin Zhang, Cong Zhang, Kaiming Cheng, Kai Li, Han Li, Haixia Tian, Yue Qiu, Peng Deng, Na Li, Chong Chen, Yaru Wang, Yi Kong, Li Chen, Jianzhan Long, Wen Xie, Guanghua Wen, Shequan Wang, Zhongjian Zhang and Tao Xu

Abstract The ICME (Integrated Computational Materials Engineering) for cemented carbides aims to combine key experiments with multi-scale simulations from nano (10^{-10}~10^{-8} m) to micro (10^{-8}~10^{-4} m) to meso (10^{-4}~10^{-2} m) and to macro (10^{-2}~10 m) during the whole R&D process of cemented carbides. Based on ICME, the framework for R&D of cemented carbides, involving end-user demand, product design and industrial application, is established. In this work, a description to our established thermodynamic and thermophysical (diffusion coefficient, interfacial energy, and thermal conductivity and so on) databases is presented, followed by simulation of microstructure evolution during sintering of cemented carbides by means of phase field method. Work is also done to investigate the correlation between microstructure and mechanical properties (crack, stress distribution, and coupled temp-displacement) by using phase field and finite element methods. The proposed ICME for cemented carbides is used to develop a few new cemented carbides (including double layer gradient cemented carbides and γ'-strengthened Co-Ni-Al binder cemented carbides), which have found industry applications.

Keywords Thermodynamic database · Thermophysical database · Cemented carbides · Phase field simulation · Finite element method

---

Y. Du (✉) · Y. Pan · W. Zhang · C. Zhang · K. Cheng · K. Li · H. Li · H. Tian · Y. Qiu · P. Deng · N. Li · C. Chen · Y. Wang · Y. Kong · L. Chen
State Key Lab of Powder Metallurgy, Central South University, Changsha, China
e-mail: yong-du@csu.edu.cn

Y. Peng
Hunan University of Technology, Zhuzhou 412007, China

P. Zhou
Hunan University of Science and Technology, Xiangtan 411201, China

J. Long · W. Xie · G. Wen · S. Wang · Z. Zhang · T. Xu
Zhuzhou Cemented Carbide Cutting Tools Inc, Zhuzhou 412007, China

© The Minerals, Metals & Materials Society 2017
P. Mason et al. (eds.), *Proceedings of the 4th World Congress on Integrated Computational Materials Engineering (ICME 2017)*,
The Minerals, Metals & Materials Series, DOI 10.1007/978-3-319-57864-4_15

Introduction

Cemented carbides are of a great importance in the manufacturing industry, which have long been used in applications such as cutting, grinding, and drilling [1]. They are hard and tough tool materials consisting of micrometer-sized tungsten carbide embedded in a metal binder phase. In order to increase the performance and extend the service lifetime of the cemented carbides, gradient cemented carbides have been developed. In the past decades, cemented carbides were mainly developed through a large degree of mechanical testing. However, there are numerous factors influencing the microstructure and properties of cemented carbides, such as alloy composition, sintering temperature and time, partial pressure and so on. These factors can only be varied in an infinite number of ways through experimental method. The need to describe the interaction of the various process conditions has led to the ICME approach, which presents the opportunity to limit the experiments to an economical level.

Computational thermodynamics, using, e.g. the Thermo-Calc and DICTRA packages, has shown to be a powerful tool for processing advanced materials in cemented carbides [2–4], which is more efficient on composition and process parameters optimization compared with expensive and time-consuming experimental methods. With the development of thermodynamic and diffusivity databases, it is possible to make technical calculations on commercial products which are multicomponent alloys. On the basis of thermodynamic database, thermodynamic calculations can give an easy access to what phases form at different temperatures and alloy concentrations during the manufacture process. In the present work, a description to our established thermodynamic and thermophysical (diffusion coefficient, interfacial energy, and thermal conductivity and so on) databases is presented, followed by simulation of microstructure evolution during sintering of cemented carbides by means of phase field method. Work is also done to investigate the correlation between microstructure and mechanical properties (crack, stress distribution, and coupled temp-displacement) by using phase field and finite element methods.

Development of the Thermodynamic and Thermophysical Databases

Thermodynamic Database

Thermodynamic database is the important part of the ICME, which can give an easy access to multi-component equilibrium state resulting from heat treatment and be combined with diffusion kinetics and phase field to simulate the grain growth, solidified microstructure and so on. In our group, a thermodynamic database which contains the major elements in cemented carbides C-W-Co-Fe-Ni-Cr-V-Ti-

Ta-Nb-Zr-Hf-Al-N-Mo was established [3]. The basic information of the database is first the unary data, i.e. pure elements, which are taken from the compilation of the Scientific Group Thermodata Europe (SGTE), and then the binary, ternary and even higher-order systems. All phases present in those binary and ternary systems are described by means of the Gibbs energy as a function of composition and temperature which enable making predictions for multi-component systems and commercial alloys. With parameters stored in database, many different models, including the substitution solution model and sublattice model have been adopted for the phases in cemented carbide systems. (Co, AlNi$_3$) is a potential material to totally or partially replace the traditional Co as an alternative binder phase for cemented carbides. The database adopts the self-consistent thermodynamic description for the entire Al-Co-Ni ternary system, which can be used to design the formation of the strengthening phase AlNi$_3$ in Co binder phase [5]. Figure 1 presents the calculated vertical section along AlNi$_3$-Co system. The ternary carbides, which show a better performance than their binary counterparts, have received extensive attention [6, 7]. The database provides an accurately thermodynamic description of the miscibility gaps of the HfC-TiC, TiC-ZrC, VC-ZrC and HfC-TiC-WC systems. Figure 2 is the calculated miscibility gaps of Hf–Ti carbides in the HfC-TiC-WC system [8].

Diffusion Database

A reliable diffusion database is critical to simulate microstructure evolution of gradient cemented carbides and cellular cemented carbides, which have better performance and longer service lifetime than traditional cemented carbides. In

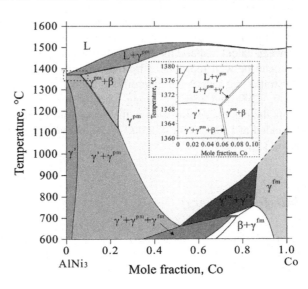

Fig. 1 Calculated vertical section along AlNi$_3$-Co

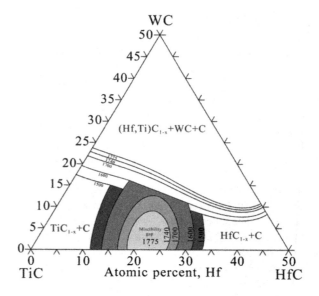

Fig. 2 Calculated miscibility gaps of Hf–Ti carbides in the HfC–TiC–WC system

2014, we established a diffusion database for multi-component C–Co–Cr–W–Ta–Ti–Nb–N cemented carbides [9]. The atomic mobility database for fcc and liquid in C-W-Co-Fe-Ni-Cr-V-Ti-Ta-Nb-Zr-Mo-Al-N cemented carbides was updated based on our new experimental data, literature data, first-principles calculation and theoretical assessment via the DICTRA (Diffusion Controlled TRAnsformation) software package. The atomic mobility parameters in liquid are theoretically calculated by the newly modified Sutherland equation, and the atomic mobility parameters in fcc phase are optimized by using the diffusivities measured in the present work and from the literature. The mobility parameters for self-diffusion and impurity diffusion in metastable fcc structure were determined through a semi-empirical method or first-principles calculations. Comprehensive comparisons between calculated and measured diffusivities indicate that most of the experimental data can be well reproduced by the currently obtained atomic mobilities. Figure 3 illustrates the calculated interdiffusion coefficients in fcc Co–W alloys at 1000, 1100, 1200, and 1300 °C along with the experimental values [10]. Figure 4 presents the calculated diffusion paths for various ternary Co–Cr–W diffusion couples annealed at 1100 °C for 432,000 s, compared with the corresponding experimental data [11].

Interfacial Energy

Interfacial energy is essential for control of grain growth and obtaining the desired microstructure in sintered poly-crystalline material. Understanding the structure, composition and energetics of internal interfaces is one of the most significant issue

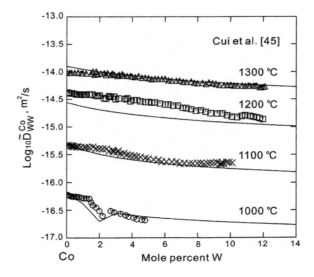

Fig. 3 Calculated interdiffusion coefficients in fcc Co–W alloys at 1000, 1100, 1200, and 1300 °C along with the experimental values [10]

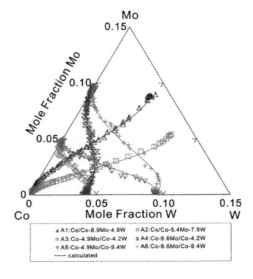

Fig. 4 Comparison between the calculated and the measured diffusion paths for the six diffusion couples annealed at 1373 K for 432.0 ks [11]

in the field of cemented carbides. A complication is that the interfacial energy is difficult to determine experimentally. During the past years, theoretical tools such as density-functional theory (DFT) have been used extensively to predict energies of various interfaces [12–14]. Using DFT calculations, Monte Carlo simulation, Calphad together with key experimental methods, the stability, segregation behavior, morphology and equilibrium interfacial phase diagram of cemented carbides were obtained. Figure 5 illustrates a typical WC/Co interfaces with

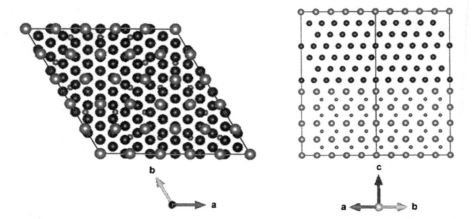

Fig. 5 The atomic structure of a WC(0001)/Co(111) interface, the atomic species are represented by different colors: Co(*blue*), W(*gray*), C(*brown*)

orientation relationship (0001)WC//(111)Co as well as [2 -1 -1 0]WC//[-1 1 0]Co. The interface energy γ was expressed as the excess Gibbs energy per unit interface area due to the presence of an interface.

Microstructure Characterization

TEM Observation

Ultrafine-grained WC-Co cemented carbides with average WC grain sizes varying from 0.2 to 0.5 μm, exhibit higher hardness and wear resistance than traditional cemented carbides. However, keeping this small grain size during the sintering process is difficult due to the existence of Ostwald-ripening mechanism. VC-Cr_3C_2 is the most effective combination of grain growth inhibitor for ultrafine-grained cemented carbides. However, the mechanism about its high effectiveness is still unclear. In a recent work [15] the inhibiting mechanism was investigated in a commercial ultrafine-grained WC-Co-VC-Cr_3C_2 cemented carbide by an objective-aberration-corrected transmission electron microscope. The results show that continuous segregation layers enveloping WC grains account for the high effectiveness of VC-Cr_3C_2 inhibitor. In detail, there are (V,W,Cr)C_x layers on (0001)WC basal facets and (Cr,W,V)C layers on {10$\bar{1}$0}WC prismatic facets (see Figs. 6 and 7). Especially, the newly observed continuous and coherent (Cr,W,V)C layers on {10$\bar{1}$0}WC prismatic facets accounts for the superiority of VC-Cr_3C_2 inhibitor compared to individual VC inhibitor.

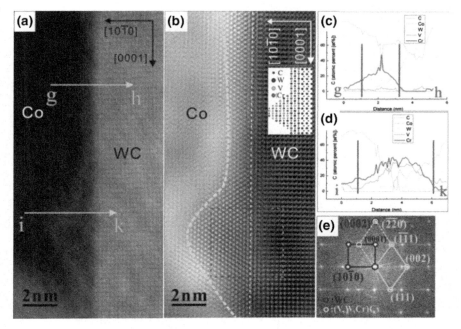

Fig. 6 TEM studies of the prismatic $(10\bar{1}0)_{WC}$ WC/Co interface co-doped with Cr and V, and viewing direction is parallel to $[1\bar{2}10]_{WC}$. **a** HAADF image of the prismatic WC/Co interface. **b** Fourier-filtered HRTEM image of the region of (**a**). **c** HREDX (high-resolution energy dispersive X-ray) line scan profile along gh shown by the *arrow* in (**a**), **d** HREDX line scan profile along ik shown by the arrow in (**a**), **e** FFT pattern of the area marked in *red dashed square* in (**b**)

Phase Field Simulation

Liquid phase sintering is a process for forming high performance, multiple-phase components from powders. The process includes very complex interactions between various mass transportation phenomena, among which the liquid phase migration represents an important one in the aspect forming gradient structure in cemented carbide.

In the present work, phase-field simulation of the liquid phase migration phenomenon during liquid phase sintering is performed in WC-Co based cemented carbide. A skeleton of 10 WC grains with particle size of 1 μm connected with SS interface is constructed and put in a box size of 128 × 128 filled with saturated liquid Co to investigate the grain separation process at 1673 K. The simulation results at 1, 5, 10 and 100 s are given in Fig. 8. It can be seen that during the separation of WC grains, the liquid phase tends to first fill in the triple junction, followed by complete particles segregation. Similar to the two-grain system, the

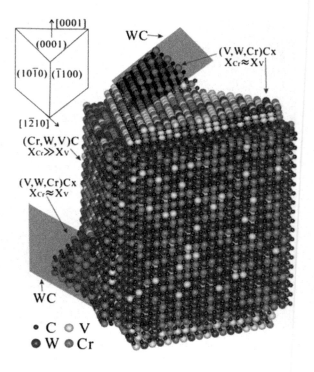

Fig. 7 Schematic of a WC grain covered by (V,W,Cr)Cx and (Cr,W,V) precipitates in atomic scale. XCr and XV represent for mole fractions of Cr and V, respectively

growth and ripening process of WC solid grains afterward take longer time. The segregation of WC grains in the liquid matrix, from another perspective, reveals the liquid binder melt absorption into the WC skeleton with no fluid flow. Quantitative analysis on the microstructure evolution as well as the binder volume change induced by this "migration" is performed as below to compare with the so called LPM process in cemented carbide system.

According to the present phase field simulation, the variation of contiguity in the WC-Co system can be analyzed quantitatively. According to the definition of contiguity, the SS and SL interface area within a 64 × 64 region located in the middle of the box in Fig. 8 is counted. The statistical results is plotted in Fig. 9 as dashed line, which is in reasonable agreement with the previous experimental result when the binder content is lower than 20 vol.%, while the computed contiguity dramatically decrease to zero at a binder content of 26 vol.%. The number of 26 vol. % denotes the critical binder content for complete separation of the WC grains in the present simulation, after which the binder content will keep growing owing to the WC grain growth. Future work considering the specific WC/WC grain boundaries ($\sigma_{SS} < 2\sigma_{SL}$) may lead to reasonable simulated contiguity of cemented carbide when the binder content is in the range of 20–60 vol.%.

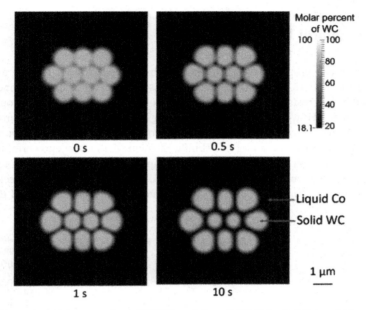

Fig. 8 Simulation of the separation of 10 WC particles at 1673 K for 0, 0.5, 1 and 10 s, box size 128 × 128. The composition of WC is denoted in *gray*, while the liquid matrix is shown in *black*

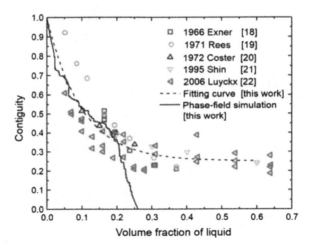

Fig. 9 Comparison between the present phase field simulated contiguity (*solid line*) and the previous experimental data

Correlation Between Microstructure and Mechanical Properties

The mechanical properties of cemented carbides are closely related to their microstructures. It is well known that hardness, TRS and fracture toughness are influenced by microstructural features such as volume fraction of phases, contiguity and grain size. However, it's still a challenging work to predict macroscopic properties from microstructure because of its complicated geometry. As a generic and flexible numerical method, FEM (Finite Element Method) is a powerful tool for investigating microstructure-property relations in cemented carbides. Both real and simulated microstructures can be used as input for FEM simulation. The mechanical behaviors of phases are described by appropriate constitutive models and damage

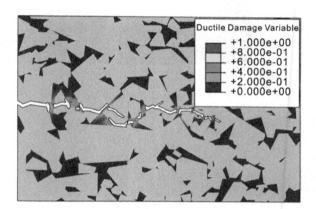

Fig. 10 Simulated crack path in cemented carbides

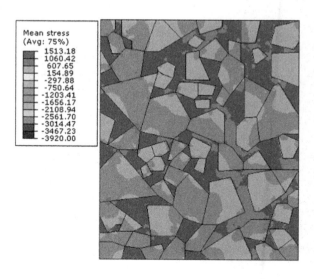

Fig. 11 Simulated thermal residual stress in cemented carbides

models. Boundary conditions are set to simulate the experiment condition. The simulation can reveal how the material responds to mechanical and/or thermal load in microscale. Macroscopic properties can also be calculated from such simulation. FEM has been used to investigate elasto-plastic behavior, strength, residual stress and crack propagation in cemented carbides. Figure 10 shows the presently simulated crack path under tensile load, and the microstructure was taken from the literature [16]. Figure 11 shows the distribution of thermal residual stress in a WC/Co composite.

Development of a Few New Cemented Carbides Through ICME

The urgent need to develop cemented carbides with excellent performance has pull industry toward ICME. With the integration of ICME into design, several brands of cemented carbides have been developed in the present work. Based on thermodynamic calculations and a concept of miscibility gap, a new type of double-layer gradient WC-Co-Ti(C,N)-ZrC cemented carbide has been prepared, as seen in Fig. 12. Based on a combination of thermodynamic calculations and calculation-designed experiment, the microstructure and concentration distribution of the different gradient layers have been systematically studied [17]. Another section, the WC-Co-Ni-Al cemented carbides with novel $\gamma + \gamma'$ Co-based binder have offered a great potential in the field of carbide rolls and hot die. Based on thermodynamic calculations, decisive experiments and microstructure characterizations, the formation mechanism for grain size of binder phase in WC-Co-Ni-Al

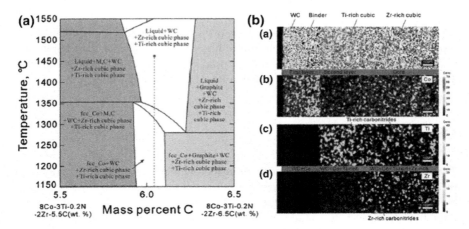

Fig. 12 a Calculated phase equilibria close to the sintering region of an alloy with the composition of WC-8Co-3Ti-2Zr-0.2 N. b SEM microstructure and distributions of elements Co, Ti and Zr of the alloy

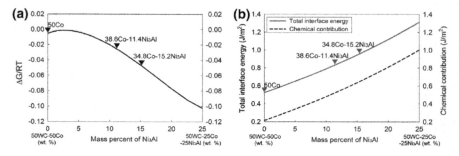

Fig. 13 a calculated driving force of the precipitation of the γ-binder phase from liquid in WC-50(Co–Ni–Al) with different binder composites at 1250 °C; **b** the solid-liquid interfacial energy in WC–50(Co–Ni–Al) with different Ni_3Al contents

Fig. 14 a EBSD maps of (**a**) WC-38.6Co-11.4Ni_3Al and **b** WC-34.8Co-15.2Ni_3Al with different binder phase grains shown in a color IPF (inverse pole figure)

cemented carbide has been clarified, as seen in Figs. 13 and 14 [18]. This work provided important information on controlling the grain size of γ′-strengthened Co-Ni-Al binder phase, which is of fundamental importance to the understanding of the performance characteristics of WC-Co-Ni-Al cemented carbides.

Conclusions

Thermodynamic and thermophysical (diffusion coefficient, interfacial energy, and thermal conductivity and so on) databases have been developed through a combination of experimental, theoretical and assessment work. The liquid phase migration process in the cellular cemented carbides was investigated through 1D numerical simulation and 2D phase field approach. The correlation between microstructure and mechanical properties has been investigated by using finite element method. An ICME framework with the validated ability to the development of a few new cemented carbides, including double-layer gradient cemented carbides, γ′-strengthened Co-Ni-Al binder cemented carbides, and so on, is demonstrated to be the effective way to develop new materials including cemented carbides.

Acknowledgements The financial support from National Natural Science Foundation of China (Grant Nos. 51371199 and 51601061) and Ministry of Industry and Information Technology of China (Grant No. 2015ZX04005008) is greatly acknowledged.

References

1. M. Rosso, G. Porto et al., Studies of graded cemented carbides components. Int. J. Refract. Metals Hard Mater. **17**, 187–192 (1999)
2. M. Ekroth, K. Frisk et al., Development of a thermodynamic database for cemented carbides for design and processing simulations. Metall. Mater. Trans. B **31B**, 615–619 (2000)
3. Y. Peng, Y. Du et al., CSUTDCC1-A thermodynamic database for multicomponent cemented carbides. Int. J. Refract. Metals Hard Mater. **42**, 57–70 (2014)
4. W. Zhang, Y. Du et al., CSUDDCC1—a diffusion database for multicomponent cemented carbides. Int. J. Refract. Metals Hard Mater. **43**, 164–180 (2014)
5. Y. Wang, P. Zhou et al., A thermodynamic description of the Al-Co-Ni system and site occupancy in Co- $AlNi_3$ composite binder phase. J. Alloy. Compd. **687**, 855–866 (2016)
6. I. Borgh, P. Hedström et al., Synthesis and phase separation of (Ti, Zr)C. Acta Mater. **66**, 209–218 (2014)
7. T. Ma, P. Hedström et al., Self-organizing nanostructured lamellar (Ti, Zr)C—a superhard mixed carbide. Int. J. Refract. Metals Hard Mater. **51**, 25–28 (2015)
8. Y. Pan, Y. Du et al., Thermodynamic description and quaternary miscibility gap of the C–Hf–Ti–W system. J. Alloy. Compd., Accepted
9. W. Zhang, Y. Du et al., CSUDDCC1-A diffusion database for multicomponent cemented carbides. Int J. Refract. Metals Hard Mater. **43**, 164–180 (2014)
10. Y.W. Cui, G.L. Xu et al., Interdiffusion and atomic mobility for face-centered cubic (FCC) Co–W alloys. Metall. Mater. Trans. A **44A**, 5–1621 (2013)
11. X. He, W. Zhang et al., Interdiffusivities and atomic mobilities in FCC Co–Mo–W alloys. CALPHAD **49**, 35–40 (2015)
12. M. Finnis, The theory of metal–ceramic interfaces. J. Phys.: Condens. Matter **8**, 5811 (1996)
13. R. Janisch, C. Elsässer, Segregated light elements at grain boundaries in niobium and molybdenum. Phys. Rev. B **67**(22), 224101 (2003)
14. W. Zhang, J. Smith et al., Influence of sulfur on the adhesion of the nickel/alumina interface. Phys. Rev. B **67**, 245414 (2003)
15. W. Guo, K. Li et al., Microstructure and composition of segregation layers at WC/Co interfaces in ultrafine-grained cemented carbides co-doped with Cr and V. Int. J. Refract. Metals Hard Mater. **58**, 68–73 (2016)
16. L.S. Sigl, H.E. Exner, Experimental study of the mechanics of fracture in WC-Co alloys. Metall. Trans. A **18**, 1299–1308 (1987)
17. W. Zhang, Y. Du et al., A new type of double-layer gradient cemented carbides: preparation and microstructure characterization. Scripta Mater. **123**, 73–76 (2016)
18. J. Long, W. Zhang et al., A new type of WC-Co-Ni-Al cemented carbide: grain size and morphology of γ′-strengthened composite binder phase. Scripta Mater. **126**, 33–36 (2016)

CSUDDCC2: An Updated Diffusion Database for Cemented Carbides

Peng Deng, Yong Du, Weibin Zhang, Cong Chen, Cong Zhang, Jinfeng Zhang, Yingbiao Peng, Peng Zhou and Weimin Chen

Abstract Cemented carbides are widely used in industry as cutting tools, wear parts, as a result of the high hardness and good toughness. A reliable diffusion database is critical to simulate microstructure evolution of cemented carbides. In 2014, we established version one of CSUDDCC1 (Central South University Diffusion Database for Cemented Carbides Version one). In this work, a description for the updated diffusion database CSUDDCC2 is presented. The atomic mobility database for fcc and liquid in C–W–Co–Fe–Ni–Cr–V–Ti–Ta–Nb–Zr–Mo–Al–N cemented carbides was established based on our new experimental data, literature data, first-principles calculation and theoretical assessment via the DICTRA (Diffusion Controlled TRAnsformation) software package. The atomic mobility parameters in liquid are theoretically calculated by the newly modified Sutherland equation, and the atomic mobility parameters in fcc phase are optimized by the diffusivities measured in the present work and from the literature. The mobility parameters for self-diffusion and impurity diffusion in metastable fcc structure were determined through a semi-empirical method or first-principles calculations. Comprehensive comparisons between calculated and measured diffusivities indicate that most of the experimental data can be well reproduced by the currently obtained atomic mobilities. Combining the thermodynamic database for cemented carbides, the diffusion database has been used to simulate the microstructure evolution during sintering of gradient cemented carbides. The simulated microstructure agrees reasonably with the experimentally observations.

Keywords Diffusion database · Atomic mobility · Cemented carbides · DICTRA simulation

P. Deng · Y. Du (✉) · W. Zhang · C. Chen · C. Zhang · J. Zhang · Y. Peng · P. Zhou · W. Chen
State Key Laboratory of Powder Metallurgy, Central South University, Changsha Shi, China
e-mail: yong-du@csu.edu.cn; yongducalphad@gmail.com

Introduction

Cemented carbides are hard and tough tool materials consisting of micrometer-sized tungsten carbide particles embedded in a metal binder phase, usually rich in Co [1]. They are widely used in applications such as cutting, grinding, and drilling [2]. In order to increase the resistance to plastic deformation or improve high-temperature performance, other carbides or carbonitrides with a cubic structure, such as TiC, (Ti, W)C, and (Ti, Ta, Nb) (C, N), have been introduced into WC–Co matrix. Having a good plasticity and toughness, the binder phase is named as soft phase. The WC and cubic phases are called the hard phase, and have a very high hardness and strength [3]. Some grain growth inhibitors such as Cr and V may also be added in small amounts. In order to prevent crack propagation from the coating into the substrate, a gradient layer, which is free of cubic phases and enriched in binder phase, is introduced between coating and substrate [4]. The formation of the gradient structure in multicomponent cemented carbides at liquid phase sintering temperature is a diffusion-controlled process, which is mainly affected by alloy compositions, sintering temperature, time and atmosphere. Knowledge of diffusivity is indispensable to understand formation mechanism, optimize technological parameters and design a new type of graded cemented carbides [5]. With the development of thermodynamic and diffusion databases, Thermo-Calc [6] and DICTRA [7] software have become a very important technique to simulate the gradient zone formation in the cemented carbides. The diffusion database CSUDDCC1 for important phases of major elements C–Co–Cr–W–Ta–Ti–Nb–N in cemented carbides has been developed in our group [5]. The simulations of different kinds of graded cemented carbides are presented to verify the validity of the diffusion database. With the development of cemented carbides, more kinds of cubic phase and binder phase are considered to improve its performance. In this work, the updated diffusion database CSUDDCC2 for fcc and liquid in C–W–Co–Fe–Ni–Cr–V–Ti–Ta–Nb–Zr–Mo–Al–N cemented carbides was established via experimental investigation and theoretical prediction. Consequently, the major purposes of the present work are: (1) to describe the development of the CSUDDCC2 database, (2) to evaluate the atomic mobilities of several binary, ternary systems by means of DICTRA applied to the experimental and theoretical diffusivities, and (3) to demonstrate two case studies (one is the grain growth simulation of cemented carbides, and the other is simulation of microstructure during sintering of gradient cemented carbides) and compare the simulations with the experimental data.

Development of CSUDDCC2

Model for Diffusion

Assuming a mono-vacancy atomic exchange mechanism for diffusion and neglecting correlation factors [8, 9], the tracer diffusivity D_i^* is related to the atomic mobility M_i by the Einstein relation:

$$D_i^* = RTM_i \tag{1}$$

For a substitutional solution phase, the interdiffusion coefficients in terms of the volume-fixed reference frame are given by the following expression [8, 9]

$$D_{kj}^n = \sum_i (\delta_{ik} - x_k) \cdot x_i \cdot M_i \cdot \left(\frac{\partial \mu_i}{\partial x_j} - \frac{\partial \mu_i}{\partial x_n} \right) \tag{2}$$

where δ_{ik} is the Kronecker delta ($\delta_{ik} = 1$ if $i = k$, otherwise $\delta_{ik} = 0$). x_i, μ_i and M_i are the mole fraction, chemical potential, and atomic mobility, respectively. The element n is chosen as the dependent element.

From the absolute reaction rate theory arguments, the atomic mobility M_i may be divided into a frequency factor M_i^0 and an activation enthalpy Q_i [8, 9]. Both M_i^0 and Q_i are in general dependent on the composition, temperature, and pressure. The M_i for disordered liquid phase and fcc phase are expressed by an equation of the form,

$$M_i = \exp\left(\frac{RT \ln M_i^0}{RT}\right) \exp\left(\frac{-Q_i}{RT}\right) \frac{1}{RT} \tag{3}$$

where R is the gas constant and T is the absolute temperature. Consequently, the composition dependence of the values of or between each endpoint of the composition space can be represented by a Redlich-Kister polynomial:

$$\Phi_i = \sum_p x_p \Phi_i^p + \sum_p \sum_{q>p} x_p x_q \left[\sum_{r=0}^m {}_2^r\Phi_i^{pq} (x_p - x_q)^r \right] + \sum_p \sum_{q>p} \sum_{v>q} x_p x_q x_v \left[v_{pqv}^s \cdot {}_3^s\Phi_i^{pqv} \right]$$

where x_p is the mole fraction of element p, Φ_i represents M_i^0 or Q_i, Φ_i^p is the value of Φ_i for i in pure p and thus represents the value of one endpoint in the composition space, ${}_2^r\Phi_i^{pq}$ and ${}_3^s\Phi_i^{pqv}$ are binary and ternary interaction parameters respectively. Each individual Φ parameter, i.e. Φ_i^p, ${}_2^r\Phi_i^{pq}$, ${}_3^s\Phi_i^{pqv}$ may be expressed as a polynomial of temperature and pressure if necessary. The parameter v_{pqv}^s is given by:

$$v_{ijk}^s = x_s + (1 - x_i - x_j - x_k)/3$$

where x_i, x_j, x_k and x_s are mole fractions of elements i, j, k and s, respectively.

Evaluation of Diffusivities for Liquid Phase and Metastable Disordered fcc Phase

Due to the lack of detailed information on liquid diffusion, Ekroth et al. [10] assumed that all elements (Co, Ti, W, Ta, Nb, C and N) have the same atomic mobility in the liquid. The activation energy was assumed to be 65,000 J/mol, and the frequency factor was chosen to be 9.24×10^{-7} m^2/s. However, this assumption is not realistic. Garcia et al. [11] assumed that the mobility of the metallic elements (W, Co, Ti, Ta and Nb) was two times slower than the atomic mobility of light non-metallic elements (C and N) by comparing the measured gradient layer thicknesses with the simulation ones. Recently, Chen et al. [12] have derived a modified Sutherland equation to predict the temperature-dependence diffusivity in liquid metals [13]. This method has been used in present work. Additionally, the ternary diffusivity of the liquid in Al–Fe–Ni system was obtained from Zhang et al. [14]. And the effect of the atomic mobilities in a liquid on microstructure and microsegregation during solidification was demonstrated with one Al–Ni binary alloy.

When the fcc states of some pure elements are metastable, it is impossible to assess the corresponding mobility parameters due to the unavailable experimental data. In this case, semi-empirical relations [15–17] or atomistic simulations [19, 20] are needed to evaluate the mobility parameters for self-diffusion in metastable fcc phase. The semi-empirical relations reviewed by Askill [20] have been applied successfully to calculate the self-diffusion coefficients of fcc-Zn and fcc-Mg. Consequently, they were thus employed in the present work to calculate the self-diffusivity of V, Ti, Zr and so on.

Critical Evaluation of Literature Data Diffusivities

Most of the mobility parameters corresponding to self-diffusion and impurity diffusion are taken from the literature. As the critical systems, the Co–W, and Co–Cr have been evaluated by Zhang et al. [5]. And considering that intermetallic compound Ni$_3$Al with fcc structure has attracted considerable attention to replace Co as new binder phase for WC-based cemented carbides due to its good mechanical properties at high temperatures [21], some new binary systems, e.g., Ni–W, and Ni–Al, are assessed based on the available experimental data. Figure 1 illustrates the calculated interdiffusion coefficients in fcc Ni–W alloys at different temperatures [22]. As a consequence, it is shown in Fig. 1 that the calculated interdiffusion coefficients according to the present atomic mobilities agree well with most of the experimental data and can reproduce much more experimental data than the results by Karunaratne et al. [23] and Campbell et al. [24]. Specially the first-principles calculations using the five jumps frequency model was utilized to evaluate the impurity diffusion coefficient of W in fcc phase.

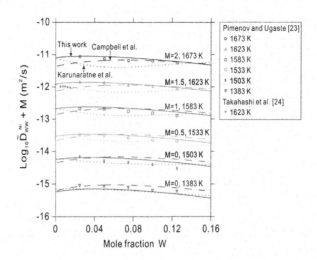

Fig. 1 Calculated interdiffusion coefficients in fcc Ni–W alloys compared with the experimental data at different temperatures [22]

Validation of the assessed atomic mobility includes not only the comparison with the experimental diffusion coefficients, but also the comparison between the predicted and the observed diffusion behaviors, such as the concentration profiles and the diffusion paths. The concentration-distance profiles in two ternary fcc/fcc-type diffusion couples of Ni–5 at.% Al/Ni–5 at.% W and pure Ni/Ni–5 at.% Al–5 at.% W annealed at 1523 K were also model-predicted in Fig. 2 [22], and compared with the experimental data measured by Hattori et al. [25]. As can be seen in the figure, the measured and calculated concentration-distance profiles are in good agreement. However, no experimental data are available in the literature for some binary and ternary systems, and the atomic mobility parameters in these systems are assessed based on the diffusivities determined by the Boltzmann–Matano method coupled with diffusion couple technique in our group. Figure 3 presents the calculated diffusion paths for various ternary Co–V–Ti diffusion couples annealed at 1100 °C for 432,000 s, compared with the corresponding experimental data [26]. It can be clearly seen from Fig. 3 that the calculated results for different Co-rich ternary diffusion couples agree well with the corresponding experimental values.

Applications of the Diffusion Database

For the cemented carbide industry, computer simulation can guide manufacturing processes and reduce the cycle for the development of new cemented carbides significantly. Many investigators have studied the gradient zone formation in the cemented carbides by computer simulation [5]. In order to validate the accuracy of the currently established diffusion database and show some applications, several

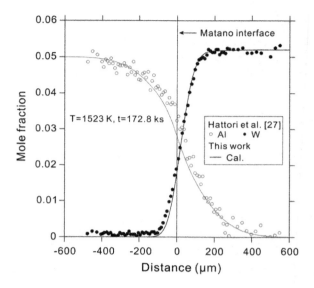

Fig. 2 Comparisons between the calculated and measured concentration-distance profiles of Ni–5 at.% Al/Ni–5 at.% W diffusion couple annealed at 1523 K for 172.8 ks [22]

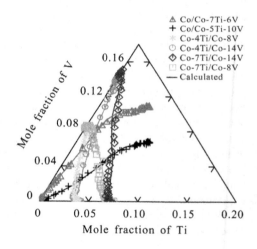

Fig. 3 Calculated diffusion paths for various ternary Co–V–Ti diffusion couples annealed at 1100 °C for 432,000 s, compared with the experimental data [26]

new types of cemented carbides were designed and sintered such as TiC–Co alloy and WC–Ti(C, N)–Ni$_3$Al grade cemented carbide. After sintering, the samples were cut, embedded in resin and polished. Scanning electron microscopy (SEM) was employed to investigate the grain size of ceramic and the microstructure of the gradient zone, the electron probe microanalysis (EPMA) and energy dispersive X-ray (EDX) were used to determine the concentration profiles of the elements.

Grain Growth Behavior of TiC–Co Alloy

The mechanical properties of Ti (C, N)-based alloys are closely related to their grain size. During the liquid phase sintering of Ti (C, N)-based alloys, the grain growth of ceramic phase occurs, which will affect the hardness, fracture toughness and cutting performance of alloys. Combining the thermodynamic database, diffusion database and the solid-liquid interfacial energies, TC-PRISMA was used to simulate grain growth behavior of TiC–Co alloys. Figure 4 shows that the TC-PRISMA simulated grain size of ceramic phase in TiC-20 vol% Co alloys sintered at 1450, 1500 and 1550 °C [27], while the grain growth experimental data were reported by Warren and Waldron [28]. Because the growth of the ceramic phase mainly depends on the dissolution of small particles, diffusion transported through the liquid phase, and precipitates on the larger particles. The high temperature state makes the liquid phase have a high diffusion coefficient, which is conducive to the growth of grain. Besides it can be found that the simulation results are in good agreement with the experimental data at 1500 and 1550 °C, despite a certain bias at 1450 °C. It is due to a reduction in volume fraction of the liquid phase as the temperature decreases, the degree of adhesion between the ceramic phase increases, thereby reducing the grain growth rate [29].

Graded WC–Ti(C, N)–Ni$_3$Al-Based Cemented Carbides

Based on thermodynamic calculation, WC–Ti(C, N)–Ni$_3$Al-based gradient cemented carbides have been prepared. Figure 5 shows SEM micrograph of WC–8Ni$_3$Al–3Ti–0.2N (wt%) and the distributions of the elements Ti, Ni, Al and

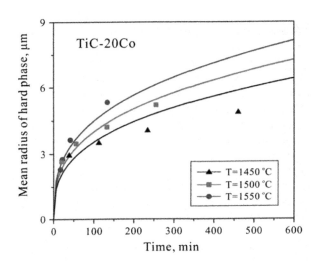

Fig. 4 The TC-PRISMA simulated grain growth behavior of ceramic phase in TiC-20 vol% Co alloys sintered at 1450, 1500 and 1550 °C [27]

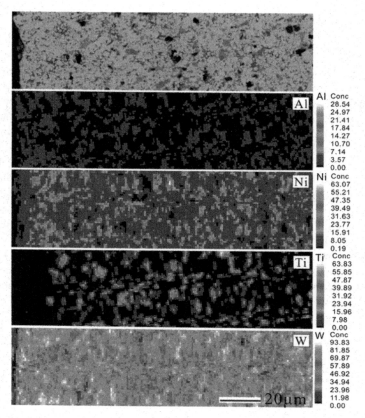

Fig. 5 The distributions of the elements Ti, Ni, Al and W in the gradient zone of the WC–8Ni3Al–3Ti–0.2N (wt%) with different binder phase vacuum sintered at 1450 °C for 2 h [30]

W in the gradient cemented carbides vacuum sintered at 1450 °C for 2 h [30]. It is obvious that the near-surface of the alloy has formed the gradient zone which is enriched in binder phase and depleted in cubic carbides. By combining the established thermodynamic and this diffusivity database, DICTRA software has been used to simulate formation of the gradient zone. Figure 6 illustrates the simulated elemental concentration profiles for Al, Ni, Ti and W in alloys, compared with the measured data [30]. The result indicates that the content of Ti is free in the near-surface zone and enrich inside the surface zone. At the near-surface zone, the contents of Ni and Al increase sharply and reached maximum value. Beneath the near-surface zone, the decrease of Ni and Al are observed, which leads to the minimum value. The change of W content in the surface layer is opposite to that of Ni and Al content. As can be seen in both figures, the updated diffusion databases can reasonably reproduce most of the experimental concentration profiles.

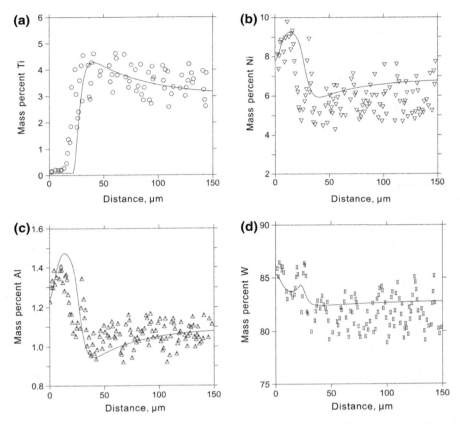

Fig. 6 Experimental measured elemental concentration distance profiles in sample WC– 8Ni3Al–3Ti–0.2N (wt%) gradient cemented carbides: **a** Ti; **b** Ni; **c** Al; **d** W [30]

Summary

An updated diffusion database for multi-component C–W–Co–Fe–Ni–Cr–V–Ti–Ta–Nb–Zr–Mo–Al–N cemented carbides has been developed through a combination of experiment, theoretical analysis and assessment. The diffusion database contains the atomic mobility parameters for different diffusing elements in liquid and fcc phase.

The new atomic mobilities of several binary, ternary systems were evaluated by means of DICTRA applied to the experimental and theoretical diffusivities. Comprehensive comparisons between calculated and measured diffusivities indicate that most of the experimental data can be well reproduced by the currently obtained atomic mobilities.

Combining the thermodynamic database, diffusion database and the solid-liquid interfacial energies, TC-PRISMA was used to simulate grain size of ceramic phase in TiC-20 vol% Co alloy sintered at different temperatures. Additionally, DICTRA software has been used to simulate formation of the gradient zone of WC–8Ni3Al–3Ti–0.2N (wt%) alloy. Good agreements between simulated and measured results indicate the powerful ability of the presently established database in cemented carbides design and process optimization.

Acknowledgements The financial support from National Natural Science Foundation of China (Grant No. 51371199) and Ministry of Industry and Information Technology of China (Grant No. 2015ZX04005008) is greatly acknowledged.

References

1. H.O. Andrén, Microstructures of cemented carbonitrides. Mater. Des. **22**, 8–491 (2001)
2. H.E. Exner, Int. Met. Rev. **24**, 149–173 (1979)
3. Y. Liu, H.B. Wang, Z.Y. Long, P.K. Liaw, J.G. Yang, B.Y. Huang, Microstructural evolution and mechanical behaviors of graded cemented carbides. Mater. Sci. Eng. A. **426**, 54–346 (2006)
4. T.E. Yang, J. Xiong, L. Sun, Z.X. Guo, D. Cao, Int. J. Miner. Metal. Mater. **18**, 709–716 (2011)
5. W.B. Zhang, Y. Du et al., CSUDDCC1—A diffusion database for multicomponent cemented carbides. Int. J. Refract. Met. Hard Mater. **43**, 164–180 (2014)
6. B. Sundman, B. Jansson, J.O. Andersson, CALPHAD **9**, 153–190 (1985)
7. J.O. Andersson, L. Hoeglund, B. Jansson, J. Agren, in *Proceedings of the International Symposium on Fundamentals and Applications of Ternary Diffusion*, 1990, pp. 153–163
8. B. Jönsson, Assessment of the mobility of carbon in fcc C–Cr–Fe–Ni alloys. Z. Metallkd. **85**, 9–502 (1994)
9. J.O. Andersson, J. Ågren, Models for numerical treatment of multicomponent diffusion in simple phases. J. Appl. Phys. **72**, 5–1350 (1992)
10. M. Ekroth, R. Frykholm, M. Lindholm, H.O. Andrén, J. Ågren, Gradient zones in WC–Ti(C, N)–Co-based cemented carbides: experimental study and computer simulations. Acta Mater. **48**, 85–2177 (2000)
11. J. Garcia, G. Lindwall, O. Prat, K. Frisk, Kinetics of formation of graded layers on cemented carbides: experimental investigations and DICTRA simulations. Int. J. Refract. Met. Hard Mater. **29**, 9–256 (2011)
12. W.M. Chen, L.J. Zhang, D.D. Liu, Y. Du, C.Y. Tan, Diffusivities and atomic mobilities of Sn–Bi and Sn–Pb melts. J. Electron. Mater. **42**, 70–1158 (2013)
13. W.B. Zhang, Y. Du, Y.B. Peng, W. Xie, G.H. Wen, S.Q. Wang, Experimental investigations and simulations of the effect of Ti and N content on formation of fcc-free surface layers in WC–Ti(C, N)–Co cemented carbides. Int. J. Refract. Met. Hard Mater. **41**, 47–638 (2013)
14. L.J. Zhang, Y. Du et al., Diffusivities of an Al–Fe–Ni melt and their effects on the microstructure during solidification. Acta Mater. **58**, 3664–3675 (2010)
15. S. Dushman, I. Langmuir: Proc. Am. Phys. Soc. **113** (1992)
16. C. Zener, J. Appl. Phys. **22**, 372 (1951)
17. R.A. Swalin, J. Appl. Phys. **27**, 554 (1956)
18. M. Matina, Y. Wang, R. Arroyave, L.Q. Chen, Z.K. Liu, C. Wolverton, Phys. Rev. Lett. **100**, 215901 (2008)
19. D.D. Zhao, Y. Kong, A.J. Wang, L.C. Zhou, S.L. Cui et al., J. Phase Equilib. Diffus. **32**, 128 (2011)

20. J. Askill, *Tracer Diffusion Data for Metals, Alloys, and Simple Oxides* (IFI, Plenum, New York, 1970)
21. M. Ahmadian, D. Wexler, T. Chandra, A. Calka, Int. J. Refract. Met. Hard Mater. **23**, 155–159 (2005)
22. C. Chen, L.J. Zhang, Y. Du et al., Diffusivities and atomic mobilities in disordered fcc and ordered $L1_2$ Ni–Al–W alloys. J. Alloys Compd. **645**, 259–268 (2015)
23. M.S.A. Karunaratne, D.C. Cox, P. Carter, R.C. Reed, *Superalloys 2000*, TMS (2000), pp. 263–272
24. C.E. Campbell, W.J. Boettinger, U.R. Kattner, Acta Mater. **50**, 775–792 (2002)
25. M. Hattori, N. Goto, Y. Murata, T. Koyama, M. Morinaga, Mater. Trans. **46**, 163–166 (2005)
26. J.F. Zhang, Paper in preparation (2017)
27. C. Zhang, Ph.D. thesis, Central South University (China), 2016 (unpublished work)
28. R. Warren, M.B. Waldron, Microstructural development during the liquid-phase sintering of cemented carbides. II. Carbide grain growth. Powder Met. **15**(30), 180–201 (1972)
29. M. Pellan, S. Lay, J.-M. Missiaen et al., Effect of Binder Composition on WC Grain Growth in cemented carbides. J. Am. Ceram. Soc. **98**(11), 3596–3601 (2015)
30. C. Chen, Ph.D. thesis, Central South University (China), 2016 (unpublished work)

Part III
Microstructure Evolution

Multi-scale Modeling of Quasi-directional Solidification of a Cast Si-Rich Eutectic Alloy

Chang Kai Wu, Kwan Skinner, Andres E. Becerra, Vasgen A. Shamamian and Salem Mosbah

Abstract Dow Corning Corporation recently examined the use of transition metal-silicon eutectics for producing melt-castable ceramic parts. These materials display good strength, wear and corrosion resistance. The near-eutectic solidification structure has significant impact on the final properties of a cast component. However, direct simulation of the cast structure at industrial scales remains a challenge. The objective of this work is to develop a multi-scale integrated solidification model that includes: density functional theory (DFT) calculations, which enable the computation of difficult-to-measure thermophysical properties; microstructural evolution simulation, which tackles nucleation eutectic growth and segregation during solidification; and casting modeling, which accounts for different boundary conditions including temperature-dependent heat transfer coefficients and geometry. The developed 3D coupled code can predict the correct morphology of the solidified composite and aid in the design and optimization of melt-cast parts based on composition and process parameters in a virtual environment. To verify the model, a mold was designed to achieve quasi-directional solidification within large regions of each casting; hypo- and hyper-eutectic Si–Cr alloys were cast into this custom mold using a vacuum tilt pour unit. Our experimental efforts focused on the quantification of the effects of process conditions on the resulting microstructure of the cast component. Local segregation was examined and compared with the model's predictions. Results are in agreement with the microstructure observed in our castings.

Keywords Solidification · Si alloy · Multi-scale modeling · Eutectic

---

C.K. Wu (✉) · K. Skinner · A.E. Becerra · V.A. Shamamian
Dow Corning Corporation, 3901 S. Saginaw Rd., Midland, MI 48686, USA
e-mail: lance.wu@dowcorning.com

S. Mosbah
Think Solidification, 2033 Gateway Place, Suite 500, San Jose, CA 95110, USA

© The Minerals, Metals & Materials Society 2017
P. Mason et al. (eds.), *Proceedings of the 4th World Congress on Integrated Computational Materials Engineering (ICME 2017)*, The Minerals, Metals & Materials Series, DOI 10.1007/978-3-319-57864-4_17

Introduction

The advent of sophisticated 3D design software integrated into predictive thermal/mechanical performance models in the automotive and aerospace industries has created opportunities for the materials processing community to develop methods that produce near net shape components of unparalleled complexity. Increased activities in the production and commercialization of parts produced by metal and ceramic injection molding, reaction bonded pressure infiltration ceramic matrix composites, and most recently, 3D printing are just some of the examples of methods entering the workplace. However, ceramic like materials and composites have lagged behind in these technologies because of their high processing temperatures, and the run to run reliability due to defects from porosity from traditional powder consolidation methods.

The over-arching goal of the project is to design melt-castable ceramic materials, and to design high-performance Si-based composite materials for structural applications. In spite of silicon's good thermal, mechanical and chemical properties (including strong acid corrosion resistance), it is one of the most brittle technical ceramics, exhibiting a fracture toughness of 0.7–0.9 MPa m$^{1/2}$. In terms of melt casting technologies, Si melts at the relatively low temperature of 1414 °C and is a strong compound former. Thus, Si and transition metals will easily form eutectics to produce brittle-brittle composite structures upon solidification of an alloy melt. Thus, the challenge is to design a polyphase microstructure that increases the toughness of the materials beyond that of the constituent phases. The required toughness of the composite can be attained through a fine spatial dispersion of one brittle phase in another. This polyphase mixture should have microstructural features sufficiently small, and thus interfacial area sufficiently high, to yield significant crack deflection. Such crack deflection is well known to yield a tough material using brittle component phases [1]. The fine distribution of the phases will be created via solidification of eutectic or near-eutectic alloys. Thus, models of the eutectic solidification process are needed to design the processing path and molds to yield these fine microstructures. The goal is to produce a castable-ceramic with a fracture toughness in the 4–6 MPa m$^{1/2}$ range.

Background

The control of microstructure, thermal stresses and casting integrity during solidification is an important concern for casting Si alloys. In order to design and cast a sound part using different solidification conditions, casting models should be used. All models require an extensive thermophysical and thermomechanical database. The nature of casting solidification is a complex process due to the broad temperature range experienced by the Si alloy and mold materials. Accurate measurements or estimations of properties contained within the thermophysical database lead to significant improvements in the prediction accuracy.

Materials and Database Generation

In this study, the complete thermophysical property database of a Si–Cr eutectic alloy is developed and used in casting simulations. The Si–Cr phase diagram is shown in Fig. 1. The most Si-rich eutectic reaction occurs at 85.24 at.% Si and at 1328 °C. It consists of the elemental silicon and chromium disilicide ($CrSi_2$) phases. Cooperative eutectic solidification results in a rod-like or fibrous silicide phase embedded in a silicon matrix. The growth is coupled by the thermal diffusion field and solute redistribution [2] in front of the growth interface. The final material and mechanical properties are strongly dependent on microstructure, especially on the diameter of silicide rods. Work related to microstructure analysis corresponding to the measured solidification rate is included, and used as a reference in developing a predictive model. The developed model is used to predict the microstructure. Predictions were verified by measurements made on cast parts.

Various properties of silicon-based alloys were calculated using ab initio molecular dynamics simulations provided in the Vienna Ab initio Simulation Package (VASP) [3]. Ionic positions were modeled via classical Nosé-Hoover dynamics for a chosen number of atoms, volume, and temperature via the NVT ensemble. Simulation cells were generally limited to 108 atoms, although some larger simulations were performed to obtain additional benchmark data. Total simulation times ranged from 4 to 10 ps, using 2.5 fs steps when calculating the

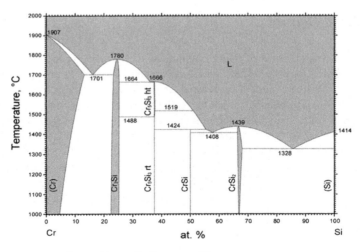

Fig. 1 Phase diagram for the binary Si–Cr system [4]

ionic trajectories via self-consistent charge densities. Correlation times were computed up to 1000 ionic steps when calculating diffusion constants. Simulations were performed with a single k-point (Γ point) when performing reciprocal space integrations. A plane wave cutoff of 320–350 eV was used in all calculations, using a PW91 (GGA) projector augmented planewave potential. Simulations were run for 10 ps at temperatures above the melting point to obtain a good starting configuration for production runs and to erase atomic configuration-based dependencies on the system energy. The positions from this mixing simulation were then used to execute shorter runs of 4 ps at various volumes. Typically five such simulations are run in order to generate enough data points when generating a fit for the zero-volume pressure, which was based on a second order polynomial fit. The simulation was then run again at the zero-volume pressure condition for 5–10 ps to generate adequate statistics for determining self diffusion constants.

Thermophysical Database

Thermal conductivity of Si–Cr eutectic is measured by Thermophysical Properties Research Laboratory (TPRL) up to 1100 °C. Temperature-dependent specific heat capacity is obtained via differential scanning calorimetry (DSC). In order to determine the melting range and latent heat, a high temperature differential scanning calorimeter (HT-DSC) was used in which energy variation due to phase transformation was observed. The measured latent heat of the Si–Cr eutectic alloy was 744.13 J/g. The measured onset and peak temperatures were 1333.1 °C and 1328.0 °C, respectively.

Solid/Liquid Density

Solid density was determined using coefficients of thermal expansion (CTE) from thermomechanical analysis (TMA) measurements. While it is straightforward to measure the density of the solid material under normal lab conditions, the determination of the liquid phase is much more challenging as these measurements must be performed with specialized equipment and at highly elevated temperatures. Recent developments in the simulation of liquids and solids using ab initio molecular dynamics (AIMD) calculations now permit atomistic simulations to be performed from first-principles calculations without the need of empirical fits or adjusted parameters. The output of these simulations include various thermal, structural and elastic properties, including the density of the material at arbitrary temperatures and pressures.

Liquid Viscosity

The shear viscosity is another quantity made available through these AIMD simulations. This value can be determined from the Einstein-Stokes equation [5]. Our viscosity value for silicon at a temperature above its melting point is within the range of published literature values (0.60–0.90 mPa s) [6]. Therefore, the AIMD data combined with the appropriate analysis gives us a tool to provide reasonable values of viscosities at arbitrary temperatures and can be used as input values for other simulations and tools. The shear viscosity of Si–Cr eutectic melt determined via AIMD was 1.6 mPa s.

Experimental—Casting

A mold was specially designed so as to obtain melt-cast cylindrical ingots with large regions exhibiting quasi-directional growth. This type of mold maximizes the axial cooling rate at the base while minimizing the cooling rate along all other directions. The main components of the mold design are shown in Fig. 2. The base of the mold was a graphite puck (shown in green) with a flat bottom surface and a concave top surface; this thin puck was placed on a water-chilled copper plate and acted as chemically inert interface for fast cooling. A hollow aluminosilicate ISA hot top (Insulation Specialties of America KVS-3584-456-3000) cylinder (shown in blue) sat on the perimeter of the graphite puck, acting as an insulating wall providing a much lower cooling rate than the chilled base. A mirror finish, aluminum sheet was also rolled into a larger cylinder and placed concentrically around the mold so as to reflect infrared radiation and further minimize the radial cooling rate. Finally, a graphite funnel was placed on top of the insulating cylinder to assist in centering the poured melt into the mold.

Ingots (900 g) at the Si–Cr eutectic composition were cast within a vacuum tilt-pour unit. For this, high-purity Si and Cr pieces were loaded into the crucible

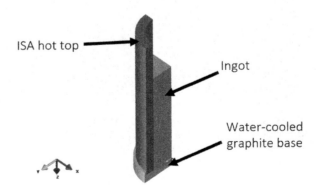

Fig. 2 Schematic showing the ingot and mold surfaces in quarter view (color online)

(an insulated SiC-coated graphite susceptor wrapped by an induction coil) and the chamber was evacuated to a base pressure below 10^{-3} Torr before heating started. The charge was melted in the crucible and superheated (typically to 1500 °C) to allow for good melt mixing, and held for 10 min. The melt was then poured smoothly into the mold, which was allowed to cool before the chamber was vented. The result of such operation is a cylindrical ingot (shown in pink) with a convex bottom surface. Due to the chilled base plate and insulating wall, a large section in the bottom half of the ingot displays quasi-directional vertical solidification. The radial solidification front does make the area of interest increasingly narrow as the distance from the bottom increases. The microstructure and composition as a function of position within the region of quasi-directional solidification could then be analyzed via scanning electron microscopy (SEM) and related techniques, as well as compared to the casting computational simulations.

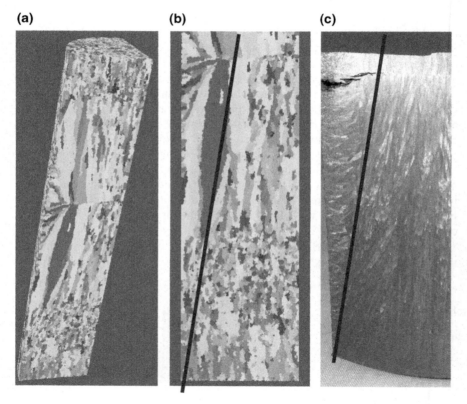

Fig. 3 Si–Cr eutectic alloy: **a** model-predicted microstructure on the quarter ingot, **b** model predicted microstructure on the cross section of the bottom half ingot and **c** cross-sectional micrograph of the bottom half ingot

Casting Simulation Details and Results

The model predictions of the Si–Cr eutectic composition, as shown in Fig. 3a, b, were compared to cast ingot. The ingot was cut in half and the cross section was polished for better visualization of the grain structure, as shown in Fig. 3c. The predicted microstructure showed good agreement with the cross-sectional micrograph regarding the location of different solidification fronts, indicated by red lines in Fig. 3b, c. Variations in color represent different grain orientations. The initial and boundary conditions of casting model are as follows:

Casting Initial and Boundary Conditions

The boundary conditions used in casting simulations were as follows:
- heat flux through the side wall = 10,000 W/m^2;
- heat flux through the open top = 150,000 W/m^2;
- heat transfer coefficient on the bottom dome = 300 W/m^{2}K;
- initial nucleation density on the wall and on the bottom dome surface = 1×10^{11} m^{-2};
- initial nucleation density on the open top = 0; and,
- initial internal nucleation density was 1×10^{10} m^{-2}.

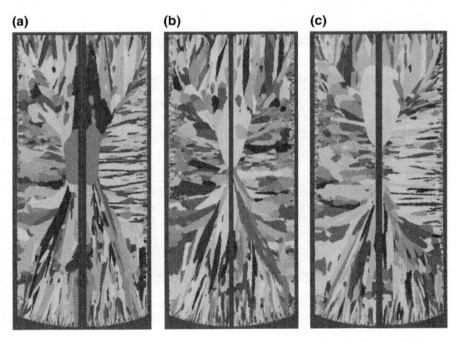

Fig. 4 Model predicted microstructure at **a** 80.24 at.%, **b** 82.24 at.% and **b** 83.24 at.% Si

Casting: Solidification Model and Predictions

Solidification Model: A representative Finite Volume, FV, is assumed to be in one of three thermodynamic states; solid, liquid and/or mixture of the two (mushy zone). The metallic alloy phase transformation from the liquid to the solid phase involves multi-scale physics that control the solute redistribution, solid/liquid phase instability, heat transfer, phases (grains) nucleation, etc. In this work, two distinct scales are considered, the macro scale, with the same order of magnitude as the

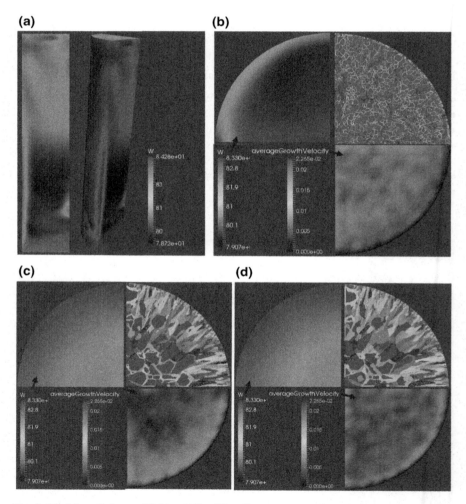

Fig. 5 Model predictions at 80.24 at.% Si. **a** W = Si concentration; **b**, **c** and **d** Cross sections at 1, 9 and 12 mm from the ingot bottom. Clockwise from *top left*: Si concentration, grain structure, average growth velocity

representative finite volume, and a meso scale of the order of magnitude of the secondary dendritic arms spacing, SDAS. The latter is used to derive a scalable tracking algorithm of the solidification grain structure and to describe grains boundaries. For each scale, an adequate mathematical formulation of the physics is used and the corresponding balance equations are solved at each discretization time step, Δt, with full coupling between the solver of the averaged macroscopic equations and computed solidification grain structure. In addition to the tracking of each grain envelope, the same Discrete Point, DP, Lagrangian based formulation is used to simulate the nucleation (primary and secondary phases), compute grains boundaries and growth kinetic. Additional details about the coupled FV-DP solidification model can be found in reference [7]. SPrime, the model implementation using the open source CFD library OpenFOAM [8] was used to carry out the simulations.

Predictions: The predicted microstructure from the model on the cross section surface of several Si–Cr hypoeutectic compositions are shown in Fig. 4a–c, respectively. In addition to grain structure, the developed model can predict segregation and average growth velocity inside a cell. The model predicted Si segregation of 80.24 at.% Si composition is shown in Fig. 5a. Meanwhile, the model-predicted grain structure, Si concentration and average growth velocity are shown in Fig. 5b–d for cross sections at different heights. Figure 5c shows results at 9 mm from the ingot bottom where the last solidified region located.

Summary and Conclusions

A multi-scale integrated solidification model was developed. The model includes density functional theory (DFT) calculations, microstructural evolution simulation and casting modeling. The developed model was validated with experimental data and used to predict the morphology, segregation and average growth velocity of the solidified Si–Cr eutectic composites.

References

1. John W. Hutchinson, Henrik M. Jensen, Models of fiber debonding and pullout in brittle composites with friction. Mech. Mater. **9**(2), 139–163 (1990)
2. H. D. Brody, Solute redistribution in dendritic solidification. Doctoral dissertation Massachusetts Institute of Technology, 1965
3. Georg Kresse, Jürgen Hafner, Ab initio molecular dynamics for liquid metals. Phys. Rev. B **47**(1), 558 (1993)
4. H. Okamoto, Si–Cr phase diagram. ASM Alloy Phase Diagrams Center (2006)
5. L. Battezzati, A.L. Greer, The viscosity of liquid metals and alloys. Acta Metall. **37**(7), 1791–1802 (1989)

6. Yuzuru Sato et al., Viscosity of molten silicon and the factors affecting measurement. J. Cryst. Growth **249**(3), 404–415 (2003)
7. J. Groh et al., Grain structure and segregation modeling using coupled FV and DP model. In *Proceedings of the 8th International Symposium on Superalloy 718 and Derivatives*, The Minerals, Metals & Materials Society, 2014
8. www.openfoam.org

Numerical Simulation of Macrosegregation in a 535 Tons Steel Ingot with a Multicomponent-Multiphase Model

Kangxin Chen, Wutao Tu and Houfa Shen

Abstract To accurately simulate the formation of macrosegregation, a major defect commonly encountered in large ingots, solidification researchers have developed various mathematical models and conducted corresponding steel ingot dissection experiments for validation. A multicomponent and multiphase solidification model was utilized to predict macrosegregation of steel ingots in this research. The model described the multi-phase flow phenomenon during solidification, with the feature of strong coupling among mass, momentum, energy, and species conservation equations. Impact factors as thermo-solutal buoyancy flow, grains sedimentation, and shrinkage-induced flow on the macroscopic scale were taken into consideration. Additionally, the interfacial concentration constraint relations were derived to close the model by solving the solidification paths in the multicomponent alloy system. A finite-volume method was employed to solve the governing equations of the model. In particular, a multi-phase SIMPLEC (semi-implicit method for pressure-linked equations-consistent) algorithm was utilized to solve the velocity-pressure coupling for the specific multiphase flow system. Finally, the model was applied to simulate the macrosegregation in a 535 tons steel ingot. The simulated results were compared with the experimental results and the predictions reproduced the classical macrosegregation patterns. Good agreement is shown generally in quantitative comparisons between experimental

K. Chen · H. Shen (✉)
School of Materials Science and Engineering, Tsinghua University,
Beijing 100084, People's Republic of China
e-mail: shen@tsinghua.edu.cn

K. Chen
e-mail: chenkx13@mails.tsinghua.edu.cn

W. Tu
SMIC Advanced Technology Research & Development (Shanghai) Corporation,
Shanghai, China

results and numerical predictions of carbon, chromium and molybdenum concentration. It is demonstrated that the multicomponent-multiphase solidification model can well predict macrosegregation in steel ingots and help optimize the ingot production process.

Keywords Macrosegregation · Multicomponent · Multiphase modeling · Steel ingots

Introduction

Large ingots are typically used to produce key components in large equipment, which are used in energy and power, metallurgical machinery, transportation, national defense and other fields. Macrosegregation, i.e. the compositional heterogeneity, is a serious defect commonly encountered in large ingots. This compositional heterogeneity usually leads to the non-uniform distribution of microstructure, harms the mechanical properties, and, at its worst, may even result in scrapping the final product. Macrosegregation results from the combined effect of compositional segregation at the microscopic scale and solid-liquid separation at the macroscopic scale. The relative motion of solid phase and liquid phase is mainly caused by such factors as thermo-solutal buoyancy flow, grains sedimentation, shrinkage-induced flow, and deformation of the solid phase skeleton [1, 2]. As macrosegregation cannot be eliminated by the subsequent heat treatment process, it is important to control the formation of macrosegregation during solidification to obtain a less segregated ingot. Due to the high cost of casting large ingots, it is unrealistic to simply use experimental methods to study macrosegregation. Numerical simulation, with the advantages of low cost and high efficiency, has become an important tool for understanding complex solidification processes and macrosegregation of large ingots.

Since the pioneering work of Flemings and Nereo [3] in 1960s, much effort has been devoted to the modeling of macrosegregation. Varieties of mathematical models have been developed to account for such mechanisms as thermal-solutal convection, free equiaxed crystal movement, solidification shrinkage-induced flow [4, 5]. The macrosegregation models commonly used can be divided into single-phase models, two-phase models and multi-phase models according to the number of basic dynamics phases used. Among them, the relatively simple single-phase models, divided into continuum models and volume average models, mainly consider the thermal-solutal buoyancy flow. The two-phase models or multi-phase models, further considering the effect of free equiaxed crystal motion or solid phase deformation on macrosegregation, however, are rather complicated. Numerous applications of macrosegregation models in steel ingots have been reported. Vannier and Combeau [6] utilized a continuum model to simulate macrosegregation of a 65 tons Fe-0.22 wt%C steel ingot, and compared the predictions with measurements of compositions in the center line. Liu et al. [7] adopted a continuum model to predict macrosegregation in a 360 tons ingot. Tu and coworkers [8] recently developed a multicomponent-multiphase model based on a two phase model previously developed in his research group, and applied the model to investigate a 36 tons Fe-0.51 wt%C-0.006 wt%S steel ingot, which was experimentally investigated by temperature recording and concentration analysis. In this experiment, the macrosegregation maps of carbon and sulphur were obtained based on 1800 component sampling points covering the entire half-section. General macrosegregation patterns in measurements have been reproduced by predictions. Good agreement is shown in quantitative comparisons between measurements and predictions of carbon and sulphur variations along selected positions. For applications in large steel ingot, single-phase models with much simplification, are still commonly used. The use of two-phase models, especially multi-phase models, is relatively rare. There are two factors accounting for this. On one hand, the development of two-phase models or multi-phase models is rather difficult, which hinders their applications to a certain extent. On the other hand, large ingots, due to their large sizes, will cost more computational time and resources with a two-phase or multi-phase model. Thus, applications of multi-phase macrosegregation models in large ingots with experiment validation are of great significance.

In this paper, the multicomponent-multiphase model developed by the authors has been adopted to predict macrosegregation in a 535 tons steel ingot in three dimensions. The ingot has been simplified as a Fe-0.24 wt%C-1.65 wt%Cr-0.39 wt%Mo quaternary alloy system for simulation. For simplicity, the model considers only the primary phase. The predictions have reproduced the classical macrosegregation patterns, including negative segregation at the bottom of the ingot and positive segregation at the top of the ingot. Finally, to quantitatively compare the results of macroscopic segregation prediction and the measured results, the predictions with measurements of compositions along the center line and three transverse sections are compared.

Mathematic Model

In current research, the air phase is further added in previous two-phase model [9] to considerate the solidification shrinkage of the ingot. Firstly, the mass, momentum, and species conservation equations for solid phase and liquid phase are each derived. Then the volume conservation equation is derived to consider the inhaled air phase due to the solidification shrinkage. Further, the energy conservation equation is derived for the solid-liquid mixture system. Besides, a multicomponent alloy system rather than a two alloy system is considered to further investigate the influence of a third or more of the alloy components on macrosegregation. Thus after adding additional solute conservation equations, the interfacial species balance is derived, further obtaining the interfacial concentration constraint relations for the multicomponent alloy system. For simplicity, the model considers only the primary

Table 1 Mathematical equations of the multicomponent-multiphase model

Mass conservation	$\frac{\partial}{\partial t}(g_s\rho_s) + \nabla \cdot (g_s\rho_s\mathbf{v}_s) = \Gamma_s$
	$\frac{\partial}{\partial t}(g_l\rho_l) + \nabla \cdot (g_l\rho_l\mathbf{v}_l) = -\Gamma_s$
Momentum conservation	$\frac{\partial}{\partial t}(g_s\rho_s\mathbf{v}_s) + \nabla \cdot (g_s\rho_s\mathbf{v}_s\mathbf{v}_s) = -g_s\nabla p + \nabla \cdot (g_s\mu_s\nabla\mathbf{v}_s) + M_s^d + g_s\rho_s\mathbf{g}$
	$\frac{\partial}{\partial t}(g_l\rho_l\mathbf{v}_l) + \nabla \cdot (g_l\rho_l\mathbf{v}_l\mathbf{v}_l) = -g_l\nabla p + \nabla \cdot (g_l\mu_l\nabla\mathbf{v}_l) - M_s^d + g_l\rho_l^b\mathbf{g}$
Volume conservation	$\Delta V_a = \left(\frac{\rho_s}{\rho_l} - 1\right)\sum (V_{\text{cell}} \cdot \Delta g_s)$
Species conservation	$\frac{\partial}{\partial t}(g_s\rho_s C_{s,i}) + \nabla \cdot (g_s\rho_s\mathbf{v}_s C_{s,i}) = \nabla \cdot (g_s\rho_s D_{s,i}\nabla C_{s,i}) + \frac{S_V\rho_s D_{s,i}}{\delta_s}(C_{s,i}^* - C_{s,i}) + C_{s,i}^*\Gamma_s$
	$\frac{\partial}{\partial t}(g_l\rho_l C_{l,i}) + \nabla \cdot (g_l\rho_l\mathbf{v}_l C_{l,i}) = \nabla \cdot (g_l\rho_l D_{l,i}\nabla C_{l,i}) + \frac{S_V\rho_l D_{l,i}}{\delta_l}(C_{l,i}^* - C_{l,i}) - C_{l,i}^*\Gamma_s$
Energy conservation	$\frac{\partial}{\partial t}\left[(g_s\rho_s c_{ps} + g_l\rho_l c_{pl})T_{s(l)}\right] + \nabla \cdot \left[(g_s\rho_s c_{ps}\mathbf{v}_s + g_l\rho_l c_{pl}\mathbf{v}_l)T_{s(l)}\right] = \nabla \cdot \left[(g_s\lambda_s + g_l\lambda_l)\nabla T_{s(l)}\right] + \Gamma_s L$
Interfacial concentration balance	$\frac{S_V\rho_s D_{s,i}}{\delta_{s,i}}(C_{s,i}^* - C_{s,i}) + C_{s,i}^*\Gamma_s + \frac{S_V\rho_l D_{l,i}}{\delta_{l,i}}(C_{l,i}^* - C_{l,i}) - C_{l,i}^*\Gamma_s = 0$

phase, neglecting the formation of the secondary phase. The model is numerically solved by a homemade FVM code based on a multi-phase SIMPLEC algorithm.

The conservation equations are summarized in Table 1. In Table 1, t is the time, g is the volume fraction, ρ is the density, Γ is the interfacial phase change rate, v is the velocity, p is the pressure, μ is the viscosity, M_s^d is the drag force coefficient between the solid and liquid, $\mathbf{g}$ is the gravitational acceleration, ΔV_a is the volume change due to solidification shrinkage per unit time, V_{cell} is the grid cell volume, Δg_s is the grid cell solid volume fraction change per unit time, C is the concentration, S_v is the interfacial area concentration, D is the mass diffusivity, δ is the interfacial solute diffusion lengths, c_p is the specific heat, T is the temperature, λ is the thermal conductivity, L is the latent heat. The subscripts "s" and "l" refer to solid phase and liquid phase, respectively.

The supplementation relations, including the descriptions of interfacial area concentration, the interfacial concentration diffusion layer length as well as the flow model, and detailed derivations of the model can be found in Refs. [8] and [9].

Results and Discussion

The multicomponent-multiphase model was applied to a 535 tons steel ingot, which was cast to produce the low pressure rotor for nuclear power. The steel ingot was 6.0 m in height and 4.0 m in mean diameter. The composition of the steel ingot with 30Cr2Ni4MoV steel grade is shown in Table 2. The molten metal was poured at 1530 °C. After being stripped out of the mold, the ingot was dissected into slices along the center plane, and the carbon, chromium and molybdenum concentrations were measured along the centerline and three transverse sections. The schematic of the ingot with the cross-section sampling positions is shown in Fig. 1.

In current simulation, the Fe-0.24 wt%C-1.65 wt%Cr-0.39 wt%Mo quaternary alloy system was taken as the simplification of the 30Cr2Ni4MoV steel ingot to predict macrosegregation of the carbon, chromium and molybdenum components. The properties of the steel ingot and mold materials used are listed in Tables 3 and 4. The thermodynamic parameters used for multicomponent alloy solidification calculation are listed in Table 5. Due to the symmetry along the centerline axis, only a quarter of the ingot system was simulated. A 3-D orthogonal grid system consisting of 104544 cells was used and the mesh size was 80 mm × 80 mm × 80 mm. A constant time step of 0.5 s was used for the simulation. The initial temperature for the mold, sleeve, and powder are 25, 25, and 1500 °C. The interfacial heat transfer coefficients of the steel ingot system are listed in Table 6.

Table 2 30Cr2Ni4MoV steel grade (wt%)	C	Cr	Ni	Mo	V	Si	P
	0.24	1.65	3.68	0.39	0.15	0.04	0.002

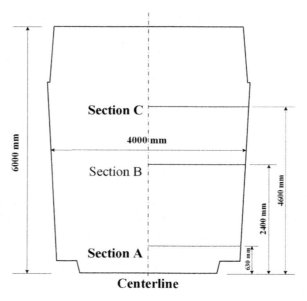

Fig. 1 Schematic of the ingot with the cross-section sampling positions

Table 3 Steel parameters [10, 11]

Property	Value	Unit
Melting temperature of pure iron	1532	°C
Steel thermal expansion coefficient	1.07×10^{-4}	K^{-1}
Liquid density	6990	$kg \cdot m^{-3}$
Solid density	7300	$kg \cdot m^{-3}$
Latent heat	2.71×10^5	$J \cdot kg^{-1}$
Liquid viscosity	4.2×10^{-3}	$kg \cdot m^{-1} \cdot s^{-1}$
Thermal conductivity	39.3	$W \cdot m^{-1} \cdot K^{-1}$
Specific heat	500	$J \cdot kg^{-1} \cdot K^{-1}$

Table 4 Thermophysical properties of the mold materials [12]

	Density ($kg \cdot m^{-3}$)	Thermal conductivity ($W \cdot m^{-1} \cdot K^{-1}$)	Specific heat ($J \cdot kg^{-1} \cdot K^{-1}$)
Mold	7000	26.3	540
Sleeve	185	1.4	1040
Powder	260	0.6	1040
Air	1.185	0.026	1000

Table 5 Thermodynamic parameters of Fe-C-Cr-Mo quaternary alloy system

Liquidus temperature (°C)	1532-7800 · C-261 · Cr-325 · Mo
Carbon solute partition coefficient	0.34
Chromium solute partition coefficient	0.76
Molybdenum solute partition coefficient	0.56
Carbon solute diffusion coefficient in liquid (m^2 · s^{-1})	2 × 10^{-8}
Chromium solute diffusion coefficient in liquid (m^2 · s^{-1})	2 × 10^{-8}
Molybdenum solute diffusion coefficient in liquid (m^2 · s^{-1})	2 × 10^{-8}
Carbon solute expansion coefficient (wt%$^{-1}$)	1.10 × 10^{-2}
Chromium solute expansion coefficient (wt%$^{-1}$)	3.97 × 10^{-3}
Molybdenum solute expansion coefficient (wt%$^{-1}$)	−1.92 × 10^{-3}

Table 6 Steel ingot system interfacial heat transfer coefficients (W · m^{-2} · K^{-1}) [12]

	Ingot	Mold	Sleeve	Powder	Air
Ingot	–	1500	2	400	5
Mold	1500	–	1000	1000	140
Sleeve	2	1000	–	1000	10
Powder	400	1000	1000	–	10
Air	5	140	10	10	–

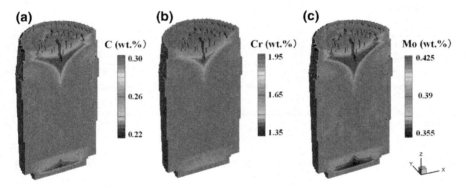

Fig. 2 **a** Predicted carbon concentration, **b** predicted chromium concentration, and **c** predicted molybdenum concentration

The predicted concentration distributions at the end of solidification are shown in Fig. 2a–c. The predicted carbon, chromium, and molybdenum concentrations all exhibit similar patterns, including negative segregation at the bottom of the ingot and positive segregation at the top of the ingot. The predicted macrosegregation patterns for the three components are proven to be qualitatively reasonable.

Quantitative comparisons between predictions and measurements of the solute concentrations along the centerline and three transverse sections A, B and C are provided in Figs. 3 and 4 to evaluate the accuracy of the macrosegregation model. As shown in Fig. 1, the centerline was chosen for its full coverage of the

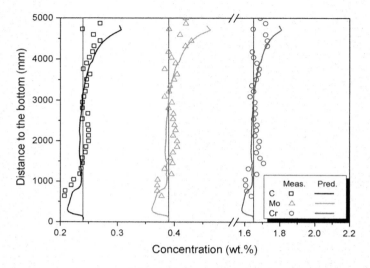

Fig. 3 Concentrations of carbon, chromium and molybdenum along the centerline

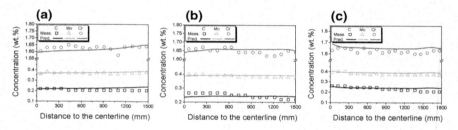

Fig. 4 Concentrations of carbon, chromium and molybdenum along **a** section A, **b** section B and **c** section C

solidification stage of the steel ingot. Section A and Section C, with the height of 630 and 4600 mm, were chosen to represent two typical serious segregation zones in the ingot, i.e. the negative segregation zone due to the sedimentation of solute-lean solid grains and the positive segregation zone formed by solute-rich melt at late stages. Section B, located at the height of 2400 mm over the bottom, represents the concerned ingot body zone. We can see that current predictions are generally in agreement with the measurements. However, for all the three species, some discrepancies arise in the positive segregation zone at the distance to the bottom higher than 3.8 m as shown in Fig. 3. Several factors may contribute to this. Firstly, we have assumed the ingot filling process is instant in the simulation, neglecting the actual multi-pouring process. This simplification ignores the initial non-uniform solute distribution after filling. Secondly, current simulation considers only four components; other components, however, may have a great impact on the solute convection during later stages of solidification [13].

Conclusions

A multicomponent-multiphase model has been applied to investigate the macrosegregation formation in a 535t steel ingot in three dimensions. Typical macrosegregation patterns are reproduced by predictions for carbon, chromium and molybdenum. Moreover, quantitative comparisons between predictions and measurements of the solute concentrations have been made. The results exhibit good agreement except for some discrepancies in the positive segregation zone at the top of the ingot. It is assumed the simplification of the multi-pouring process and ignorance of other components than the Fe-C-Cr-Mo quaternary alloy system may be the main factors.

Acknowledgements This work was financially supported by the NSFC-Liaoning Joint Fund (U1508215) and the project to strengthen industrial development at the grass-roots level of MIIT China (TC160A310/21).

References

1. E.J. Pickering, Macrosegregation in steel ingots: the applicability of modelling and characterisation techniques. ISIJ Int. **53**(6), 935–949 (2013)
2. C. Beckermann, Modelling of macrosegregation: applications and future needs. Int. Mater. Rev. **47**(5), 243–261 (2002)
3. M.C. Flemings, G.E. Nereo, Macrosegregation: part I. Trans. Metall. Soc. AIME **239**(9), 1449–1461 (1967)
4. J. Ni, C. Beckermann, A volume-averaged two-phase model for transport phenomena during solidification. Metall. Trans. B **22**(3), 349–361 (1991)
5. M. Wu, A. Ludwig, A three-phase model for mixed columnar-equiaxed solidification. Metallur. Mater. Trans. A **37**(5), 1613–1631 (2006)
6. I. Vannier, H. Combeau, G. Lesoult, Numerical model for prediction of the final segregation pattern of bearing steel ingots. Mater. Sci. Eng. A **173**(1–2), 317–321 (1993)
7. D.R. Liu et al., Numerical simulation of macrosegregation in large multiconcentration poured steel ingot. Int. J. Cast Met. Res. **23**(6), 354–363 (2010)
8. W.T. Tu et al., Three-dimensional simulation of macrosegregation in a 36-ton steel ingot using a multicomponent multiphase model. JOM **68**(12), 3116–3125 (2016)
9. W.S. Li, H.F. Shen, B.C. Liu, Numerical simulation of macrosegregation in steel ingots using a two-phase model. Int. J. Miner., Metallurg. Mater. **19**(9), 787–794 (2012)
10. H. Combeau et al., Prediction of macrosegregation in steel ingots: influence of the motion and the morphology of equiaxed grains. Metallur. Mater. Trans. B **40**(3), 289–304 (2009)
11. L. Thuinet, H. Combeau, Prediction of macrosegregation during the solidification involving a peritectic transformation for multicomponent steels. J. Mater. Sci. **39**(24), 7213–7219 (2003)
12. W.S. Li, Numerical simulation of macrosegregation in large steel ingots based on two-phase model. Ph.D. thesis, Tsinghua University, 2012, pp. 59–103
13. M.C. Schneider, C. Beckermann, Formation of macrosegregation by multicomponent thermosolutal convection during the solidification of steel. Metallur. Mater. Trans. A **9**(26), 2373–2388 (1995)

Validation of CAFE Model with Experimental Macroscopic Grain Structures in a 36-Ton Steel Ingot

Jing'an Yang, Zhenhu Duan, Houfa Shen and Baicheng Liu

Abstract In order to recognize macroscopic grain structures evolution within large heavy casting, a 36-ton steel ingot has been experimentally investigated. Thirteen thermocouples have been used to record temperature variations during solidification of the ingot to ensure a reliable simulation of temperature field. Half of the ingot tail in the longitudinal section has been etched to obtain as-cast macrostructure. Fine equiaxed grains are found in the ingot tail periphery, then slender columnar grains next to them, finally widely spread coarse equiaxed grains in the ingot tail center. Then, simulation of macroscopic grain structure is processed by a three dimensional Cellular Automaton Finite Element (CAFE) module of ProCAST software. The nucleation algorithm is based on an instantaneous nucleation model considering a Gaussian distribution of nucleation sites proposed by Rappaz. The growth algorithm is based on the growth of an octahedron bounded by (111) faces and the growth kinetics law is given by the model of Kurz et al. The microscopic CA and macroscopic FE calculation are coupled where the temperature of each cell is simply interpolated from the temperature of the FE nodes using a unique solidification path at the macroscopic scale. Simulation parameters of CAFE about Gaussian nucleation and growth kinetics have been adjusted so that the macroscopic grain structures correlate with the as-cast macrostructure experiment.

Keywords Steel ingot · Macroscopic grain structures · Thermocouples · CAFE · Simulation

---

J. Yang (✉) · H. Shen (✉) · B. Liu
School of Materials Science and Engineering, Tsinghua University,
Beijing 100084, People's Republic of China
e-mail: yja12@mails.tsinghua.edu.cn

H. Shen
e-mail: shen@tsinghua.edu.cn

Z. Duan
School of Engineering and Design, Lishui University, Zhejiang 323000,
People's Republic of China

© The Minerals, Metals & Materials Society 2017
P. Mason et al. (eds.), *Proceedings of the 4th World Congress on Integrated Computational Materials Engineering (ICME 2017)*,
The Minerals, Metals & Materials Series, DOI 10.1007/978-3-319-57864-4_19

Introduction

Ingots usually require a second procedure of shaping, such as cold/hot working, cutting, or milling to produce a useful final product. As-cast macroscopic grain structures have a great effect on the subsequent processing. Many studies [1–4] have been made on as-cast grain structures evolution by changing cooling conditions or altering the melt composition. Nuri et al. [5] reported on the solidification grain structures of steels added with rare-earth metals, finding that the growth direction of dendrite arm is less oriented, primary dendritic arm spacing is narrower and primary dendrite arm length is shorter. Tyurin et al. [6] studied the macroscopic grain structures of long cylindrical ingots, finding that horizontal crystallization basically consists of a trans-crystallization accompanied by certain smaller elements.

On the other hand, many efforts [7–11] have been made on the modelling and simulation of grain structures evolution. In 1990s, Rappaz and Gandin [12, 13] proposed a 2D Cellular Automaton (CA) technique for the simulation of dendritic grain formation during solidification. The CA model takes into account the heterogeneous nucleation, the growth kinetics and the preferential growth directions of the dendrites. This CA algorithm, which applies to non-uniform temperature situations, is fully coupled to an enthalpy-based Finite Element (FE) heat flow calculation. Despite the new insight brought by the 2D CAFE model into the modeling of solidification structures, a 3D CAFE model had been developed in order to quantitatively predict real as-cast structures. Gandin et al. [14] presented a 3D CAFE model for the prediction of dendritic grain structures formed during solidification. The microscopic CA and macroscopic FE calculation are coupled and special dynamic allocation techniques have been designed in order to minimize the computation costs and memory size associated with a very large number of cells. The 3D CAFE algorithm was further programmed into CALCOSOFT software, and then incorporated into ProCAST software, which has been a convenient tool for researchers.

In this paper, the CAFE module in ProCAST was used to investigate the macroscopic grain structures of a 36-ton steel ingot. The simulated results were compared with the etched samples. Besides, the temperature variations considering air gap formation contrasted with temperature measurements had been discussed.

Experimental Process

A 36-ton steel ingot was cast in CITIC Heavy Industries Co., Ltd., P.R. China. The ingot is a big top and small bottom octagonal $45^{\#}$ ordinary carbon steel ingot with 3434 mm in height and 1500 mm in mean diameter. The riser is 1700 mm in height, embedded with a layer of 140 mm thick refractory bricks. Molten steel was poured from the mold bottom and about 50 kg exothermic powder and thermal insulating agent were used to protect pouring molten steel. Most thermal insulating agent was divided into several bags and hung in the mold centerline, finally forming a layer of about 200 mm anti-piping compound on the melt top. The geometry characteristic of ingot and mold system is shown in Fig. 1a and the steel melt composition is listed in Table 1.

The pouring temperature was 1560 °C, the pouring time was 35 min and the stripping time was 10 h. Thirteen thermocouples were arranged to monitor the temperature variations as shown in Fig. 1b. They were divided into A, B, C three groups. Thermocouples of A group (A1–A5) were immersed into the melt and the ones of B group (B1–B4) were at the very inner face of the mold or insulating tiles. For the nine thermocouples of A and B groups required long-term contact with the molten steel, S-type thermocouples of good high temperature stability were used. Thermocouples of C group were K-type and located within the mold or insulating tiles. A5 was inserted into the melt vertically in the ingot centerline under the anti-piping compound within the depth of 200 mm. The other twelve thermocouples were horizontally located at four different heights (800, 1600, 2600 and 3000 mm) over the mold bottom, in which the thermocouples of B group were along the ingot interface with mold or insulating tiles and the ones of A and C groups were 30 mm away from the interface respectively as illustrated in Fig. 1b. All the thermocouples had been carefully calibrated before experiment and the data was recorded by a 24 channel paperless recorder at the frequency of 1 Hz.

After stripping out, the ingot was sectioned by flame cutting parallel to the central longitudinal plane. Then, a center plate of about 600 mm thickness was fetched out and sent to the heat treatment plant for stress relief annealing, holding for 24 h at 500 °C. Subsequently, a part of the plate containing ingot tail was sliced down and further thinned to about 30 mm by milling machine to remove heat

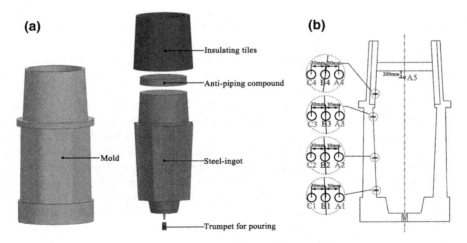

Fig. 1 a Geometry characteristic of ingot and mold system, b schematic of thirteen temperature measurement positions

Table 1 Composition of steel ingot

Elements	C	Si	Mn	S	P	Ni	Mo	Fe
(Wt Pct)	0.51	0.26	0.65	0.006	0.017	0.08	0.016	Bal.

affected zone. Considering the symmetry, half of the thinned plate was etched to obtain as-cast macroscopic grain structure. Due to the limitation of the pickling tank, it was divided into two pieces. Additionally, the surfaces roughness of them prepared for etching were made under Ra 3.2 by grinding machine. Hot acid etching standard of GB226-1991 (Etch test for macrostructure and defect of steels) was carried out in the experimental study on macroscopic grain structure of the $45^{\#}$ steel ingot.

Experimental Results

Temperature Variations

Temperature variations of the thirteen measurement positions are shown in Fig. 2. Figure 2a displays temperature variations along the ingot lateral surface recorded by the thermocouples A1–3, which is close to the inner wall of the mold. A failure of recording is noticed for A3 after two hours. The thermocouples A1–3 were horizontally fixed and probed into the ingot body. During the ingot solidification, they bear forces continuously from three aspects, such as horizontal pulling force toward ingot center due to the ingot solidification shrinkage, horizontal pushing force outward ingot center due to thermal expansion of the mold and vertical shear force due to contraction of the ingot in vertical direction. Additionally, they are under long-term high temperature state. Therefore, the thermocouples A1–3 have great risk in being distorted along the ingot lateral surface and A3 is the most distorted one, failing to record the temperature variation at that position after enough distortion.

It can be seen from the partial enlarged view in Fig. 2a that the temperature records of A1, A2 and A3 increase sharply to their maximum values one by one and then decrease with time. As a consequence of the different heights for A1–3, their maximum temperature has been up to 1323 °C, 1439 °C and 1390 °C respectively. Moreover, the temperature decreasing rate of A2 is lower than that of A1 and A3. At the initial stage of ingot solidification, the gap width due to ingot solidification shrinkage and mold expansion at A2 position is larger than that at A1 position, which lead to greater thermal resistance. When the molten steel rise to A3 position, it has been cooled to a certain extent. So the maximum temperature of A3 is 49 °C lower than that of A2. Owing to the release of latent heat, there are some temperature rise in the ingot cooling process.

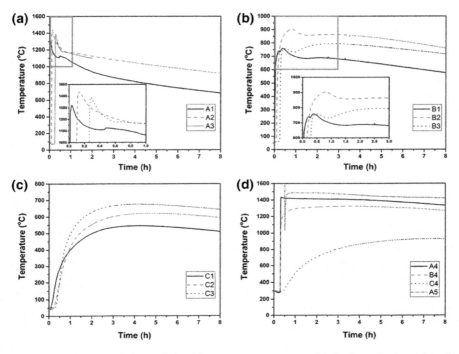

Fig. 2 Temperature variations of the thirteen measurements. **a** A1–3 along the ingot lateral surface, **b** B1–3 at the very inner face of the mold, **c** C1–3 within the mold 30 mm away from the mold inner face, **d** A4, B4, C4 and A5 at the riser part of ingot and mold system

Figure 2b displays temperature variations at the very inner face of the mold recorded by the thermocouples B1–3. They show similar trends as the records of A1–3, including rapid rise at the initial stage, highest maximum temperature sequence and temperature rise due to the release of latent heat. As depicted in the partial enlarged view in Fig. 2b, there is a small fluctuation of B1 recorded at the initial stage. The possible reason is that the solidified shell formed at the early stage of ingot solidification has not built up enough strength, resulting in the instability of air gap. In addition, large amount of latent heat from the ingot riser makes the temperature rise of B3 more obvious than those of B1 and B2.

Figure 2c displays temperature variations within the mold 30 mm away from the mold inner face recorded by the thermocouples C1–3. They are in general increasing trends corresponding with the heating of mold during the ingot filling. Additionally, the higher position of the thermocouples is, the higher maximum temperature it can reach.

Figure 2d displays temperature variations at the riser part of ingot and mold system recorded by the thermocouples A4, B4, C4 and A5. Due to good insulation capacity of the insulation tiles, the temperature of A4, which probes into the melt for 30 mm, increases sharply to a maximum value of 1433 °C and decreases at a very low speed. The similar trend is found in the temperature of B4, which is at the very inner face of insulation tiles. The temperature curve trend of C4 is much like that of C1–3, but it reaches a much higher maximum value. As for temperature variation of A5 in the ingot hot top, it increases sharply to 1488 °C and decreases at a very low speed as A4.

Macroscopic Grain Structures

Figure 3 shows macroscopic grain structures of the $45^{\#}$ steel ingot tail according to hot acid etching standard of GB226-1991. The two digital photos were stitched together as shown in Fig. 3a. Columnar-equiaxed transition (CET) positions are depicted by a solid curve at the longitudinal section shown in Fig. 3a. It can be seen that columnar grains growth directions are parallel to the heat extraction directions.

Figure 3b is a detailed view of the dotted box indicated in Fig. 3a, which is a horizontal bar from the ingot tail. As shown in Fig. 3b, fine equiaxed grains (b1) are found in the ingot periphery, then slender columnar grains (b2) next to them, finally widely spread coarse equiaxed grains (b3) in the ingot center.

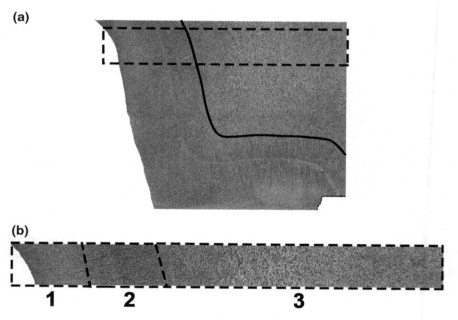

Fig. 3 a Macroscopic grain structures of the ingot tail. **b** Detailed view of the dotted box indicated in (a)

Numerical Simulation

A 3D FEM mesh including the ingot and mold system (Fig. 1a) with 2,414,710 elements was generated for the numerical simulation based on ProCAST software. The temperature, stress, flow and CAFE module had been used. Besides, the thermophysical properties, thermo-mechanical properties and KGT parameters were calculated by the ProCAST database based on the chemical composition of $45^{\#}$ steel ingot. Since coupling calculation of flow and stress is difficult to converge, the bottom pouring process was first calculated with temperature and flow. When the filling was completed, temperature field information was extracted to the second case and thermo-mechanical coupling calculation with proper boundary conditions was built to continue the calculation until complete solidification of the ingot. Finally, CAFE module was called base on the second temperature field to simulate as-cast macroscopic grain structures.

Temperature variations of four positions were chosen to evaluate temperature field simulation as shown in Fig. 4. As in the thermo-mechanical calculation filling time was skipped, the curves trends of experiment and simulation at initial stage did not agree very well. However, they shared similar trends at the later stage. In the numerical simulation results, the temperature at the very inner face of mold went lower than the experimental one, while the temperature in the mold went higher. This may be caused by underestimate of gap width, leading to quicker heat extraction of the ingot.

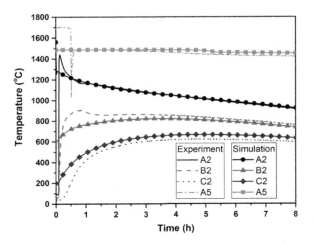

Fig. 4 Temperature variations comparison of experiment and simulation at the position of A2, B2, C2 and A5

Gap Width Modeling

During thermo-mechanical calculation, the gap model assumes that heat conduction through the air gap is progressively replacing the heat transfer coefficient due to the contact. When there is a contact, the heat transfer coefficient is increased as a function of the pressure,

$$h = h_0 \cdot \left(1 + \frac{P}{A}\right) \quad (1)$$

where h is the adjusted heat transfer coefficient, h_0 is the initial value of the heat transfer coefficient, P is the contact pressure and A is an empirical constant to account for contact pressure. When the gap is formed, the heat transfer coefficient is defined as

$$h = \frac{1}{1/h_0 + R_{gap}} \quad \text{with} \quad R_{gap} = \frac{1}{k/gap + h_{rad}} \quad (2)$$

where R_{gap} is thermal resistance of air gap, k is conductivity of air, gap is air gap width and h_{rad} is radiative equivalent heat transfer coefficient.

Figure 5 shows simulation results of gap width along the ingot lateral surface at the position of H1–4. The position of H1–4 is in accordance with that of A1–4. As can be seen from the partial enlarged view in Fig. 5, the gap width of H2 position is larger than that of H1 position at initial stage. Besides, the gap width of H3 position is the largest among that of the four positions at initial stage. It just explains why temperature of A2 and A3 positions can reach higher maximum value than that of A1 position and the maximum temperature rise was found at B3 position. H4 position is at the interface of ingot riser and insulating tiles. Owing to the long-term

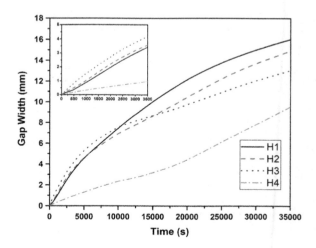

Fig. 5 Gap width simulation along the ingot lateral surface at the position of H1–4

high temperature state of ingot riser and low ductility of insulating tiles, the gap width of H4 has been the smallest one.

CAFE Results

The nucleation algorithm is based on an instantaneous nucleation model considering a Gaussian distribution of nucleation sites proposed by Rappaz [12]. It is

$$\frac{dn}{d(\Delta T)} = \frac{n_{max}}{\sqrt{2\pi}\Delta T_\sigma} \exp\left[-\frac{1}{2}\left(\frac{\Delta T - \Delta T_\mu}{\Delta T_\sigma}\right)^2\right] \quad (3)$$

where ΔT_μ is the mean undercooling, ΔT_σ is the undercooling standard deviation of the Gaussian distribution, n_{max} is the maximum nucleation sites in unit volume. The total density of grains n at a given undercooling is the integral of the distribution as Eq. (4).

$$n(\Delta T) = \int_0^{\Delta T} \frac{dn}{d(\Delta T')} d(\Delta T') \quad (4)$$

As the most of equiaxed grains in the ingot periphery is under 1 mm, so the maximum nucleation site is taken as 10^9–10^{10} m^{-3}. And the undercooling standard deviation is taken as 1 K, for the large difference of fine equiaxed grains, slender column grains and coarse equiaxed grains.

Besides, the growth algorithm is based on the growth of an octahedron bounded by (111) faces and the growth kinetics law is given by the model of Kurz et al. [13]. A cubic fit polynomial is used for the growth rate of the dendrite tips, it is

$$v(\Delta T) = a_2 \Delta T^2 + a_3 \Delta T^3 \quad (5)$$

where v is the growth rate, a_2 and a_3 are constants relating to the phase diagram of alloys and the diffusion coefficient of the alloy elements. The two parameters can be calculated by CompuTherm database of ProCAST and the values are 2.2776×10^{-7} ms^{-1} k^{-2} and 4.9501×10^{-8} ms^{-1} k^{-3} respectively.

For simplicity, volume and surface nucleation are assumed taking the same value of Gaussian distribution parameters. Subsequently, the three parameters of Gaussian distribution, i.e. the mean undercooling, ΔT_μ, the undercooling standard deviation, ΔT_σ, and the maximum nucleation sites in unit volume, n_{max}, are tried by various combinations aiming at reasonable CET positions similar to the experimental results. Table 2 lists seven groups of parameters for the simulation.

The microscopic CA and macroscopic FE calculation are coupled where the temperature of each cell is simply interpolated from the temperature of the FE nodes using a unique solidification path at the macroscopic scale. The simulation results of

Table 2 Surface and volume nucleation parameters

	Surface nucleation			Volume nucleation		
	ΔT_μ	ΔT_σ	n_{max}	ΔT_μ	ΔT_σ	n_{max}
1	8	1	7.8×10^{10}	8	1	7.8×10^{10}
2	9	1	7.8×10^{10}	9	1	7.8×10^{10}
3	10	1	7.8×10^{10}	10	1	7.8×10^{10}
4	9	0.1	7.8×10^{10}	9	0.1	7.8×10^{10}
5	9	10	7.8×10^{10}	9	10	7.8×10^{10}
6	9	1	7.8×10^{9}	9	1	7.8×10^{9}
7	9	1	7.8×10^{11}	9	1	7.8×10^{11}

the seven groups of parameters are exhibited in Fig. 6b–h. And Fig. 6a demonstrates CET positions in the ingot tail by a solid curve for comparison. Figure 6b–d correspond with group 1–3 in Table 2, where ΔT_σ and n_{max} are fixed and the mean undercooling, ΔT_μ, changes from 8 to 10. It can be seen that the greater the mean undercooling, the thicker the columnar grains zone. Figure 6e and f correspond with group 4 and 5 in Table 2, where the undercooling standard deviation, ΔT_σ,

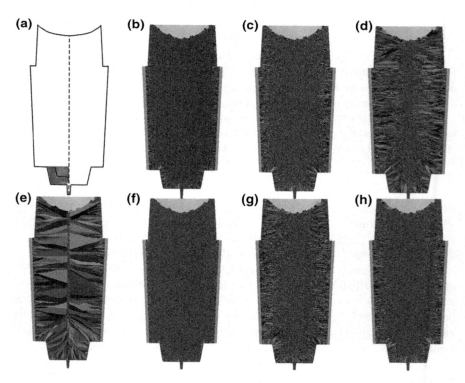

Fig. 6 a CET positions in the ingot tail, (**b–h**) the simulation results of the seven groups of parameters in Table 2

changes. In the Gaussian distribution, 99.7% of the data is distributed within 3 standard deviations. Therefore, the smaller the standard deviation is, the more nuclei form around mean undercooling. Correspondingly, the less nuclei form far from mean undercooling and grow into bigger grains as shown in Fig. 6e. When the standard deviation is large enough, nuclei will be distributed very evenly and grow into grains of similar size as almost all equiaxed grains displayed in Fig. 6f. Figure 6g and h correspond with group 6 and 7 in Table 2, where the maximum nucleation sites in unit volume, n_{max}, changes. It can be seen that this parameter is not so sensitive to the final results as the other two. There is a slight trend that the greater the maximum nucleation sites, the smaller the grains and the thinner the columnar grains zone. Compared with experimental results, parameters of group 2 in Table 2 are most reasonable in predicting CET positions.

Figure 7 shows the CAFE results of the ingot longitudinal section by the parameters of group 2 in Table 2 compared with etched macroscopic grain structures of the ingot tail. It well reproduces CET positions as indicated by a solid curve in Fig. 7b. However, the predicted columnar grains are greater in width compared to the slender columnar grains in experiment and fine equiaxed grains in the ingot periphery are not so conspicuous in the simulation results.

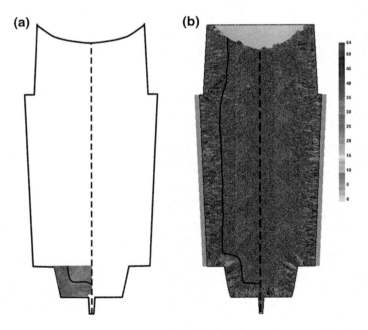

Fig. 7 CAFE results of the ingot longitudinal section. **a** Etched macroscopic grain structures, **b** CAFE results

Conclusions

A 36-ton steel ingot has been cast and sliced to study the as-cast macroscopic grain structures. Twelve horizontal thermocouples at four different heights had clearly recorded the temperature variations near the ingot and mold or insulation tiles interface. The last thermocouples was vertically immersed into melt and recorded the riser center temperature variation. This is a recommendable experimental technique in continuous temperature measurement during ingot solidification for engineering practice in developing useful ICME tools.

The behavior of temperature variations near the ingot and mold or insulation tiles interface can be explained by air gap formation as the thermo-mechanical calculation shows. At the initial stage of ingot solidification, the higher position at the ingot and mold interface, the larger the air gap.

The ingot as-cast macroscopic grain structures are mainly consisted of fine equiaxed grains, slender columnar grains and coarse equiaxed grains from surface to core. CET positions is clear and distinguishable. CET prediction of the CAFE result is in good agreement with the experiment. Besides, the CAFE module of ProCAST has been a convenience tool in prediction of macroscopic grain structures. It can be integrated into more powerful ICME tools for improving manufacturing process after more efforts should be done to acquire higher accuracy.

Acknowledgements This work was financially supported by the NSFC-Liaoning Joint Fund (U1508215) and the project to strengthen industrial development at the grass-roots level of MIIT China (TC160A310/21).

References

1. A.L. Greer, A.M. Bunn, A. Tronche, P.V. Evans, D.J. Bristow, Modelling of inoculation of metallic melts: application to grain refinement of aluminium by Al–Ti–B. Acta Mater. **48**(11), 2823–2835 (2000)
2. V.M. Schastlivtsev, T.I. Tabatchikova, I.L. Yakovleva, S.Y. Klyueva, Effect of structure and nonmetallic inclusions on the intercrystalline fracture of cast steel. Phys. Met. Metall. **114**(2), 180–189 (2013)
3. X.B. Qi, Y. Chen, X.H. Kang, D.Z. Li, Q. Du. An analytical approach for predicting as-cast grain size of inoculated aluminum alloys. Acta Mater. **99**, 337–346 (2015)
4. Suyitno, V.I. Savran, L. Katgerman, D.G. Eskin, Effects of alloy composition and casting speed on structure formation and hot tearing during direct-chill casting of Al-Cu alloys. Metallur. Mater. Trans. A, **35**(11), pp. 3551–3561 (2004)
5. Y. Nuri, T. Ohashi, T. Hiromoto, O. Kitamura, Solidification macrostructure of ingots and continuously cast slabs treated with rare earth metal. Trans. Iron Steel Inst. Japan **22**(6), 408–416 (1982)
6. V.A. Tyurin, Y.V. Lukanin, A.V. Morozov, Ingot molds for obtaining long cylindrical ingots and features of the macrostructure of the metal. Metallurgist **56**(9–10), 742–747 (2013)
7. S.C. Flood, J.D. Hunt, Columnar and equiaxed growth-I. A model of a columnar front with a temperature dependent velocity. J. Cryst. Growth **82**(3), 543–551 (1987)

8. S.C. Flood, J.D. Hunt, Columnar and equiaxed growth-II. Equiaxed growth ahead of a columnar front. J. Cryst. Growth **82**(3), 552–560 (1987)
9. M.F. Zhu, C.P. Hong, A modified cellular automaton model for the simulation of dendritic growth in solidification of alloys. ISIJ Int. **41**(5), 436–445 (2001)
10. A.I. Ciobanas, Y. Fautrelle, Ensemble averaged multiphase Eulerian model for columnar/equiaxed solidification of a binary alloy: I. The mathematical model. J. Phys. D Appl. Phys. **40**(12), 3733–3762 (2007)
11. S. Vernede, M. Rappaz, A simple and efficient model for mesoscale solidification simulation of globular grain structures. Acta Mater. **55**(5), 1703–1710 (2007)
12. M. Rappaz, C.A. Gandin, Probabilistic modelling of microstructure formation in solidification processes. Acta Metall. Mater. **41**(2), 345–360 (1993)
13. C.A. Gandin, M. Rappaz, A coupled finite element-cellular automaton model for the prediction of dendritic grain structures in solidification processes. Acta Metall. Mater. **42**(7), 2233–2246 (1994)
14. C.A. Gandin, J.L. Desbiolles, M. Rappaz, P. Thevoz, A three-dimensional cellular automation-finite element model for the prediction of solidification grain structures. Metallur. Mater. Trans. A **30**(12), 3153–3165 (1999)

Analysis of Localized Plastic Strain in Heterogeneous Cast Iron Microstructures Using 3D Finite Element Simulations

Kent Salomonsson and Jakob Olofsson

Abstract The design and production of light structures in cast iron with high static and fatigue performance is of major interest in e.g. the automotive area. Since the casting process inevitably leads to heterogeneous solidification conditions and variations in microstructural features and material properties, the effects on multiple scale levels needs to be considered in the determination of the local fatigue performance. In the current work, microstructural features of different cast irons are captured by use of micro X-ray tomography, and 3D finite element models generated. The details of the 3D microstructure differ from the commonly used 2D representations in that the actual geometry is captured and that there is not a need to compensate for 3D-effects. The first objective with the present study is to try and highlight certain aspects at the micro scale that might be the underlying cause of fatigue crack initiation, and ultimately crack propagation, under fatigue loading for cast iron alloys. The second objective is to incorporate the gained knowledge about the microstructural behavior into multi-scale simulations at a structural length scale, including the local damage level obtained in the heterogeneous structure subjected to fatigue load.

Keywords Cast iron · Microstructure · X-ray tomography · Characterization

Introduction

The ever-increasing demands for lower emissions in the automotive industry require lighter components to be manufactured. Materials such as polymeric materials, aluminum and magnesium alloys have increased in use. However, there is a strong connection with lowered density and lowered strength. Thus, attention has also been turned towards the use of optimizing stronger materials topologically

K. Salomonsson (✉) · J. Olofsson
Jönköping University, School of Engineering, Materials and Manufacturing,
P.O. Box 1026, SE-551 10 Jönköping, Sweden
e-mail: kent.salomonsson@ju.se

to reduce weight. In this context, cast iron has been, and still is, beneficial to use when constructing components for the automotive industry since its manufacturability has several strong benefits such as e.g. high machinability, high heat resistance, good wear resistance. Different cast iron alloys have different benefits. For example, ductile iron or spherical graphite iron (SGI) has high strength and low thermal conductivity, whereas grey iron or lamellar graphite iron (LGI) has high thermal conductivity and lower strength. Compacted graphite (CGI) is thus of interest for components where e.g. both strength and thermal conductivity is sought after, e.g. engine blocks.

Determination of the mechanical properties for different cast iron alloys has gained interest over the past decade and specifically locally varying mechanical properties that result from e.g. the geometrical features of the component, chemical composition, cooling rates etc. [1]. Locally varying mechanical properties is traditionally not taken into account when performing strength analyses for e.g. automotive components [2]. It was shown by [3] that there are considerable differences when analyzing a component using isotropic homogenous properties compared to the use of locally varying material properties. The difference can be as high as 90 MPa locally [3]. The microstructures for SGI, LGI and CGI contain complex combinations of different phases such as e.g. graphite morphology, fractions of ferrite and/or pearlite which govern the mechanical performance of the material [1]. Thus, depending on the geometry of the component these parameters will change and as a result, the mechanical performance will change locally [3].

In the field of shape analysis, it is common practice to specify one global measure of the graphite particle and one concerning the morphology to determine different microstructural features [4]. Several parameters have been defined over the years that serve the purpose of defining particular properties specifically for compacted and lamellar graphite inclusions [5]. Several methods have been developed that propose to characterize microstructures, see e.g. [6–8]. The proposed methods use two-dimensional images in the characterization framework. The images provide a nodule count which is converted to the three-dimensional nodule count by determination of the area fraction of graphite and an assumption of average graphite diameter to yield the volume count. The assumption is that all the graphite nodules are of equal size and perfectly spherical for e.g. SGI. In the work by [9], they refined the expression and used a parametrized model to account for different sized and shaped nodules. The characterization of different cast iron alloys is frequently determined by the value of nodularity where values range between 65 and 90% for SGI and lower values correspond to CGI (10–25% nodularity) and LGI (0–5% nodularity), respectively.

Nevertheless, it has been shown that it is not the grade % of graphite that determines the shift in ultimate tensile strength (UTS) observed for different SGI alloys, but the grade % of pearlite and ferrite [10]. Interestingly, [10] showed that an increase in nodularity increases the UTS for SGI, whereas an increase of vermicularity decreases the UTS for CGI. It has been observed in e.g. [11] for SGI that underlying cause of this effect is due to the shape of graphite inclusions. In particular, high local strains are observed experimentally in [11] by use of digital image correlation.

In the present paper, we analyze SGI, LGI and CGI microstructures by use of three-dimensional finite element models to try and highlight particular shapes and distances that localize plastic strain well below homogenized levels of strain at the structural scale. The three-dimensional numerical models modeled as representative volume elements with virtually the same size, i.e. $0.7 \times 0.7 \times 0.7$ mm. The paper is organized as follows. Firstly, we elaborate on the generation of the three-dimensional finite element models. Secondly, we characterize the microstructures by use of characterization parameters for two-dimensional approaches. The numerical models are presented next and the paper is finalized by the results and concluding remarks.

Generation of Microstructures

By use of micro X-ray tomography, it is possible to produce two-dimensional slices of a specimen. The size of the specimen is generally determined by the sought-after resolution one is aiming for. In the present work, the aim was to distinguish different phase morphologies. Pictures of different microstructures are depicted in Fig. 1.

In order to distinguish between different phases in the microstructure it is suggested that an image analysis is performed prior to segmentation of the images. Figure 2 illustrates an example for a SGI microstructure where the original micrograph is seen together with its corresponding probability density which is generated by use of the Fiji/ImageJ plugin Weka segmentation. The extent of the ferrite phase surrounding the graphite is clearly visible by observing Fig. 2.

The resolution in which one can determine the extent of different phases is related to the bit-depth as well. For the case of 8-bit pictures, the image intensity limited to values between 0 and 255, whereas for the 16-bit depth images between 0 and 4095. Thus, details in intensity shift can be captured more easily for the 16-bit images.

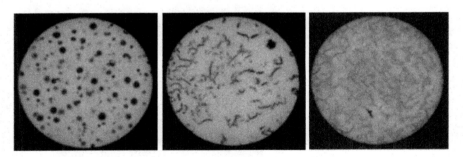

Fig. 1 Pictures taken from the X-ray tomograph. *Left* SGI, *center* CGI and *right* LGI

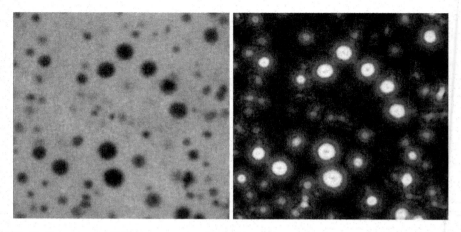

Fig. 2 Example of a SGI microstructure (*left*), probability density of intensity (*right*)

Once the images have been pre-analyzed using image analysis software, each image is segmented to establish the extent of each microstructural phase. The segmented images are then interpolated to yield a three-dimensional representation of the microstructure, see Fig. 3 and e.g. [12] for a fairly recent study on three-dimensional analysis using micro μCT to analyze SGI.

Fig. 3 Three-dimensional representation of the interpolated segmented images

Numerical Model

The size of the representative volumes that were selected for SGI and LGI in the present study are about $0.7 \times 0.7 \times 0.7$ mm^3, whereas a cylinder is used for the CGI, see Fig. 4.

The numerical models contain different number of finite elements due phase morphology. Specifically, for the SGI microstructure it was essential to resolve the ferrite phase boundary surrounding the graphite inclusions. Thus, the number of elements used for SGI is approximately 5 million linear tetrahedral elements. The cylinder representing the CGI microstructure contains approximately 2.5 million linear tetrahedral elements and the final microstructure, LGI, contains approximately 4 million linear tetrahedral elements. For simplicity, the interfaces connecting different phases share nodes.

It should be noted that the carbide inclusions were considered virtually rigid in comparison to the surrounding phase pearlite and we, thus, chose to model carbide as elastic and homogenous with an artificially stiff Young's modulus of 300 GPa. The graphite was considered to be elastic and homogenous and since the Young's modulus for graphite varies significantly in literature, we chose to use 25 GPa. The ferrite and pearlite phases were modeled as elasto-plastic materials using a rate in-dependent Johnson-Cook material model. The material properties for the different phases and the different microstructures are presented in Table 1.

In order to analyze and load the microstructures, a displacement field is extracted from a structural analysis of an engine block mount for simplicity and to yield

Fig. 4 Three-dimensional representations of SGI (*left*), CGI (*center*) and LGI (*right*)

Table 1 Material parameters used in the simulations. Young's modulus and Johnson-Cook parameters		E [GPa]	A [MPa]	B [MPa]	n [−]
	Graphite	25			
	Ferrite [13]	197	175	571	0.35
	Pearlite [13]	206	750	593	0.33
	Carbide	300			

"actual" loading. The selected volume, from which the displacement field is extracted, can be seen in Fig. 5. This is also a region which was strained the most in the structural simulation.

The loading of the microstructures was in form of a controlled displacement vector, $u \approx (5.2, -3.5, 6.9)$ µm, of the top surface (positive z-direction according to Fig. 6) and the bottom surface is clamped. It should be noted that all the microstructures were loaded by the same displacement vector on the top surface (Fig. 6).

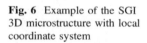
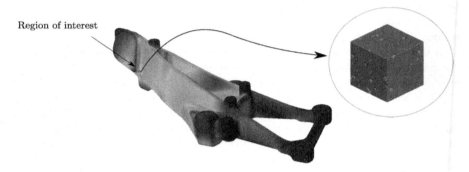

Fig. 5 Strained engine block mount with the selected region of interest

Fig. 6 Example of the SGI 3D microstructure with local coordinate system

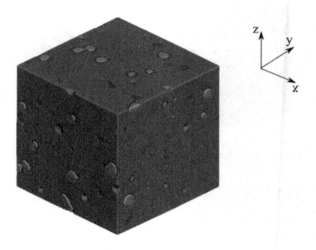

Results

The intention of the present study is to highlight differences between different microstructures and compare the effective localized plastic strain development. Figure 7 shows the effective plastic strain for the three microstructures as a result of the controlled displacement vector. Clearly, the strains for the CGI microstructure are not directly comparable due to a the difference in the shape of the representative volume. Nevertheless, the SGI and the LGI microstructures are comparable and it can be observed that the SGI microstructure displays localized plastic strains which are not seen for the LGI microstructure. The ferrite phase surrounding the graphite nodules have a lower yield stress than the pearlite and the graphite inclusions act similar to voids. Thus, stress concentrations are expected around all graphite nodules that are loaded, which leads to the increased plastic strain primarily in the ferrite phase. For the LGI microstructure, the graphite inclusions are in form of large connected flakes which instead act as dampers that reduce the stress concentration. This effect can also be observed for the CGI with its vermicular graphite inclusions. Spherical inclusions are also present in CGI which will result in similar stress states as those observed for SGI.

Considering ultimate tensile stress levels, SGI will generally have the highest value followed by CGI and then LGI [14]. This is directly related to the localization of stress concentrators such as graphite inclusions and their shapes, respectively, see e.g. [15, 16]. Notice also that three-dimensional effects such as e.g. graphite nodules, will increase the strain locally, point A for SGI in Fig. 7, which is not possible to predict in a two-dimensional analysis of the corresponding microstructure, see Fig. 8.

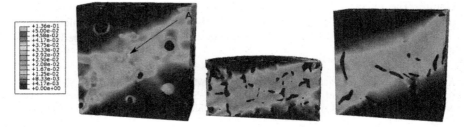

Fig. 7 Contour plots of a center cross section showing the equivalent plastic strain for the loaded microstructures; from *left* to *right*, SGI, CGI and LGI, respectively

Fig. 8 Undeformed corresponding 2D representation of the 3D microstructure for SGI shown to the *left* in Fig. 7

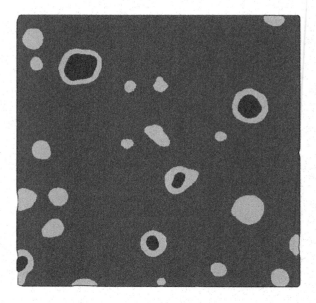

Conclusions

In the present paper, three-dimensional finite element models have been developed in order to better understand the effects of graphite inclusion morphology under similar loading conditions. The localization of plastic strain has been proven to be more pronounced for SGI than for the other two microstructures, CGI, and LGI. It has also been concluded that numerical analysis in two dimensions are not sufficient when analyzing actual microstructural features, which is common observed in literature. The effects originating from the three-dimensional representation of the microstructure is most likely the cause of mismatch between e.g. micro digital image correlation (μDIC) and finite element simulations trying to predict strain fields locally [11].

Acknowledgements The authors would like to acknowledge Professor Ragnvald Mathiesen for supplying the μCT-scans of the cast iron alloys. The Swedish Knowledge Foundation is also acknowledged for financial support of the research profile CompCAST at Jönköping University.

References

1. J.R. Davis, *Cast irons*, (ASM International, Materials Park, Ohio, 1996)
2. J. Olofsson, I.L. Svensson, Incorporating predicted local mechanical behaviour of cast components into finite element simulations. Mater. Design. 34, 494–500 (2012)
3. J. Olofsson, I.L. Svensson, The effects of local variations in mechanical behaviour—numerical investigation of a ductile iron component. Mater. Design, 43, 264–271 (2013)

4. U. Riebel, V. Kofler, F. Löffler, Shape characterization of crystal and crystal agglomerates. Part. Part. Syst. Charact. 8, 48–54 (1991)
5. Y. Tanaka, H. Saito, I. Tokura, K. Ikawa, Relationship between some physico-mechanical properties and numerical indexes of graphite shape in compacted-vermicular graphite cast irons. Trans. ASME **104**, 60–65 (1982)
6. H. Ledbetter, S. Datta, Cast iron elastic constants: effect of graphite aspect ratio. Z. Metallkd. **83**(3), 195–198 (1992)
7. S.H. Pundale, R.J. Rogers, G.R. Nadkarni, Finite element modelling of elastic modulus in ductile irons: effect of graphite morphology. AFS Trans. **98–102**, 99–105 (2000)
8. K.M. Pedersen, N.S. Tiedje, Graphite nodule count and size distribution in thin-walled ductile cast iron. Mater. Charact. **59**, 1111–1221 (2008)
9. T. Owadano, Graphite nodule number in spheroidal graphite and malleable cast irons. Imono/Jpn Foundrymen's Soc **45**, 193–197 (1973)
10. C. Fragassa, N. Radovic, A. Pavlovic, G. Minak, Comparison of mechanical properties in compacted and spheroidal graphite irons. Tribol. Ind. **38**(1), 49–59 (2016)
11. K. Kasvayee, K. Salomonsson, E. Ghassemali, A. Jarfors, Microstructural strain distribution in ductile iron; comparison between finite element simulation and digital image correlation measurements. Mater. Sci. Eng. A **655**, 27–35 (2015)
12. G. Fischer, J. Nellesen, N.B. Anar, K. Ehrig, H. Riesemeier, W. Tillmann, 3D analysis of micro-deformation in VHCF-loaded nodular cast iron by & #x03BC;CT. Mater. Sci. Eng. A **577**, 202–209 (2013)
13. M. Abouridouane, F. Klocke, D. Lung, Microstructure-based 3D FE modeling for micro cutting ferritic-pearlitic carbon steels. In *Proceedings of ASME 2014 Manufacturing Science and Engineering Conference MSEC2014*, Detroit, Michigan, USA. MSEC2014-4011 9–13 June 2014
14. T. Sjögren, I.L. Svensson, Modelling the Effect of Graphite Morphology on the Modulus of Elasticity in Cast Irons. In *International Journal of Cast Metals Research*, **17**(5), 271–279 (2004)
15. V.I. Litovka, N.I. Bekh, O.I. Shinskii, N.I. Tarasevich, P. Yakovlev, G.A. Kosnikov, Effects of composition and structure on the state, fatigue and damping properties of high-strength cast iron. Strength Mater. **27**(8), 448–453 (1995)
16. Y.B. Zhang, T. Andriollo, S. Fœster, W. Liu, J. Hattel, R.I. Barabash, Three-dimensional local residual stress and orientation gradients near graphite nodules in ductile cast iron. Acta Mater. **121**, 173–180 (2016)

An Integrated Solidification and Heat Treatment Model for Predicting Mechanical Properties of Cast Aluminum Alloy Component

Chang Kai Wu and Salem Mosbah

Abstract In this work, a newly developed modeling tool is presented which computes the local mechanical properties of cast and precipitation hardening heat treated aluminum alloy component. The integrated model simulates both casting and heat treating processes, and it computes the local hardness, yield strength and ultimate tensile strength, that developed in the casting during each step. Both alloy solidification and precipitation hardening heat treatment steps are simulated. The solidification model takes into account grains nucleation and the mushy zone front undercooling to predict the growth of the dendritic and eutectic microstructures. The predicted secondary dendrite arm spacing (SDAS) map is used to calculate the local strengths in the subsequent heat treatment steps. The heat treating model takes into account quenching and aging steps. The integrated model uses an extensive database that was developed specifically for the A356 alloy under consideration. The database includes temperature dependent mechanical, physical, and thermal properties of the alloy.

Keywords ICME · Quench factor · Solidification · Lagrangian tracking · Grain structure

Introduction

There is a substantial increase in the market demand for aluminum casting alloys due to their low density, high corrosion resistance and good ductility. The melt-cast process is most often used for making complex shapes that would be difficult or uneconomical using other methods, such as forging, extrusion and machining. For A365 cast alloy, common casting processes include sand casting, permanent mold

---

C.K. Wu
Dow Performance Silicones, 3901 S. Saginaw Rd., Midland, MI 48686, USA

S. Mosbah (✉)
Think Solidification, 3401 Iron Point Dr. Unit 572, San Jose, CA 95134, USA
e-mail: salem.mosbah@live.com

casting and investment casting, and the control of microstructure and casting integrity during solidification is key in choosing which process to use. Numerical simulation of solidification has improved our understanding of casting processes significantly over the last few decades and the study of the casting process design effect on the as cast product quality. In particular, as reported elsewhere [1], understanding the grain structure features, such as secondary dendrite arm spacing (SDAS) map which will determine the effectiveness of any subsequent heat treatment processes, requires the modeling of the effect of the bulk flow, mushy zone and the coupling with the grain structure [2]. A brief review of the current solidification models capabilities can be found in [1, 2] and shows that the current available tools can only be applied to small geometries. Indeed, despite the fact that these tools provide an important insight of the solidification process and incorporate some important physical phenomena, as reported in [3] for CA-FE based models, the required computational resources would render such approaches impractical to daily R&D use. Typically, aluminum alloy cast parts are subsequently strengthened by precipitation hardening heat treatment. It consists of three steps: (1) solutionizing, (2) quenching, and (3) aging. Obviously, these processing steps involve significant thermal changes that may be different from location to location in the casting.

The objective of this research is to develop and verify an integrated model that enables predicting the local physical and mechanical property in aluminum alloy castings in response to both solidification and precipitation hardening heat treatment. First, the solidification model used in this paper is briefly described. Following, the precipitation hardening heat treatment model is explained in details. Finally, the model required input parameters are listed, and the model predictions are compared to the experimental measurements.

Background

Solidification Model

The solidification model consists of two solvers: Finite Volume (FV) model to solve the macroscopic equations at the macro scale and a Lagrangian based tracking solver to compute the solidification grain structure. A brief description of the two models is given as follows and the details can be found in references [4, 5]. The model has been validated using the solidification benchmark developed by Bellet et al. [6] and the reader is referred to the same reference for further details about the benchmark. The scheme is based on a set of linear transformation/creation of a virtual particle which tracks the dendritic tips evolution as the solidification progresses. Both the FV equations and the solidification grains envelope tracking are implemented using the open source C++ libraries OpenFOAM.[1]

---

[1]OpenFOAM Foundation | OPENFOAM and OpenCFD are registered trademarks of OpenCFD Ltd.

Precipitation Hardening Heat Treatment Model

The heat treatment model is created based on coupling the Quench Factor Analysis (QFA) [7, 8] and the Shercliff-Ashby aging model [9]. Coupling detail is described in elsewhere [10]. QFA assumes that the precipitation transformation follows the Johnson-Mehl-Avarmi-Kolmogorov (JMAK) equation [11]. For continuous transformations, the isothermal holding time (t) in the JMAK equation can be replaced by the cumulative Quench Factor (Q) [12]. Since the first QFA model published, there are several modifications and improvements have been made over years. Rometsch [13] suggested that the development of strength in a precipitation hardened metallic component is proportional to the square root of the volume fraction of precipitate, so that instead of using the Avrami exponent (n), he proposed the square root should be introduced.

Materials and Database Generation

Commercial aluminum casting alloy A356 with 0.2% with TiB (5:1) grain refiner addition was used to develop and demonstrate the procedures for obtaining the necessary database and modeling solidification and heat treatment.

Solidification Database

The bulk chemistry is reduced to a pseudo binary mixture with a nominal 7 wt% Silicon composition. All the solidification simulation input parameters are listed in Table 1. In this model, a simple instantaneous nucleation law is considered which characterizes the volume number density of grains seeds as a function of the nucleation undercooling, ΔTn at randomly selected positions within the fluid volume and in contact with the mold boundaries. Similarly, the crystallographic orientation defined by the three Euler angles is randomly generated. At the boundaries, the nucleation undercooling is assumed to be zero, a few degree undercooling is considered for the volume nucleation. Solidification model predicted local growth velocity was converted to SDAS values based on Jackson-Hunt's eutectic growth model [14]. The available literature [15] was used as a database, and the determined constant was 119261.8.

Heat Treatment Database

The database developed specifically for the A356 alloy includes physical properties, mechanical properties, thermal conductivity, a quenching and aging database, and

Table 1 Solidification simulation parameters and physical properties used in the model [19]

Symbol	Parameter description	Value	Units
T_M	Melting temperature	663.5	°K
T_E	Eutectic temperature	577.0	°K
M	Liquidus slope	−6.5	°K/wt%
k	Segregation coefficient	0.13	–
w_E	Eutectic composition	13.31	wt%
C_p	Heat capacity	$C_p(T)$	J/kg °K
L	Latent heat of fusion	400844.0	J/kg
κ	Thermal conductivity	κ(T)	W/m °K
ρ_0	Reference density	2370.0	Kg m^{-3}
T_0	Reference temperature	660.2	°K
μ	Dynamic viscosity	0.001	Pa s
λ_2	Secondary dendrite arm spacing	250.0	μm
T_{init}	Initial temperature	660.2	°K
T_{ext}	External temperature	25.0	°K
w_0	Si nominal composition	7.0	wt%

Table 2 Heat treatment simulation parameters used in each model

HT Parameters	Hardness model	YS model	UTS model
K_1	−0.00501		
K_2	2.87E-10	6.36E-24	3.02E-10
K_3 (J/mol °K)	5043	30522	3758
K_4 (°K)	813		
K_5 (J/mol)	78143.72		
σ_i (MPa)	114.6	46.3	143.2
σ_q (MPa)	463.4	80.1	247.7
Qa (J/mol)	78143.72		
Pp (s/°K)	3.34E-8		
Ts (°C)	330		
S_0 max (MPa)	1238.2	331	247.6
K	0.4496	0.4013	0.6327

heat transfer coefficients for the quenching step of the precipitation strengthening heat treatment, all as functions of temperature. Measurements were made on samples cut from bars cast in the cast iron ASTM standard tensile bar mold [16], and the determined SDAS at this location was roughly 25 microns. Thermal conductivity was directly measured from a thermal gradient induced in the material as per ASTM standard E1225 [17]. The time dependent quenching heat transfer coefficients (HTCs) was measured by using the lumped parameter analysis [10]. Other required alloy properties, such as density, specific heat, etc., were obtained from JMatPro Software.[2] The required kinetic quenching and aging parameters are

---

[2]Developed and marketed by Sente Software Ltd., Surrey Technology Centre, 40 Occam Road, GU2 7YG, United Kingdom.

measured and calibrated based on each property, as shown in Table 2. Detail calibrating procedures are described in elsewhere [10]. Heat treated yield strength and ultimate tensile strength are assumed to vary linearly with local SDAS, and the available literature [18] was used to determine the slope. The heat treated dataset was generated on cast samples with SDAS of 25 microns, thus, this value was used as the origin. The heat treated hardness is assumed to be independent of local SDAS variation.

Measured and Model Predicted Results

A356 alloy cast in the ASTM standard cast iron tensile bar mold was chosen to demonstrate the model capabilities and verify the accuracy of its predictions. Few bars were cast with a k-type thermocouple permanently inserted in it at gage length location to measure temperature profiles during solidification and subsequent water quenching. The computer-generated renditions of the tensile bar geometry are shown in Fig. 1, and the thermocouple location is indicated by the circle. Only a quarter of the tensile bar geometry is modeled due to its symmetry. Casting models were created base on bars cast in two different mold materials, cast iron and sand, by applying different casting HTCs on the alloy surfaces. Heat treatment models were created based on bars solutionized at 538 °C for 12 h and then quenched in

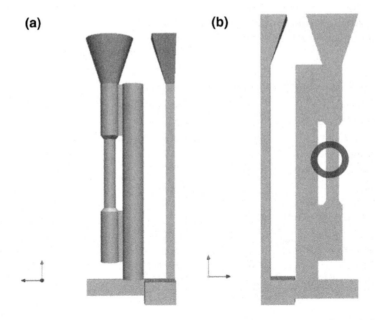

Fig. 1 The ASTM standard tensile quarter model of different views, **a** and **b** *circle* indicates thermocouple location

Table 3 Measured properties of tensile bar cast in a cast iron mold

Measured values	Cast iron mold	Standard deviation
SDAS (μm)	25	N/A
Mold temperature (°C)	462	N/A
Solidification rate (°C/s)	24.3	N/A
As cast, yield strength (MPa)	101.8	1.2
As cast, ultimate tensile strength (MPa)	169.4	8.9
As cast, hardness (HRB)	22	0.5
T6, yield strength (MPa)	282.73	2.5
T6, ultimate tensile strength (MPa)	344.27	3.6
T6, hardness (HRB)	42.58	1.5

80 °C water. Subsequently, two standard aging conditions, namely T6 and T7, were simulated. T6 aging was done by holding the part at 155 °C for 4 h. T7 aging was done by holding the part at 227 °C for 8 h. The measured values are shown in Table 3. The boundary conditions used and predicted properties at thermocouple location are shown in Table 4. Compared to measured values of bars cast in cast iron mold followed by T6 treatment, predicted hardness, yield and ultimate tensile strength are all within 5% difference. The solidification rates were calculated as an average value between 760 and 700 °C. Screen captures from the computer simulations are shown in Figs. 2, 3, 4 and Fig. 5 for different properties, respectively. The computer-predicted yield strength and ultimate tensile strength values vary from location to location within the casting due to the unique local predicted SDAS values and quenching rates at each location.

Table 4 Boundary conditions used and model predicted values of location 1

Predicted values	Cast iron mold	Sand mold
Mold temperature used (°C)	462	25
HTC used (W/m² °C)	900	150
Solidification rate (°C/s)	26.3	11.5
Growth velocity (μm/s)	287	194
SDAS (μm)	20.4	24.8
T6, yield strength (MPa)	269.4	265.7
T6, ultimate tensile strength (MPa)	342.7	336.8
T6, hardness (HRB)	44.4	44.4
T7, yield strength (MPa)	213.1	209.4
T7, ultimate tensile strength (MPa)	315.9	309.6
T7, hardness (HRB)	22.6	22.6

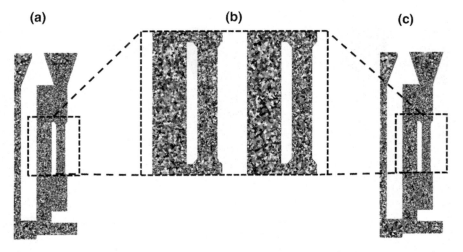

Fig. 2 The predicted grain structure cross section where each gray level represents a given orientation of A356 alloy cast in **a** cast iron mold **b** zoom out at thermocouple location and **c** sand mold. The average SDAS value is different between the cast and sand mold. This is in line with the difference notes in Table 4

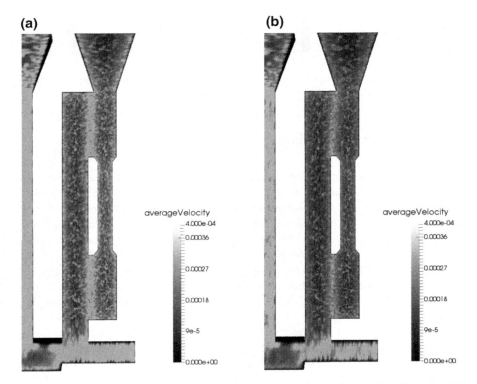

Fig. 3 The model predicted growth velocity of A356 alloy cast in **a** cast iron mold and **b** sand mold

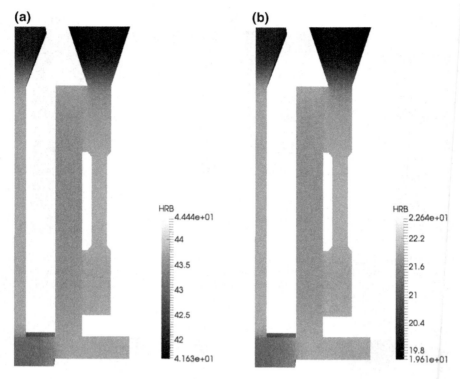

Fig. 4 The model predicted HRB hardness of A356 alloy **a** cast in cast iron mold followed by T6 treatment and **b** cast in sand mold followed by T7 treatment

Summary and Conclusions

An integrated solidification and precipitation hardening heat treatment model has been developed. Casting and heat treatment CFD simulations were performed to understand and predict the response of cast aluminum alloy component including quenching and aging. The model calculates the quenching temperature profile using a measured database of time dependent quenching HTCs. Two cases were considered based on cast iron mold and sand mold solidifications followed by the standard T6 and T7 precipitation hardening heat treatments. The predicted values of T6 heat treated bars were validated by comparing them to measurements of temperature profiles, SDAS and corresponding properties made using processing conditions identical to those used in the simulations. A good agreement with measurements made on T6 heat-treated bars demonstrates the computational utility of the cooling rate, grain structure, grain size and orientation, SDAS and properties predictions, and was successfully applied to the cast iron mold cast tensile bars. The comprehensive modeling approach presented in this paper is unique in its ability to

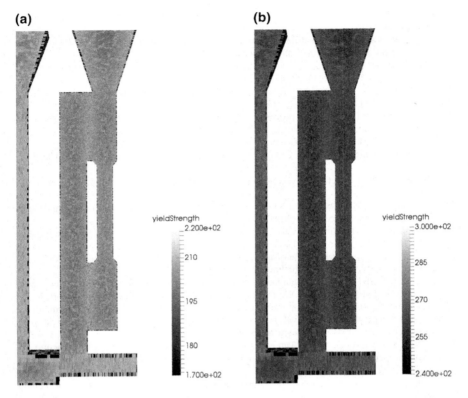

Fig. 5 The model predicted yield strength of A356 alloy **a** cast in cast iron mold followed by T7 treatment and **b** cast in sand mold followed by T6 treatment

simulate grain structure, size and orientation at the entire casting scale, and to capture variations in part properties by incorporating local SDAS and quenching cooling rates.

References

1. J. Dantzig, M. Rappaz, Solidification. (EPFL press, 2009)
2. T.L. Finn, M.G. Chu, W.D. Bennon, Micro/macro scale phenomena in solidification. ASME, HTD **218**, 17–26 (1992)
3. CA Gandin et al, A Three-dimensional cellular automation-finite element model for the prediction of solidification grain structures. Metall. Mater. Trans. A **30**(12), 3153–3165 (1999)
4. S. Mosbah, Grain structure and segregation modeling using coupled FV and DP model. Paper presented at the 8th International Symposium on Superalloy 718 and Derivatives, Pittsburgh, PA (2014)

5. S. Mosbah, Three dimensional model of solidification grain structure and segregation (in preparation, 2015)
6. M. Bellet et al, Call for contributions to a numerical benchmark problem for 2D columnar solidification of binary alloys. Int. J. Therm. Sci. **48**(11), 2013–2016 (2009)
7. J.W. Evancho, J.T. Staley, Kinetics of precipitation in aluminum alloys during continuous cooling. Metall. Trans. **5**(1), 43–47 (1974)
8. J.T. Staley, R.D. Doherty, A.P. Jaworski, Improved Model to Predict Properties of. Metall. Trans. A, 24 (11), 2417–2427 (1993)
9. H.R. Shercliff, M.F. Ashby, A process model for age hardening of aluminium alloys—I. The model. Acta Metall. Mater. **38**(10), 1789–1802 (1990)
10. C. Wu, M.M. Makhlouf, Ph.D. Thesis, Worcester Polytechnic Institute, 2012
11. M. Avrami, Kinetics of phase change. I general theory. J. Chem. Phys. **7**(12), 1103–1112 (1939)
12. J.T. Staley, Quench factor analysis of aluminium alloys. Mater. Sci. Technol. **3**(11), 923–935 (1987)
13. P.A. Rometsch, M.J. Starink, P.J. Gregson, Improvements in quench factor modelling. Mater. Sci. Eng., A **339**(1), 255–264 (2003)
14. K.A. Jackson, J.D. Hunt, Lamellar and rod eutectic growth. AIME Met. Soc. Trans. **236**, 1129–1142 (1966)
15. S.J. Hong et al, Effect of microstructural variables on tensile behaviour of A356 cast aluminium alloy. Mater. Sci. Technol. **23**(7), 810–814 (2007)
16. ASTM B108/B108 M-15, Standard specification for aluminum-alloy permanent mold castings. ASTM International, West Conshohocken, PA (2015). http://www.astm.org
17. ASTM E1225–04, Standard test method for thermal conductivity of solids by means of the guarded-comparative-longitudinal heat flow technique. ASTM International, West Conshohocken, PA (2004). http://www.astm.org
18. G. Ran, J. Zhou, Q.G. Wang. The effect of hot isostatic pressing on the microstructure and tensile properties of an unmodified A356-T6 Cast aluminum alloy. J. Alloys Comp., **421**(1), 80–86 (2006)
19. T. Carozzani, Ch.-A. Gandin, H. Digonnet. Optimized parallel computing for cellular automaton–finite element modeling of solidification grain structures. Modell. Simul. Mater. Sci. Eng., **22**, 015012, 21 (2014)

Linked Heat Treatment and Bending Simulation of Aluminium Tailored Heat Treated Profiles

Hannes Fröck, Matthias Graser, Michael Reich, Michael Lechner, Marion Merklein and Olaf Kessler

Abstract Precipitation hardening aluminium alloys enable tailoring of mechanical properties through the dissolution of strength-increasing precipitates during a local short-term heat treatment. Tailor Heat Treated Profiles (THTP) are aluminium extrusion profiles with locally different material properties, specifically optimised for succeeding bending processes. Softened areas need to be generated next to hardened areas to optimise the material flow during the forming process. To determine the optimised layout of softened and hardened areas, a process chain simulation consisting of the simulation of the short-term heat treatment and the subsequent forming process seems purposeful. The numerical modelling of short-term heat treatment requires a coupled computation of thermal and mechanical simulation with particular focus on the evaluation of microstructure and consequently on the change of mechanical properties. The dissolution and precipitation behaviour during heating and cooling of aluminium profiles 6060 T4 is investigated using differential scanning calorimetry. Thermo-mechanical analysis is applied for evaluation of the mechanical properties. This behaviour should be described in a material model with the software LS DYNA. The heat treatment simulation provides a distribution of mechanical properties along the profile, which is an important input parameter for the following forming simulation. In order to avoid a loss of information between the heat treatment simulation and forming simulation, both linked simulations are performed with the software LS DYNA.

H. Fröck (✉) · M. Reich · O. Kessler
Chair of Materials Science, Faculty of Mechanical Engineering and Marine Technology, University of Rostock, Albert Einstein-Str. 2, 18059 Rostock, Germany
e-mail: hannes.froeck@uni-rostock.de

M. Graser · M. Lechner · M. Merklein
Institute of Manufacturing Technology, Friedrich-Alexander-Universität Erlangen-Nürnberg, Egerlandstraße 11–13, 91058 Erlangen, Germany

O. Kessler
Faculty of Interdisciplinary Research, Department Life Light & Matter
(Research Competence Centre CALOR), University of Rostock,
Albert-Einstein-Str. 25, 18059 Rostock, Germany

© The Minerals, Metals & Materials Society 2017
P. Mason et al. (eds.), *Proceedings of the 4th World Congress on Integrated Computational Materials Engineering (ICME 2017)*,
The Minerals, Metals & Materials Series, DOI 10.1007/978-3-319-57864-4_22

Keywords Aluminium alloy · EN AW-6060 · Tailor heat treated profiles (THTP) · Differential scanning calorimetry

Introduction

Lightweight structures made of precipitation hardening aluminium alloys, like the Al-Mg-Si-alloys, are applied in various industrial areas, such as automotive or aviation industry. This is due to the good weldability, corrosion resistance and above all the good relationship of strength to density. The consistent use of lightweight structures, like aluminium extrusion profiles, in the automotive industry leads to a reduction in vehicle mass and thus to a decrease in energy consumption. Precipitation hardening aluminium profiles provide the possibility to generate rigid and light structures, which are required in various lightweight constructions. However, in comparison to mild steels, the cold formability of precipitation hardening aluminium alloys in their higher strength states is low, and causing unacceptable cross-sectional deformations during the bending process of small radii. Thus, complex parts are hardly producible at low temperatures.

To improve the cold formability of aluminium sheet components, the local short-term heat treatment has been established in recent years, to achieve a local softening of the material due to a dissolution of precipitations. To optimise the material flow during the forming process, softened areas need to be generated next to hardened areas. The so called Tailor Heat Treated Blanks (THTB) are of interest for example in the automotive industry [1]. Hence, the experiences made with the sheet metal materials can be transferred to THTP. To determine the optimised layout of softened and hardened areas, a process chain simulation consisting of the simulation of the short-term heat treatment and the subsequent forming process seems purposeful.

The simulation of the short-term heat treatment provides a distribution of mechanical properties and residual stresses along the profile, which are determining input parameters for the subsequent forming simulation. The material modelling of the aluminium alloy during short-term heat treatment poses a major challenge for the simulation chain. The mechanical properties during the heat treatment depend on a number of influencing factors. Due to the dissolution of strength-increasing precipitates, the mechanical properties of the heating step differ from these of the cooling step. Furthermore, the dissolution of the precipitates depends on the maximum temperature during the heat treatment, which results in different mechanical properties in dependence of the achieved maximum temperature. A further factor of influence is the actual temperature during the heat treatment. These dependencies should be depicted in a material model.

In order to provide a reliable data base, the dissolution and precipitation behaviour during heating of the aluminium alloy EN AW–6060 T4 is investigated in a wide dynamic range of heating rates, using the differential scanning calorimetry (DSC). While different short-term heat treatments, tensile tests were carried out during the heating and cooling, as well as on different actual temperatures and maximum temperatures, to map the course of the mechanical properties during various short-term heat treatments.

Tailor Heat Treated Profiles

The high stiffness of aluminium extrusion profiles combined with the low density makes aluminium extrusion profiles an ideal design element for frame structures of automobiles or aircrafts. However, aluminium has an up to 20% lower limiting drawing ratio in comparison to mild steels [2]. This leads to a material failure during cold forming which is expressed in local thinning and cracks. Various methods for profile forming at low temperatures, such as the usage of mandrels, have been developed [3]. However, these methods are often inflexible with respect to geometry and require high tooling costs or long process times. Another approach to improve formability of aluminium extrusion profiles are locally different material properties. The high strength of precipitation hardening aluminium alloys is mainly based on the obstruction of the dislocation glide by precipitates. During a short–term heat treatment, the precipitates are dissolved in the aluminium matrix, fewer obstacles hinder the dislocation glide and the material is softened. The material is only locally heat treated, mainly in areas that are not directly located in the critical forming zone [4]. Therefore, the material flow in direction of the forming area is enhanced, occurring stresses are decreased and as a consequence material failure is prevented. Subsequently, the strength increase again after the forming process through a paint–bake cycle.

The local short-term heat treatment can be carried out with a laser, whereby the heat treatment can be integrated flexibly into the production chain. In order to achieve the optimum heat treatment layout for the forming process, a deep understanding of the progressing precipitation and dissolution reactions and the associated mechanical properties during the heat treatment is necessary.

Thermo-Mechanical Behaviour of EN AW-6060 T4 During Short-Term Heat Treatment

Thermal Analysis by Differential Scanning Calorimetry

The mechanical properties of precipitation hardening aluminium alloys are decisively influenced by the precipitates in the aluminium matrix. It has been shown in previous work that the mechanical properties of the investigated aluminium alloy correlate with the progressing dissolution and precipitation reactions [5].

The differential scanning calorimetry (DSC) has been established, to record in situ the dissolution and precipitation behaviour of aluminium alloys in a wide

Table 1 Mass fraction of alloying elements of the investigated material in %

Aluminium	Mass fraction in %							
Alloy	Si	Fe	Cu	Mn	Mg	Cr	Zn	Ti
EN AW-6060 T4	0.40	0.22	0.07	0.14	0.56	0.02	0.02	0.02
DIN EN 573-3	0.3–0.6	0.1–0.3	≤0.1	≤0.1	0.35–0.6	≤0.05	≤0.15	≤0.1

range of heating and cooling rates [6]. To investigate the dissolution and precipitation behaviour of the aluminium alloy EN AW–6060 in the natural aged state T4, heating curves with heating rates of 0.01 up to 5 Ks^{-1} from 20 to 600 °C were recorded. To measure the wide range of heating rates, two different types of DSC-devices were used. The slow rates between 0.01 and 0.1 Ks^{-1} were carried out in a heat-flux DSC of the Calved-type (Setaram Sensys evo DSC) and the faster heating rates from 0.5 up to 5 Ks^{-1} were performed in a power-compensated DSC (PerkinElmer Pyris Diamond DSC). More experimental details on the DSC experiments can be found in [7].

As base material a hollow quadratic extrusion profile with the dimensions of 20 × 20 × 2 mm was used. The chemical composition of the investigated alloy is given in Table 1. Samples of 99.9995% pure aluminium were used as reference material.

Figure 1 displays selected heating curves of heating rates from 0.03 up to 5 Ks^{-1}. It can be seen, that endothermic and exothermic reactions alternate. The dissolution of precipitates proceeds endothermic, shown in all figures as an exceeding deviation from the zero level. However, the formation of precipitates acts exothermic, shown as a deviation below the zero level (dotted line). Since only the resulting heat flow is measured, the single peaks cannot be separated accurately, due to the overlapping of the various individual reactions. However, many individual peaks can be detected, particularly during slow heating. In recent years, this

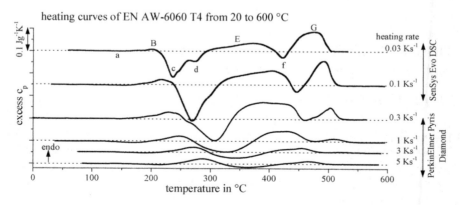

Fig. 1 Selected heating curves of EN AW-6060 T4 recorded with heating rates from 0.03 to 5 Ks^{-1} from 20 to 600 °C

alloy system has been extensively studied such as certain reactions can be attributed to the individual peaks [8].

As can be seen from Fig. 1, the continuous heating curve starts with a slight exothermic peak (a) at low temperatures, which is interpreted as the formation of clusters [9]. The following endothermic peak (B) is considered as the dissolution of cluster and GP–zones [10]. The dissolution of clusters and GP–zones is the main factor for the softening of the material [7] and is therefore of particular importance for the case of application [11]. Subsequently, two exothermic peaks are recorded. Peak (c) is interpreted as the formation of the metastable phase β'', whereas the peak (d) is regarded to the formation of the metastable phase β' [12]. The following endothermic peak (E) is construed as the dissolution of the previous formed metastable phases β'' and β' [13]. Thereafter, the equilibrium phase β (Mg_2Si) is formed in the exothermic peak (f) [14]. The endothermic peak (G) displays the dissolution of the ß-phase as well as all other secondary phases which have not been dissolved completely before. An extensive description of the individual reactions may be found in [7].

It can also be seen in Fig. 1 that the reactions shift to higher temperatures and are increasingly suppressed with an increasing heating rate. This behaviour is also discussed in detail in [7].

The zero crossings of the individual peaks were determined from all recorded heating curves. Due to the overlapping of the individual reactions, the zero-crossings are not necessarily the correct start respectively end temperatures of the expiring reactions. However, since the resulting heat flow is measured, a more precise reaction separation is not yet possible. In this way, the determinable start and end temperatures of the reactions were evaluated from all heating curves and converted into a continuous heating dissolution diagram, see Fig. 2.

Due to equipment restrictions, it is not yet possible to record directly heating curves with heating rates higher than 5 Ks^{-1} of aluminium alloys from the initial state. Nevertheless, a typical laser short-term heat treatment has heating rates up to several 100 Ks^{-1}, no soaking time at the maximum temperature and with a free

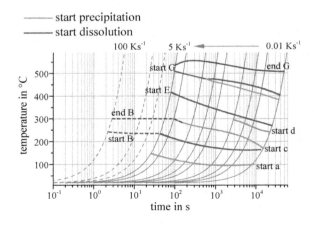

Fig. 2 Continuous heating dissolution diagram of EN AW-6060 T4

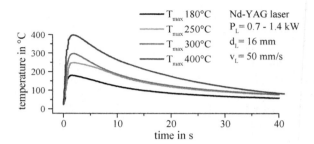

Fig. 3 Recorded temperature profiles of different laser heat-treatments [11]

convection to the environment, a subsequent cooling with some 10 Ks^{-1}. Figure 3 shows the time-temperature profile of four laser heat treatments with different performance settings.

In order to fill the gap between the measured slower heating rates and the application-typical high heating rates, the continuous heating dissolution diagram was extrapolated to high heating rates with the aid of indirect DSC measurements at high heating rates [15]. The extrapolated reaction temperatures are shown in Fig. 2 with dashed lines. For the production of THTP, it is essential that softened areas are next to hardened areas. In order to achieve this, on the one hand the heat treatment should be sufficiently fast, on the other hand, the maximum temperature should be selected to be high enough to soften the material. With these arrangements, the heat affected zone can be kept as small as possible for creating different strength areas.

Thermo-Mechanical Analysis During Short-Term Heat Treatment

In order to establish a data base for the thermo-mechanical simulation of this process, the mechanical properties were recorded during four different relevant short-term heat treatments according to Fig. 3. For this purpose, the recorded time-temperature profiles were imitated in the quenching and deformation dilatometer type Bähr 805 A/D. This device combines the advantages of a quenching dilatometer and a mechanical testing machine [16]. The device requires a special sample geometry which does not correspond to a standard tensile test sample, as shown in Fig. 4. However, comparative tensile tests were carried out before with the dilatometer and standard sample geometry, which show that the results of both are comparable [11].

The schematic measurement plan is shown in Fig. 4. In order to measure the mechanical properties during the short-term heat treatment, the heat treatment was interrupted at different temperatures/times. Directly after, tensile tests were carried out at these temperatures, see the arrows in Fig. 4. Three tensile tests were performed for each parameter set. Figure 5 displays the tensile and yield strengths during the various short-term heat treatments.

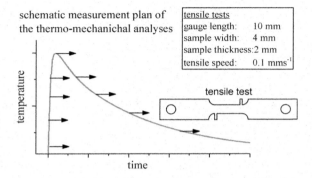

Fig. 4 Schematic measurement plan of the thermo-mechanical analysis, with major parameter of the tensile test

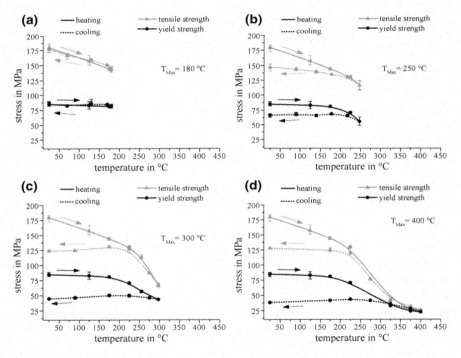

Fig. 5 Mechanical properties (yield and tensile strength) during various short-term heat treatments

Despite the heating rate, most of the softening of the material takes place during heating up to the maximum temperature. For this reason, the heating step is of particular interest. In order to show both, the fast heating step and the slow cooling step, the mechanical properties are shown as a function of the temperature. The results of the heating step are shown in the figures with a solid line, while the results of the cooling step are represented by a dotted line. The mechanical properties shown refer to the presented time/temperature profiles in Fig. 3.

Figure 5 A displays the tensile and yield strength during a short-term heat treatment with a maximum temperature of 180 °C. It becomes clear that the tensile strength decreases slightly during heating to the maximum temperature, and increases again with a decreasing temperature, during cooling. The tensile strength after the heat treatment is on the same level as before. As Fig. 2 has shown, up to the maximum temperature of 180 °C, no GP-zones have been dissolved at high heating rates. Therefore the precipitation state should be the same after the heat treatment as before, such that the mechanical properties are comparable. The yield strength is almost unaffected during this heat treatment.

In a short-term heat treatment with the maximum temperature of 250 °C, the tensile strength decreases during heating, as shown in Fig. 5b. The tensile strength decreases from 180 MPa in the initial state to 116 MPa at 250 °C. During cooling, the strength increases again up to 147 MPa. The initial strength is not reached again after the heat treatment. As illustrated in Fig. 2, the dissolution of GP–zones starts in the temperature range around 240 °C. Due to the change of the precipitation state caused by the dissolution of GP–zones, the material exhibits different mechanical properties during heating and cooling. The yield strength is also reduced from 85 MPa in the initial state to 55 MPa at 250 °C and then increases again to 66 MPa after the heat treatment.

A comparable behaviour is shown with the maximum temperature of 300 °C, see Fig. 5c. The tensile strength drops to 67 MPa at 300 °C and rises to 124 MPa after the heat treatment. During this short-term heat treatment, the precipitation state in the matrix has even changed stronger as indicated by the reduction in strength and as shown in Fig. 2. At a maximum temperature of 300 °C the dissolution of GP–zones is almost completed. The yield strength decreases to 44 MPa at 300 °C and amounts 44 MPa after heat treatment.

A similar course of the mechanical properties can be observed during a short-term heat treatment with the maximum temperature of 400 °C, as shown in Fig. 5d. Due to the high temperatures and the dissolution of the strength-increasing precipitates, the tensile strength at 400 °C amounts only 26 MPa. By cooling, the tensile strength rises to 128 MPa after the heat treatment. At the maximum temperature of 400 °C, the GP zones are dissolved. The yield strength decreases from 85 MPa in the initial state to 22 MPa at 400 °C and increases only slightly to 38 MPa after the heat treatment.

A data basis for the complex mechanical behaviour of the alloy EN AW–6060 T4 in THTP was acquired through the thermal analysis by DSC in combination with the thermo-mechanical analysis by tensile tests. Through these investigations the main influencing factors on the mechanical properties were identified. These are the heating rate as well as the cooling rate [17]. The main influencing factor, however, is the temperature. On the one hand, the actual temperature is important for the mechanical properties during the heat treatment, on the other hand the maximum temperature which the material has undergone during the heat treatment has a very important influence on the mechanical properties during and after the cooling.

Simulation of a Short-Term Heat Treatment

The aim of heat treatment simulation is the coupled thermo-mechanical simulation, taking into account the change of the mechanical properties due to phase transformations. To set up a linked heat treatment and bending simulation of aluminium THTP, both the thermo–mechanical simulation and the bending simulation are performed in the FEM–program LS-DYNA. Through the use of the same FEM-program, the results of the thermo-mechanical simulation can be transferred into the subsequent bending simulation without loss of information.

In the simulation setup, a hollow quadratic extrusion profile with the dimensions of $20 \times 20 \times 2$ mm was discretised as volume model with solid elements. A discretisation with shell elements does not appear to be a guide, since measurements have shown that the laser heat treatment on the profile surface results in a temperature gradient even over the thin material thicknesses of 2 mm [18]. In order to be able to represent this gradient, a discretisation with solid elements is necessary. Since the optimum strength layout has to be found for the use case, simplifications cannot be utilised by existing symmetries of the initial geometry since it could not be ruled out that these symmetries will also be reflected in the used heat treatment simulation.

The boundary conditions for thermo–mechanical simulation are an initial temperature of the profile of 25 °C, and a free convection to the environment was considered with a heat transfer coefficient of $\alpha = 30$ Wm^{-2} K^{-1}.

To simulate the moving heat input, different options are available. The moving laser spot produces a temperature field that is comparable to a temperature distribution during welding. Thereby, it is possible to apply the heat via an implemented LS-DYNA welding keyword [19]. A double ellipsoid heat source is implemented, the so-called Goldak source, which can be controlled by a few parameters [20]. By using this heat source, the heat can simply applied, but no user-defined laser spot geometries can be specified.

The more detailed way is the own description of the power distribution of the used laser via a self-defined function. This function is afterwards transferred via a flux heat to the self-defined heat source. Thus, any radiant heat sources can be described.

For the present simulation of heat treatment, a round base area with a diameter of 16 mm and a Gaussian power distribution with an output of 1.28 kW has been implemented for the laser spot. Figure 6 displays the simulated temperature distribution in an aluminium profile during a laser short-term heat treatment with the above-mentioned parameters. The simulated temperature profiles were compared to

Fig. 6 Simulated temperature distribution in an aluminium profile during a laser short-term heat treatment

measured temperature profiles at selected positions. The heat treatment simulation could be verified by a high consistency of the temperature profiles.

For the FEM calculation, no material model is available, which constitutes the influencing variables in the short-term heat treatment of precipitation hardening aluminium alloys. No material model has yet been implemented which defines the mechanical properties during a heat treatment as a function of the current temperature, the heat treatment history and the heating or cooling step (Fig. 5). However, this is necessary in future THTP simulations.

For the subsequent bending simulation, mechanical properties must be assigned to the material as a function of the maximum temperature reached during the heat treatment. Areas which must be softened for an improvement of the forming process could be identified by this method. Further, the residual stresses and deformations from the heat treatment simulation are necessary as input variables for the subsequent forming simulation.

Outlook and Summary

In order to display the stresses, deformations and mechanical properties in aluminium THTP simulation, it is necessary to use a suitable thermo-mechanical material model. This should depend on the current temperature, the heat treatment history and the heating or cooling step and is not available until now. However, in recent publications, a new material model, which may possibly close this gap, was presented [21].

Based on the present study a pool of stress/strain-curves during different short-term heat treatments of aluminium alloy 6060 T4 could be recorded. Due to the combination of DSC and thermo-mechanical analysis the changes of the mechanical properties can also be explained on the microstructural level.

It has been shown that the tensile strength during heating decreases with increasing temperature. However, the yield strength remains relatively unaffected up to a temperature of 200 °C. At higher temperatures strength increasing precipitates dissolve and the tensile as well as the yield strength decrease. During cooling, the tensile strength and yield strength increase again with decreasing temperature. Depending on the maximum temperature, the strengths after the short-term heat treatment are lower than the initial strengths. This softening is to be used specifically for the improvement of forming of aluminium profiles, which allows the production of THTP.

Acknowledgements The authors gratefully acknowledge funding of this work by the German Research Foundation (DFG KE616/22-1, DFG ME2043/45-1).

References

1. M. Merklein, M. Johannes, M. Lechner et al., A review on tailored blanks—production, applications and evaluation. J. Mater. Process. Technol. **214**(2), 151–164 (2014)
2. F. Ostermann, *Anwendungstechnologie Aluminium*, 3rd edn. (Springer 2014)
3. F. Vollertsen, A. Sprenger, J. Kraus, et al. Extrusion, channel, and profile bending: a review. J. Mater. Process. Technol. **87**, 1–27 (1999)
4. M. Geiger, M. Merklein, U. Vogt, Aluminum tailored heat treated blanks. Prod. Eng. Res. Devel. **3**(4–5), 401–410 (2009)
5. H. Fröck, M. Graser, B. Milkereit et al., Precipitation behaviour and mechanical properties during short-term heat treatment for tailor heat treated profiles (THTP) of aluminium alloy 6060 T4. MSF **877**, 400–406 (2016)
6. J. Osten, B. Milkereit, C. Schick et al., Dissolution and precipitation behaviour during continuous heating of Al-Mg-Si alloys in a wide range of heating rates: materials. Mater **8**(5), 2830–2848 (2015)
7. H. Fröck, M. Graser, M. Reich et al., Influence of short-term heat treatment on the microstructure and mechanical properties of EN AW-6060 T4 extrusion profiles: part A. Prod. Eng. Res. Dev. **10**(4–5), 383–389 (2016)
8. M. Takeda, F. Ohkubo, T. Shirai et al., Precipitation behaviour of Al–Mg–Si ternary alloys. Mater. Sci. Forum **217–222**, 815–820 (1996)
9. S.N. Kim, J.H. Kim, H. Tezuka et al., Formation behavior of nanoclusters in Al–Mg–Si alloys with different Mg and Si concentration. Mater. Trans. **54**(03), 297–303 (2013)
10. M. Murayama, K. Hono, Pre-precipitate clusters and precipitation processes in Al-Mg-Si alloys. Acta Mater. **47**(5), 1537–1548 (1999)
11. M. Graser, H. Fröck, M. Lechner et al., Influence of short-term heat treatment on the microstructure and mechanical properties of EN AW-6060 T4 extrusion profiles—part B. Prod. Eng. Res. Dev. **10**(4–5), 391–398 (2016)
12. C.-S. Tsao, C.Y. Chen, U.-S. Jeng et al., Precipitation kinetics and transformation of metastable phases in Al–Mg–Si alloys. Acta Mater. **54**(17), 4621–4631 (2006)
13. G.A. Edwards, K. Stiller, G.L. Dunlop et al., The precipitation sequence in Al-Mg-Si alloys. Acta Mater. **46**(11), 3893–3904 (1998)
14. L.C. Doan, Y. Ohmori, K. Nakai, Precipitation and dissolution reactions in a 6061 aluminum alloy. Mater. Trans. **41**(2), 300–305 (2000)
15. D. Zohrabyan, B. Milkereit, O. Kessler et al., Precipitation enthalpy during cooling of aluminum alloys obtained from calorimetric reheating experiments. Thermochim. Acta **529**, 51–58 (2012)
16. M. Reich, O. Keßler, Quenching simulation of aluminum alloys including mechanical properties of the undercooled states. Matls Perf Char. **1**(1), 104632 (2012)

17. B. Milkereit, O. Kessler, C. Schick, *Continuous Cooling Precipitation Diagrams of Aluminium-Magnesium-Silicon Alloys*, ed. by J. Hirsch, B. Skrotzki, G. Gottstein (eds), vol 2. (DGM; WILEY-VCH Weinheim), pp. 1232–1237 (2008)
18. M. Merklein, M. Lechner, M. Graser Influence of a short-term heat treatment on the formability and ageing characteristics of aluminum profiles, in Lasers in Manufacturing Conference (2015)
19. T. Klöppel, T. Loose, Recent developments for thermo-mechanically coupled simulations in LS-DYNA with focus on welding processes, in European LS-DYNA Conference
20. J. Goldak, A. Chakravarti, M. Bibby, A new finite element model for welding heat sources. MTB **15**(2), 299–305 (1984)
21. T. Loose, T. Klöppel, An LS-DYNA material model for the consistent simulation of welding forming and heat treatment, in 11th International Seminar Numerical Analysis of Weldability (2015)

Numerical Simulation of Meso-Micro Structure in Ni-Based Superalloy During Liquid Metal Cooling Process

Xuewei Yan, Wei Li, Lei Yao, Xin Xue, Yanbin Wang, Gang Zhao, Juntao Li, Qingyan Xu and Baicheng Liu

Abstract Ni-based superalloys are the preferred material to manufacture turbine blades for their high temperature strength, microstructural stability and corrosion resistance. As a new method, liquid-metal cooling (LMC) process is prospective used in manufacturing large-size turbines blades. Unfortunately, there are many casting defects during LMC directional solidification, such as stray grain, freckle, cracking. Moreover, the trial and error method is time and money cost and lead to a long R&D cycle. As a powerful tool, numerical simulation can be used to study LMC directional solidification processes, to predict final microstructures and optimize process parameters. Mathematical models of microstructure nucleation and growth were established based on the cellular automaton-finite difference

---

X. Yan (✉) · Q. Xu (✉) · B. Liu
Key Laboratory for Advanced Materials Processing Technology, Ministry of Education, School of Materials Science and Engineering, Tsinghua University, Beijing 100084, China
e-mail: yanxuewei2013@gmail.com

Q. Xu
e-mail: scjxqy@tsinghua.edu.cn

B. Liu
e-mail: liubc@mail.tsinghua.edu.cn

W. Li · L. Yao · X. Xue · Y. Wang · G. Zhao · J. Li
Beijing CISRI-GAONA Materials & Technology Co. LTD, Beijing 100081, China
e-mail: 13911591161@139.com

L. Yao
e-mail: 58485360@qq.com

X. Xue
e-mail: xuetiger@126.com

Y. Wang
e-mail: 137804898@qq.com

G. Zhao
e-mail: 13910603539@139.com

J. Li
e-mail: juntaoli168@163.com

© The Minerals, Metals & Materials Society 2017
P. Mason et al. (eds.), *Proceedings of the 4th World Congress on Integrated Computational Materials Engineering (ICME 2017)*, The Minerals, Metals & Materials Series, DOI 10.1007/978-3-319-57864-4_23

(CA-FD) method to simulate meso-scale grain and micro dendrite growth behavior and morphology. Simulated and experimental results were compared in this work, and they agreed very well with each other. Meso-scale grain evolution and micro dendritic distribution at a large scale were investigated in detail, and the results indicated that grain numbers reduced with the increase of height of the casting, and stray grain will be relatively easy to produce in the platform. In addition, secondary dendrite arms were very tiny at the bottom of the casting, and they will coarsen as the he height of the cross section increased.

Keywords Numerical simulation · Meso-scale grain · Micro dendrite · LMC process

Introduction

Nickel-based superalloys are widely used in modern advanced aero and power industry because of their excellent high temperature mechanical properties and corrosion resistance [1]. As the key components of turbine engines, high-performance turbine blades are usually produced by directional solidification (DS) process [2]. Since the superalloy properties are largely dependent on microstructure, columnar grains or single crystal (SC) structure are expected to obtain during DS process [3]. Nowadays, the more common DS process is Bridgman or high-rate solidification (HRS) process, which provides a vertical temperature gradient by a water cooling chill-plate and a withdrawal unit. This process has been highly optimized for the production of aero-engine scale components, which represent the majority of production. However, there are several issues when the HRS process is scaled to produce larger components, such as deformation and crack of mold shell, chemical reaction in the interface between ceramic and liquid metal, and freckles, stray grain of castings [4, 5]. Recently, more efforts have focused on the development of novel DS techniques utilizing high thermal gradient and high cooling rate in order to solve these issues, such as liquid metal cooling (LMC) process [6] or gas cooling casting (GCC) process [7]. In LMC process, the ceramic mold containing the metal melt is withdrawn from a hot zone through a baffle opening and simultaneously immersed into a liquid metal coolant with low melting temperature, e.g. stannum (Fig. 1). The benefits of LMC process have been identified, especially for large cross-section components requiring significant heat extraction to maintain directional solidification. Although the LMC process can achieve thermal gradient over $10 \text{ K} \cdot \text{mm}^{-1}$, it will require significant development before put into production [8].

Over the past few decades, considerable progress has been made in understanding microstructure growth in LMC process from both experimental and theoretical studies. Elliott et al. [9] analyzed microstructure morphology at different withdrawal rates in large cross-sectional castings, and the results showed that LMC process exhibited higher gradients at all withdrawal rates and higher thermal gradients

Fig. 1 Schematic diagram of LMC furnace

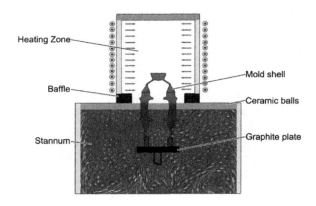

resulted in a refined microstructure measurable by the finer dendrite-arm spacing. Kermanpur et al. [10] investigated the influence factors of grain structure in land-based turbine blade through experiment, and used a cellular automaton (CA) coupled with finite-element (FE) model to simulate the grain growth in LMC process. CA is a powerful method for predicting microstructure growth presented by Rappaz and Gandin [11] in the 1990s. Subsequently, there have been many studies on the simulation of DS dendrite growth and their evolution behavior using CA method. Recently, Zhang and Xu [12] presented a two-type directional dendrite growth model to realize the multi-scale simulation based on the CA-FD model considering macro DS parameters. However, most of studies mentioned above are 2D or pseudo 3D simulation for confining to calculation capability. Although some techniques such as the parallel computing and adaptive mesh refinement methods have been developed in order to enhance computational efficiency, large scale or 3D dendrite growth simulation are still a challenge, as they require significant computational resources.

In this study, mathematical models of microstructure nucleation and growth were established to simulate grain and dendrite growth in LMC process. Simulation parameters were calculated by dedicated software JMatPro, meso-scale grain growth and micro dendrite morphologies were simulated by the coupled CA-FD model. The verification experiment was also carried out by using a simplified geometry specimen, and the simulation results can predict the experimental results very well. Large scale 2D dendritic distribution were investigated by experimental and simulation, the dendrites growth and stray grain defect were also analyzed.

Mathematical Models and Experimental Details

Microstructure Growth Model

In CA model, the liquid cells in the growth directions <100> were captured preferentially, and the cells in the direction which is vertical to the prior growth directions were also captured. These two capturing rules were the basic process of CA model, which describe the dendrite growth and coarsening. Moreover, some special cells should be captured in this step to modify the grains directions and keep the <100> directions as the prior growth direction.

The KGT model [13] is based on the assumption that the dendrite tip has an ideal parabolic shape and is advancing with a steady state velocity and the kinetic undercooling is neglected. However, the DS dendrite growth is always influenced by solute distribution, the KGT model is no longer appropriate to describe the dendrite growth. Therefore, the solute diffusion is calculated by Eq. (1) in the bulk of the liquid and the solid phase without considering nature and forced convection influence.

$$\frac{\partial C_i}{\partial t} = \nabla \cdot (D_i \nabla C_i) + C_i(1-k_0)\frac{\partial f_s}{\partial t} \quad (1)$$

where C is the composition with its subscript i denoting solid or liquid, D is the solute diffusion coefficient and k_0 is the equilibrium partition coefficient. The last terms on the right hand denotes the amounts of solute rejected due to the increment of solid fraction at the S/L interface. At the interface of the liquid and solid, the partitioning of solute in the growing cell is determined by Eq. (2).

$$C_s = k_0 C_L \quad (2)$$

where C_S and C_L are the average solute concentrations of the solid and liquid, respectively, in the solid at the liquid/solid interface, k_0 is the solute partition coefficient. The local equilibrium composition C_L at the interface is obtained by Eq. (3).

$$C_L^* = C_0 + \frac{1}{m_L}\left(T^* - T_L + \Gamma \kappa f(\theta_i)\right) \quad (3)$$

where C_0 is the original solute concentration in the liquid, m_L is the liquidus slope, T^* is the actual interface equilibrium temperature, T_L is the equilibrium liquidus temperature at initial solute composition, Γ is the Gibbs-Thomson coefficient, κ is curvature of S/L interface and $f(\theta_i)$ is a anisotropy function can be described by Eqs. (4) and (5).

$$\kappa = \frac{1}{a_m}\left\{1 - \frac{2}{N+1}\left[f_s + \sum_{i=1}^{N} f_s(i)\right]\right\} \quad (4)$$

$$f(\theta_i) = \prod_{i=l,m,n}[1 + \varepsilon \cos(\delta_i \theta_i)] \quad (5)$$

where N is the number of neighboring cells, a_m is the micro grid step length, f_s is the solid fraction, $f_s(i)$ is the solid fraction of i cell, ε is the anisotropy coefficient, δ_i is the anisotropic modulus, θ_i is the interface anisotropy angle, and l, m, n are the coordinates.

Simulation Parameters and Experimental Methods

In this study, the superalloy employed is DZ24 (Ni-8.5 Cr-5.4 Al-4.5 Ti-3.1 Mo-13.5 Co-1.4 W-0.15 C, wt%). Since the DD6 superalloy is a multi-component alloy, a binary approximation Ni-X alloy was performed to simulate the solidification process variables on the development of the dendritic structure. Some equivalent parameters of Ni-X alloy were calculated using the professional metal performance calculation software JMatPro as shown in Table 1.

The pouring experiment was carried out in a 10 kg HRS and LMC amphibious DS furnace. The ceramic mold shell and a simplified geometry casing used in the experiment are shown in Fig. 2. The withdrawal rate was 15 mm·min^{-1} and the pour temperature was 1500 °C for the LMC casting experiment. The liquid-metal bath container held approximately 1200 kg of tin, and the temperature was set to 250 °C which controlled by a recirculating thermal oil system. After solidification, the casting was cleaned and lightly polished to remove mold material and was sectioned in different position for microstructural analysis (Fig. 2b). Following metallographic preparation, polished surfaces were etched to reveal the dendritic structure by using the solution containing 70 mL CH_3CH_2OH, 30 mL HCl and 2.5 g $CuCl_2$. Dendrite morphology of the castings with different processes was observed by using LEICA DM6000 M optical microscope.

Table 1 The thermophysical parameters of DZ24 superalloy

Parameters	Values
Liquidus temperature (K)	1619
Solidus temperature (K)	1505
Thermal conductivity (kJ·m^{-1}·s^{-1}·K^{-1})	0.0336
Density (kg·m^{-3})	7950
Specific heat (kJ·kg^{-1}·K^{-1})	0.863
Latent heat (kJ·kg^{-1})	90
Liquidus slope	−3.73
Solute partition coefficient	0.562
Concentration (wt%)	32.15
Anisotropy coefficient ε	0.03
Gibbs-Thomson coefficient Γ (K·m)	3.05×10^{-7}
Liquid diffusion coefficient D_L (m^2·s^{-1})	1.73×10^{-9}
Solid diffusion coefficient D_s (m^2·s^{-1})	1.16×10^{-12}

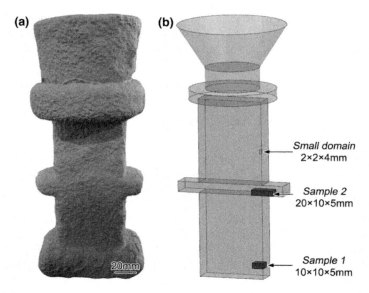

Fig. 2 a Ceramic mold shell, b corresponding casing and samples

Results and Discussion

Simulation of Grain Growth

Figure 3 shows the meso-scale grain growth in different stages during LMC process, which reflects the competitive growth of the DS grains, and different colors represent different columnar grains, and the liquid metal is set to be achromatic in order to present the morphology and location of the microstructure clearly. In the present case, the 3D grain structure simulated with the CAFD model and nucleation

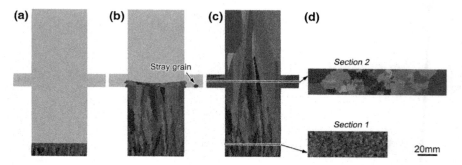

Fig. 3 Simulation results of grain structure: **a** $t = 59$ s, $f_s = 2.3\%$; **b** $t = 672$ s, $f_s = 19.6\%$; **c** $t = 1482$ s, $f_s = 82.1\%$; **d** cross-section in different position

takes place at the bottom of the starter block, and after that these tiny nuclei grow in the opposite direction of the heat flow. These grains have the [001] crystal directions pointing to all directions, but only those whose [001] preferential growth orientation best aligned with that of heat flow could get the fastest growth rate. As the best-aligned grains continually grow and become coarse (Fig. 3a), the grains which were not well aligned with respect to the maximum gradient of the temperature field would grow at a much slower speed and may be restrained and submerged. It can be seen in Fig. 3b, stray grain has formed at the platform of casting. To some extent, this is due to a hot spot exists at the inner corner of the cross-sectional transition, resulting from the poor local cooling condition. This hot spot hinders single-crystal growth from the casting portion into the platform. In order to be accepted, such DS casting must have a well-oriented columnar grain structure (parallel to the blade axis) without any stray grain and with a minimum number of grains in the casting. Therefore, the casting manufactured by this process will not be accepted, even though its high-temperature mechanical properties are superior to conventionally cast ones. Figure 3c shows that the grain boundaries in the casting are not well oriented along the casting axis, and they seem to grow toward the center of the casting. This is mainly because the withdrawal rate is too high, mushy zone interface becomes concave, which is higher on the edges side and lower in the center, and will lead to the incline growths of grains on the edge sides and the transverse coarsening of grains. In the real production, it is necessary to adopt a higher withdrawal rate to improve the productivity. But too high withdrawal rate will lead to the declining grains and instable growth, and the property of the grain structure will be hard to control [14]. However, low withdrawal rate not only decreases the productivity but also causes coarsened grains and uneven sizes of different columnar grains. In total, choosing a proper value of withdrawal rate may get better grain structure for the DS castings. Figure 4d shows the grain morphologies of different transversal sections which are the same place

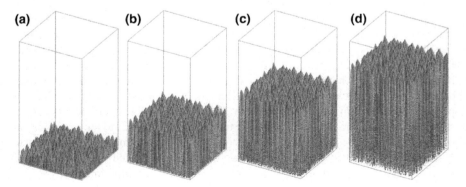

Fig. 4 Simulation results of 3D dendrite growth of the small domain: **a** $f_s = 10\%$, **b** $f_s = 30\%$, **c** $f_s = 60\%$, **d** $f_s = 80\%$

with sample 1 and sample 2. A large number of grains were observed at section 1 in Fig. 3d and the size of grains were very small, while the grain numbers decreased in section 2 in Fig. 3d and an obvious stray grain was found in the platform.

Dendrite Morphology in 3D

The dendrites are formed at the moving solid-liquid interfaces, and the simulated dendrite growth behavior are shown in Fig. 4. In this 3D case, a domain of 400 × 400 × 800 cells (small domain in Fig. 2b) was used, and the cell size is 5 μm. At the bottom of the calculation domain, hundreds of solid seeds with a preferential crystallographic orientation (0°, 0°, 0°) parallel to the axis of the (x, y, z) coordinate system were placed, and the position of these grain seeds were randomly distributed (Fig. 4a), then they grew in the unidirectional temperature distribution. Initially, these grain seeds showed a severely competitive growth, and the dendrite arm spacing was adjusted accordingly. The uneven fluctuation of temperature and components led to grains with preferred orientation growing quickly, and other ones being blocked and eventually eliminated. The distribution of heat and solute then settled into a dynamic balance, which would lead to a stable growth of dendrites (Fig. 4c). Finally, the dendrites were stable growth and the dendrite arm spacing was equally distributed, as shown in Fig. 4d. The dendrite tip morphology characteristics, determined using the detailed numerical analysis of the wavelengths of instabilities along the sides of a dendrite by Langer and Müller-Krumbhaar [15], followed the predicted scaling law:

$$\lambda_2/R = 2.1 \pm 0.03 \qquad (9)$$

where λ_2 is the secondary dendrite arm spacing, R is the steady-state dendrite tip radius.

Figure 5a shows a SEM micrograph in the top view which was obtained from the upper end of the casting, where the solidifying material has run out of interdendritic melt material and free standing dendrites can be seen without further metallographic preparation and the detailed description can be seen in reference [16]. It was observed that four secondary dendrite arms emanate from the central primary dendrite, which show an un-strict fourfold symmetric, and the tertiary dendrite has not grown to a significant length. In addition, it suggests that the primary dendrite arm spacing is close to 300 μm. Figure 5b is the corresponding simulation result. The simulated dendritic morphology is similar to the experimental result, especially the secondary dendrite arms. Thus, this model was shown to describe the experimental process very well, and the simulated dendrite "cruciate flower" morphology and arm spacing agreed well with the experiments.

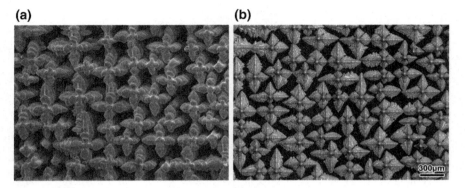

Fig. 5 Dendritic morphologies in the top view: **a** Experimental result [12], **b** simulation result

Dendrite Distribution at Large Scale

The dendrites are the substructure of the grains, which are the strategic link between materials processing and materials behavior. Therefore, dendrites growth and distribution control is essential for any processing activity. However, the scale of the dendrite is very small and the growth stage is ordinary unobservable which will cause much hardship to study the dendrite. Although the simulation method can describe the process of the dendrite growth, over a large scale of simulation is still a challenge due to the computation bound. In this work, parallel algorithm and modified self-adaption mesh refinement method was used to simulate the dendrite growth at the macro scale. Figure 6a shows the experimental result of transverse section of sample 1 in the top surface, and Fig. 6b is the corresponding simulation result, the calculation domain contains the whole section which reaches about 100 mm^2. In the simulation result, different grains with different Euler angles are presented in various colors. Each grain consists of dendrites with the same Euler angles as a sub-structure. It is clearly to see the simulation and experimental results are in good agreement with each other. In these figures, many grains with a large amount of dendrites competitively grow and the secondary dendrite arms are tiny due to the great thermal gradient. Figure 6c, d are the experimental and simulation results of transverse section in the upper of sample 2, and calculation domain was about 200 mm^2. It is obvious found that as the height of the cross section increased, the number of columnar grains decreased. When dendrites with different orientations encountered each other at the boundary of columnar grains, some branches were well developed in the free space, whereas some appeared competitive in a limited space. As the DS height increased, the dendrite arm spacing also increased and the dendritic morphologies became more complicated. A stay grain was also found in these dendritic morphology graphs which verify the accuracy of the simulation result in Fig. 3. Thus, it is acceptable to use the modified microstructure model to predict the dendrite growth based on the coupling simulation of the grains structure.

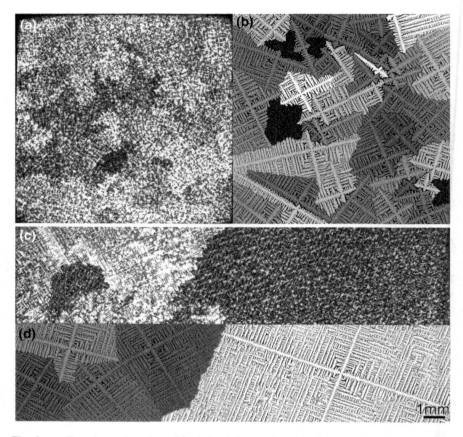

Fig. 6 **a, c** Experimental results and **b, d** simulation results of dendrite morphologies in different transverse sections: **a, b** for sample 1 and **c, d** for sample 2

Conclusions

(1) The mathematical models of microstructure nucleation and growth were built based on the CA-FD method. The grain structure evolution and dendrite growth behavior and distribution could be predicted by using the multi-scale coupling model.
(2) Meso-scale grain structure for the whole casting was simulated. The grains growing preferentially toward the centerline of the casting due to the too high withdrawal rate, and a stray grain was found in the platform.
(3) 3D dendrite growth behavior was simulated and corresponding model validations were performed, they were in good agreement. 2D large scale dendrite distribution was also simulated by parallel algorithm and modified self-adaption mesh refinement method, and the evolution law of dendrite arm was analyzed by experimental and simulation results.

Acknowledgements The authors gratefully acknowledge the financial support of the National Basic Research Program of China (Grant No. 2011CB706801), the National Natural Science Foundation of China (Grant Nos. 51374137 and 51171089), and the National Science and Technology Major Projects (Grant Nos. 2012ZX04012-011 and 2011ZX04014-052).

References

1. F.R.N. Nabarro, The superiority of superalloys. Mater. Sci. Eng. A **184**, 167–171 (1994)
2. B.H. Kear, E.R. Thompson, Aircraft gas turbine materials and processes. Science **208**, 847–856 (1980)
3. M. Rappaz, Modeling and characterization of grain structures and defects in solidification. Curr. Opin. Solid State Mater. Sci. **20**, 37–45 (2016)
4. D. Pan, Q.Y. Xu, B.C Liu, Modeling of grain selection during directional solidification of single crystal superalloy turbine blade castings. JOM, **62**(5), 30–34 (2010)
5. H. Zhang, Q.Y. Xu, N. Tang et al., Numerical simulation of microstructure evolution during directional solidification process in directional solidified (DS) turbine blades. Sci. China Technol. Sci. **54**, 3191–3202 (2011)
6. M.M. Franke, R.M. Hilbinger, A. Lohmuller et al., The effect of liquid metal cooling on thermal gradients in directional solidification of superalloys: thermal analysis. J. Mater. Process. Technol. **213**, 2081–2088 (2013)
7. F. Wang, D.X. Ma, J. Zhang et al., A high thermal gradient directional solidification method for growing superalloy single crystals. J. Mater. Process. Technol. **214**, 3112–3121 (2014)
8. X.W. Yan, N. Tang, X.F. Liu et al., Modeling and simulation of directional solidification by LMC process for nickel base superalloy casting. Acta Metall. Sin. **51**, 1288–1296 (2015)
9. A.J. Elliott, S. Tin, W.T. King et al., Directional solidification of large superalloy castings with radiation and liquid-metal cooling: a comparative assessment. Metall. Mater. Trans. A **35A**, 3221–3231 (2004)
10. A. Kermanpur, N. Varahram, P. Davami et al., Thermal and grain-structure simulation in a land-based turbine blade directionally solidified with the liquid metal cooling process. Metall. Mater. Trans. B **31B**, 1293–1304 (2000)
11. M. Rappaz, C.A. Gandin, Probabilistic modelling of microstructure formation in solidification process. Acta Metall. Mater. **41**, 345–360 (1993)
12. H. Zhang, Q.Y. Xu, Multi-scale simulation of directional dendrites growth in superalloys. J. Mater. Process. Technol. **238**, 132–141 (2016)
13. W. Kurz, B. Giovanola, R. Trivedi, Theory of microstructural development during rapid solidification. Acta Metall. **34**, 823–830 (1986)
14. R. Chen, Q.Y. Xu, B.C. Liu, Cellular automaton simulation of three-dimensional dendrite growth in Al-7Si-Mg ternary aluminum alloys. Comput. Mater. Sci. **105**, 90–100 (2015)
15. J.S. Langer, H. Müller-Krumbhaar, Theory of dendritic growth-I. Elements of a stability analysis. Acta Metall. **26**, 1681–1978 (1978)
16. H. Zhang, Q.Y. Xu, Z.X. Shi et al., Numerical simulation of dendrite grain growth of DD6 superalloy during directional solidification process. Acta Metall. Sin. **50**, 345–354 (2014)

Part IV
Phase Field Modeling

Multiscale Simulation of α-Mg Dendrite Growth via 3D Phase Field Modeling and Ab Initio First Principle Calculations

Jinglian Du, Zhipeng Guo, Manhong Yang and Shoumei Xiong

Abstract Based on synchrotron X-ray tomography and electron backscattered diffraction techniques, recent studies revealed that the α-Mg dendrite exhibited an 18-primary-branch morphology in 3D, of which six grew along $<11\bar{2}0>$ on the basal plane, whereas the other twelve along $<11\bar{2}3>$ on non-basal planes. To describe this growth behaviour and simulate the morphology of the α-Mg dendrite in 3D, an anisotropy function based on cubic harmonics was developed and coupled into a 3D phase field model previously developed by the current authors. Results showed that this anisotropy function, together with the phase field model could perfectly describe the 18-primary-branch dendrite morphology for the magnesium alloys. The growth tendency or orientation selection of the 18-primary-branch morphology was further investigated by performing ab initio first principle calculations based on the hexagonal symmetry structure. It was showed that those crystallographic planes normal to the preferred growth directions of α-Mg dendrite were characterized by higher surface energy than these of others, i.e. coinciding with the 18-primary-branch dendritic morphology. Apart from agreement with experiment results and providing great insights in understanding dendrite growth behaviour, such multiscale computing scheme could also be employed as a standard tool for studying general pattern formation behaviours in solidification.

Keywords Magnesium alloys · Orientation selection · Dendritic morphology · Phase field simulation · Ab initio calculations

---

J. Du · Z. Guo · M. Yang · S. Xiong (✉)
School of Materials Science and Engineering, Tsinghua University, Beijing 100084, China
e-mail: smxiong@tsinghua.edu.cn

J. Du · Z. Guo · M. Yang · S. Xiong
Key Laboratory for Advanced Materials Processing Technology, Ministry of Education, Beijing, China

Introduction

Magnesium alloys are widely applied in aerospace, automobile and electronic communication industries due to superior properties such as lightweight and high specific strength [1–3]. Die casting is the primary manufacturing technique of magnesium alloys for structural applications [4]. During the casting process, alloy melts solidify dendritically and form the so-called α-Mg dendrites, whose size, orientation and distribution will exert profound influence on the practical performance of magnesium alloy products [5, 6]. Therefore, acquiring key information on α-Mg dendritic microstructure is essential for improving the properties of magnesium alloys.

Compared with aluminum alloys with cubic symmetry [6, 7], the dendritic orientation selection and growth morphology of magnesium alloys are much more complicated because of their hexagonal symmetry atomic structure [8, 9]. It has been generally accepted that α-Mg dendrites exhibit typical six-fold symmetry pattern in two-dimensional (2D) scale [10–12]. However, the 3D microstructure of α-Mg dendrites, which is very important to understand the properties of magnesium alloys, has not been identified until the rapid advancement of experimental technology [13]. With synchrotron X-ray tomography and EBSD techniques, it was found recently that the preferred growth orientations of α-Mg dendrites were $<11\bar{2}0>$ in the basal plane and $<11\bar{2}3>$ in non-basal planes, resulting in an 18-primary-branch morphology in 3D [14, 15]. For the best knowledge of the current authors, the underlying mechanism determining such growth pattern of α-Mg dendrite was still unclear. Since atomic structure and surface energy play important role in driving the growth pattern of crystals [16–18], it is necessary to investigate the growth dependency of α-Mg dendrites from the significance of these factors.

In this work, the dendritic orientation selection of magnesium alloys was investigated by 3D phase field simulations and ab initio first principle calculations. Based on the experimental results on the 18-primary-branch morphology, an anisotropy function was developed and coupled into a 3D phase field model. Results indicated that this anisotropy function, together with the phase field model could perfectly describe the 3D growth morphology of α-Mg dendrites. By performing first principle calculations, it was found that these specific crystallographic planes perpendicular to which lie the dendritic preferred orientations of magnesium alloys always exhibit higher surface energy than other planes, which agrees well with our experimental results. The present investigation achieves an aggregate picture with fascinating 3D morphology of α-Mg dendrites, and thus provides great insights in understanding the dendritic orientation selection of magnesium alloys.

Phase Field Model

The governing equations of the phase field (PF) model adopted in this study were expressed as [19]:

$$\tau \frac{\partial \phi}{\partial t} = \nabla \cdot (W(\vec{n})^2 \nabla \phi) + \frac{\partial}{\partial x}\left(|\nabla \phi|^2 W(\vec{n}) \frac{\partial W(\vec{n})}{\partial \phi_x}\right) + \frac{\partial}{\partial y}\left(|\nabla \phi|^2 W(\vec{n}) \frac{\partial W(\vec{n})}{\partial \phi_y}\right)$$
$$+ \frac{\partial}{\partial z}\left(|\nabla \phi|^2 W(\vec{n}) \frac{\partial W(\vec{n})}{\partial \phi_z}\right) + \phi(1-\phi^2) - \lambda(1-\phi^2)^2(\theta + kU) \tag{1}$$

$$\left(\frac{1+k}{2} - \frac{1-k}{2}\phi\right)\frac{\partial U}{\partial t} = \nabla \cdot \left(D\frac{1-\phi}{2}\nabla U - \vec{J}_{at}\right) + \frac{1}{2}[1+(1-k)U]\frac{\partial \phi}{\partial t} \tag{2}$$

$$\frac{\partial \theta}{\partial t} = \alpha \nabla^2 \theta + \frac{1}{2}\frac{L/c_p}{\Delta T_0}\frac{\partial \phi}{\partial t} \tag{3}$$

In the above equations, τ is the relaxation time, ϕ is the phase field, $W(\vec{n})$ is the anisotropic width of the diffuse interface with $\vec{n}$ the unit normal vector of the solid/liquid interface, k is the partition coefficient, D and α are the solute and thermal diffusivities respectively, L is the heat of fusion per unit volume and c_p is the specific heat of alloy. λ is the scaling parameter and given as:

$$\lambda = \frac{15}{16}\frac{RT_M(1-k)}{v_0 H|m|}\Delta T_0 \tag{4}$$

where R is the gas constant, T_M is the melting temperature of the solvent, v_0 is the molar volume, H is the energy barrier of the double well potential, and m is the liquidus slope in the phase diagram. $\vec{J}_{at}$ is the 'anti-trapping' current which is defined as:

$$\vec{J}_{at} = -\frac{W}{\sqrt{2}}\frac{c/c_\infty}{[1+k-(1-k)\phi]}\frac{\partial \phi}{\partial t}\frac{\nabla \phi}{|\nabla \phi|} \tag{5}$$

U is the dimensionless solute concentration and given as:

$$U = \frac{\frac{2c/c_\infty}{1+k-(1-k)\phi} - 1}{1-k} \tag{6}$$

where c is the solute concentration, c_∞ is the initial solute concentration, and ΔT_0 is the equilibrium freezing temperature range and denoted as:

$$\Delta T_0 = \frac{|m|c_\infty(1-k)}{k} \tag{7}$$

θ is the dimensionless temperature and denoted as:

$$\theta = \frac{T - T_M - mc_\infty}{\Delta T_0} \tag{8}$$

The crystal anisotropy was introduced by $\tau = \tau_0 A(\vec{n})^2$ and $W(\vec{n}) = W_0 A(\vec{n})$ with $A(\vec{n})$ the anisotropy function, $\phi_x = \partial\phi/\partial x$, $\phi_y = \partial\phi/\partial y$ and $\phi_z = \partial\phi/\partial z$ [20].

Computational Methods

The first-principles calculations were performed within the framework of density functional theory, as implemented in the Vienne Ab initio Simulation Package (VASP) [21]. The exchange and correlation interactions were described in local density localization (LDA) methods [22]. The interactions between ions and valence electrons were modeled by the projector-augmented wave (PAW) potentials [23]. A plane wave cutoff energy of 450 eV was used for pure Mg, and 420 eV for binary Mg–Al/Ba/Sn alloys after performing relevant convergence tests. Brillouin zone integration was modeled by the Monkhorst-Pack k-point mesh [24], and the k-point separation in the Brillouin zone of the reciprocal space was set as 0.01 Å^{-1} with respect to each surface unitcell. The total energy was converged to 5×10^{-7} eV/atom with respect to electronic, ionic and unitcell degrees of freedom.

The surface energy [25, 26], which is defined as the energy required to form a unit area of surface, was used to analyze the growth tendency or orientation selection of α-Mg dendrites, and the calculation formula was expressed as:

$$E_{surf} = (E_n - nE_b)/(2A) \tag{9}$$

where E_n is the total energy per primitive surface unitcell, E_b is the total energy per primitive bulk unitcell, n is the number of layers in the slab model, and A is the area per primitive surface unitcell, respectively.

The surface structure was simulated by the slab model composed of a periodic boundary condition and a vacuum region [27]. Meanwhile, the top atomic layers were relaxed whereas the bottom ones were fixed. Based on the experimental results [14], different surface models corresponding to low index as well as high index crystallographic planes of magnesium and its binary dilute alloys were obtained. The atomic structure of binary magnesium alloys was simulated by solid solution models, where certain number of solvent atoms were substituted randomly by the solute atoms in a (4 × 4×2) supercell of magnesium. The surface energy values were satisfactorily converged to <0.001 eV/Å² with respect to slab thickness, vacuum thickness and number of relaxed atomic layers of the slab model. Initial benchmark calculations were performed on bulk magnesium to guarantee the accuracy of the computational method. Results indicated that the optimized lattice parameters for bulk magnesium were in good agreement with the experimental values [28], which confirmed that the computational scheme is reliable.

Results and Discussion

The Anisotropy Function

In light of our experimental results on the 18-primary-branch morphology of α-Mg dendrite, an anisotropy function was developed upon the spherical harmonics to describe the 3D dendritic growth pattern of magnesium alloys [14]:

$$\gamma(\theta,\varphi) = \gamma_0 \left[1 + \varepsilon_1 \left(3n_z^2 - 1\right)^2 + \varepsilon_2 \left(n_x^3 - 3n_x n_y^2\right)^2 \left(9n_z^2 - 1 + \varepsilon_3\right)^2 \right] \quad (10)$$

where θ and φ are spherical coordinate angles used to denote the unit normal vector $\vec{n}$ of the solid/liquid interface, with $n_x = \sin\theta \cos\varphi$, $n_y = \sin\theta \sin\varphi$ and $n_z = \cos\theta$, as depicted in Fig. 1a, b. ε_1, ε_2 and ε_3 are the anisotropic factors along different directions, with ε_1 relating to dendritic growth along the principal axial direction, ε_2 to dendritic growth along the 18-primary-branch pattern, and ε_3 to the growth direction within the basal plane. The graphic illustration for this anisotropy function describing the 3D dendritic morphology of magnesium alloys was presented in Fig. 1c.

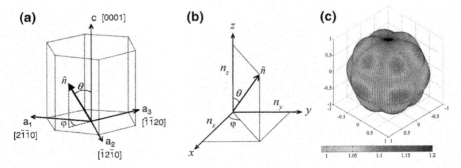

Fig. 1 Schematic illustration for the coordinate systems **a** and **b** adopted in this work. **c** shows the graphic illustration of the anisotropy function used to describe the 18-primary-branch morphology of α-Mg dendrites, with $\varepsilon_1 = -0.02$, $\varepsilon_2 = 0.15$ and $\varepsilon_3 = 0.15$

The Phase Field Simulation

The according anisotropy function was then coupled into the phase field (PF) model by using a Para-AMR algorithm previously developed by Guo and *co-workers* [19]. The simulation was performed using a hierarchical mesh with 5 levels of grids, and the domain size was set as 819.2 × 819.2 × 819.2, other relevant parameters were set as $\theta = -0.10$, $k = 0.15$ and $\lambda = 30$.

The PF simulation results on the growth process of an equal-axial α-Mg dendrite were shown in Fig. 2, together with the corresponding coordinate system, where z-axis is along the <0001> direction. Figure 2a presents the dendritic growth morphology at the initial solidification stage. The dendrite exhibits six-fold symmetry features viewed from z-axis, and there are three layers with each layer having six branches along the view from x-/y-axis, i.e. α-Mg dendrite will grow along 18-primary branches. With the growth proceeding, the secondary branches appear with the same growth direction as the primary branches, as shown in Fig. 2b–f, respectively.

To validate the reliability of the present model, the PF simulation results were compared with the X-ray tomography experiments on the 3D dendritic morphology of magnesium alloys, as shown in Fig. 3a, b. Futher comparisons of 3D dendritic morphology between PF simulation and experimental results along different view perspectives were shown in Fig. 3c–e. The α-Mg dendrite exhibits an six-fold symmetry feature normal to <0001>, and there are twelve branches normal to $< 10\bar{1}0 >$ direction. Although the dendritic morphology presents differences along

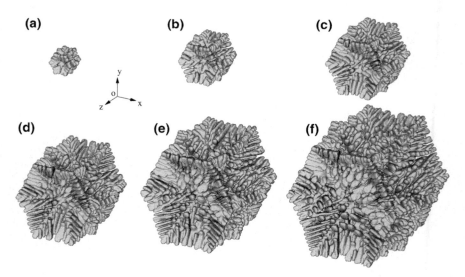

Fig. 2 Phase field simulation of the α-Mg dendrite evolution at time steps of, **a** 5000, **b** 10000, **c** 15000, **d** 20000, **e** 25000, and **f** 30000

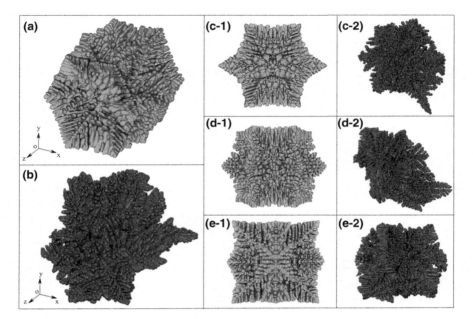

Fig. 3 Comparisions of 3D dendritic morphology between **a** PF simulation and **b** X-ray tomography experiments of Mg-30 wt% Sn alloy. The view directions for the corresponding comparisons were **c** $<0001>$, **d** $<10\bar{1}0>$ and **e** $<11\bar{2}0>$, respectively

different view directions, the results showed that the PF simulation on the growth pattern of α-Mg dendrite was in good agreement with the X-ray tomography experimental results on the 18-primary-branch morphology.

The First-Principle Calculations

The surface energy was investigated to further understand the dendritic orientation selection of magnesium alloys. Based on the experimental results [14], the crystallographic planes including $\{0001\}$, $\{10\bar{1}0\}$, $\{10\bar{1}1\}$, $\{11\bar{2}0\}$, and $\{11\bar{2}k\}$ with k changing from 1 to 9, were analyzed in this work. For the *hcp* structure, the crystallographic plane perpendicular to $<11\bar{2}x>$ crystallographic direction is denoted as $\{11\bar{2}k\}$, and k formulates with the *c/a*-ratio and x via $k = 2x(c/a)^2/3$. By using the ideal *c/a*-ratio of 1.62 for pure magnesium and its dilute alloys, it was deduced that the crystallographic plane perpendicular to $<11\bar{2}3>$ is $\{11\bar{2}5\}$, and that perpendicular to the $<22\bar{4}5>$ as reported by Pettersen [11] is $\{11\bar{2}4\}$.

According to Eq. (9), the surface energy of relevant crystallographic planes was calculated by performing first principle calculations. As shown in Fig. 4a, b, the

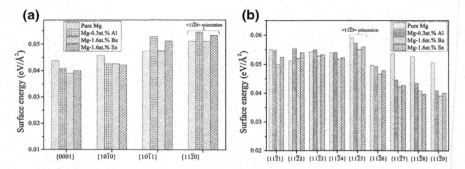

Fig. 4 Surface energy of pure magnesium and binary dilute magnesium alloys. **a** High symmetry surface orientations, and **b** high index surface orientations, showing that the surface energy of $\{11\bar{2}0\}$ and $\{11\bar{2}5\}$ corresponding to dendritic preferred growth directions $<11\bar{2}0>$ and $<11\bar{2}3>$ is higher than the other surface orientations

surface energy of those high symmetrical crystallographic planes, including $\{0001\}$, $\{10\bar{1}0\}$, $\{10\bar{1}1\}$ and $\{11\bar{2}0\}$ was lower than the high index planes $\{11\bar{2}k\}$. This indicates that those high symmetrical crystallographic planes were energetically more favorable than others. It was confirmed that crystals usually favor to grow along those orientations with higher surface energy, appearing as the preferred growth directions [18]. Accordingly, the fact that in the basal plane, the surface energy corresponding to $<11\bar{2}0>$ was higher than that to $<10\bar{1}0>$ indicated that the α-Mg dendrite would prefer to grow along $<11\bar{2}0>$ direction. Analogously, the high surface energy of the $\{11\bar{2}5\}$ plane implied that its normal direction $<11\bar{2}3>$ would be another preferred orientation for α-Mg dendritic growth in non-basal planes. Although Pettersen reported that the preferred growth direction of α-Mg dendrites in the non-basal plane was $<22\bar{4}5>$, our calculated results showed that the surface energy of $<11\bar{2}3>$ orientation was higher than that of $<22\bar{4}5>$. This suggested that the preferred growth direction of α-Mg dendrite in non-basal planes was $<11\bar{2}3>$. Our ab initio prediction for dendritic orientation selection based on surface energy was in good accordance with the experimental findings on the preferred growth directions of α-Mg dendrites.

Conclusion

The dendritic orientation selection of magnesium alloys was investigated by 3D phase field simulations and ab initio first principle calculations. Based on the 18-primary-branch dendritic morphology, an anisotropy function was developed and coupled into the PF model to perfectly describe the 3D dendritic morphology of magnesium alloys. First principle calculations on the orientation-dependent surface energy indicate that those preferred growth directions $<11\bar{2}0>$ and $<11\bar{2}3>$ of α-Mg dendrites were characterized by higher surface energy than others. The simulation results coincided well with the experimental findings on the dendritic orientation selection of magnesium alloys.

Acknowledgements The work was financially supported by the National Key Research and Development Program of China (No. 2016YFB0301001), the Tsinghua University Initiative Scientific Research Program (20151080370) and the UK Royal Society Newton International Fellowship Scheme. The authors would like to thank the supports from Shanghai Synchrotron Radiation Facility for the provision of beam time and the National Laboratory for Information Science and Technology in Tsinghua University for access to supercomputing facilities.

References

1. S. Sandlöbes, M. Friák, S. Zaefferer, A. Dick, S. Yi, D. Letzig, The relation between ductility and stacking fault energies in Mg and Mg–Y alloys. Acta Mater. **60**, 3011–3021 (2012)
2. X. Li, S.M. Xiong, Z. Guo, On the porosity induced by externally solidified crystals in high-pressure die-cast of AM60B alloy and its effect on crack initiation and propagation. Mater. Sci. Eng. A **633**, 35–41 (2015)
3. S. Sandlöbes, Z. Pei, M. Friák, L.F. Zhu, F. Wang, S. Zaefferer, Ductility improvement of Mg alloys by solid solution: Ab initio modeling, synthesis and mechanical properties. Acta Mater. **70**, 92–104 (2014)
4. B.S. Wang, S.M. Xiong, Effects of shot speed and biscuit thickness on externally solidified crystals of high-pressure diet cast AM60B magnesium alloy. Trans. Nonferr. Metals Soc. China. **21**, 767–772 (2011)
5. D. Casari, W.U. Mirihanage, K.V. Falch, I.G. Ringdalen, J. Friis, R. Schmid-Fetzer, α-Mg primary phase formation and dendritic morphology transition in solidification of a Mg-Nd-Gd-Zn-Zr casting alloy. Acta Mater. **116**, 177–187 (2016)
6. T. Haxhimali, A. Karma, F. Gonzales, M. Rappaz, Orientation selection in dendritic evolution. Nat. Mater. **5**, 660–664 (2006)
7. Q. Xu, W. Feng, B. Liu, S. Xiong, Numerical simulation of dendrite growth of aluminum alloy. Acta Metall. Sin. **38**, 799–803 (2002)
8. J. Eiken, Phase-field simulations of dendritic orientation selection in Mg-alloys with hexagonal anisotropy. Mater. Sci. Forum **649**, 199–204 (2010)
9. Z.G. Xia, D.Y. Sun, M. Asta, J.J. Hoyt. Molecular dynamics calculations of the crystal-melt interfacial mobility for hexagonal close-packed Mg. Phys. Rev. B. 75:012103(1–4) (2007)
10. K. Pettersen, N. Ryum, Crystallography of directionally solidified magnesium alloy AZ91. Metall. Trans. A **20A**, 847–852 (1989)
11. K. Pettersen, O. Lohne, N. Ryum, Dendritic solidification of magnesium alloy AZ91. Metall. Trans. A **21A**, 221–230 (1990)
12. M. Wang, T. Jing, B. Liu, Phase-field simulations of dendrite morphologies and selected evolution of primary α-Mg phases during the solidification of Mg-rich Mg-Al-based alloys. Scripta Mater. **61**, 777–780 (2009)
13. M.Y. Wang, J.J. Williams, L. Jiang, F. De Carlo, T. Jing, N. Chawla, Dendritic morphology of α-Mg during the solidification of Mg-based alloys: 3D experimental characterization by X-ray synchrotron tomography and phase-field simulations. Scripta Mater. **65**, 855–858 (2011)
14. M. Yang, S.M. Xiong, Z. Guo, Characterisation of the 3-D dendrite morphology of magnesium alloys using synchrotron X-ray tomography and 3-D phase-field modelling. Acta Mater. **92**, 8–17 (2015)

15. M. Yang, S.M. Xiong, Z. Guo, Effect of different solute additions on dendrite morphology and orientation selection in cast binary magnesium alloys. Acta Mater. **112**, 261–272 (2016)
16. Y. Luo, R. Qin, Surface energy and its anisotropy of hexagonal close-packed metals. Surf. Sci. **630**, 195–201 (2014)
17. L. Zhang, X. Liu, C. Geng, H. Fang, Z. Lian, X. Wang, Hexagonal crown-capped zinc oxide micro rods: hydrothermal growth and formation mechanism. Inorg. Chem. **52**, 10167–10175 (2013)
18. J.V.D. Planken, A. Deruyttere, A scanning electron microscope study of vapour grown magnesium. J. Cryst. Growth **11**, 273–279 (1971)
19. Z. Guo, S.M. Xiong, On solving the 3-D phase field equations by employing a parallel-adaptive mesh refinement (Para-AMR) algorithm. Comput. Phys. Commun. **190**, 89–97 (2015)
20. Z. Guo, J. Mi, S. Xiong, P.S. Grant, Phase field simulation of binary alloy dendrite growth under thermal- and forced-flow fields: an implementation of the parallel-multigrid approach. Metall. Mater. Trans. B **44**, 924–937 (2013)
21. G. Kresse, M. Marsman, J. Furthüller. VASP the guide. http://cmsmpiUnivie.ac.at/VASP/
22. C. Stampf, C.G. Van de Walle, Density-functional calculations for III-V nitrides using the local-density approximation and the generalized gradient approximation. Phys. Rev. B **59**, 5521–5535 (1999)
23. P.E. Blöchl, Projector augmented-wave method. Phys. Rev. B **50**, 17953–17979 (1994)
24. H.J. Monkhorst, J.D. Pack, Special points for Brillouin-zone integrations. Phys. Rev. B **13**, 5188–5192 (1976)
25. H.L. Skriver, N.M. Rosengaard, Surface energy and work function of elemental metals. Phys. Rev. B **46**, 7157–7168 (1992)
26. D.P. Ji, Q. Zhu, S.Q. Wang, Detailed first-principles studies on surface energy and work function of hexagonal metals. Surf. Sci. **651**, 137–146 (2016)
27. J. Du, B. Wen, R. Melnik, Mechanism of hydrogen production via water splitting on 3C-SiC's different surfaces: a first-principles study. Comput. Mater. Sci. **95**, 451–455 (2014)
28. P. Villars, L.D. Calvert, *Pearson's Handbook of Crystallographic Data for Intermetallic Phases* (ASM International, Materials Park, OH, 1997)

Macro- and Micro-Simulation and Experiment Study on Microstructure and Mechanical Properties of Squeeze Casting Wheel of Magnesium Alloy

Shan Shang, Bin Hu, Zhiqiang Han, Weihua Sun and Alan A. Luo

Abstract The macro- and micro-simulation based on a coupled thermo-mechanical simulation method using ANSYS® and phase field modeling with pressure effects were carried out for squeeze casting wheel of AT72 alloy. The mechanical properties at different positions of the wheel and under different pressures were analyzed by the macro- and micro-simulation and experimental results, and the corresponding strengthening mechanism was discussed. Firstly, the mechanical properties in spoke are better than those in rim due to higher integrity associated with more forced feeding including more liquid flow feeding and almost all of the plastic deformation feeding in spoke. Furthermore, the mechanical properties increase with pressure due to the enhanced forced feeding shown by the macro-simulation results and the more developed dendrite arms, finer dendrites and more solutes in dendrites under higher pressure indicated by the micro-simulation and experimental results. As analyzed, the mechanical properties are improved by applied pressure according to the strengthening mechanism, including strengthening associated with high integrity, fine-grain strengthening and solution strengthening.

S. Shang (✉) · Z. Han (✉)
Key Laboratory for Advanced Materials Processing Technology,
Ministry of Education, School of Materials Science and Engineering,
Tsinghua University, Beijing 100084, China
e-mail: shangs13@mails.tsinghua.edu.cn

Z. Han
e-mail: zqhan@tsinghua.edu.cn

B. Hu
General Motors China Science Laboratory, Shanghai 201206, China

W. Sun · A.A. Luo
Department of Materials Science and Engineering, The Ohio State University,
Columbus, OH 43210, USA

A.A. Luo (✉)
Department of Integrated Systems Engineering, The Ohio State University,
Columbus, OH 43210, USA
e-mail: luo.445@osu.edu

Keywords Macro- and Micro-simulation · Phase field · Squeeze casting magnesium alloy · Strengthening mechanism

Introduction

With the requirement of lightweight in automotive industry, magnesium alloys are supposed to be promising automotive materials, with low density and high strength and stiffness. Squeeze casting is an advanced near-net-shape materials processing technology, in which liquid metal solidifies under applied pressure, making products with high casting integrity and excellent mechanical properties. It is of significant importance to understand the mechanism how squeeze casting process especially pressure influence the final microstructure and mechanical properties [1].

Macro numerical simulation and analysis of squeeze casting has been conducted in many literatures by the control-volume-finite-difference approach [2] or the finite-element method based on ABAQUS® [3] or ANSYS® software [4]. As to micro-simulation, phase field modeling is a powerful tool to simulate microstructure evolution during phase transformation, which are thermodynamically consistent based on energy description of alloy systems [5], making it feasible to take the effects of pressure on thermodynamics and kinetics of dendritic growth into account. Macro- and micro-simulation plays a great role in predicting the location-specific microstructure and location-specific mechanical properties of the cast products in an integrated computational materials engineering (ICME) framework [6, 7].

In this work, the macro- and micro-simulation on squeeze casting wheel of AT72 alloy was conducted based on the coupled thermo-mechanical simulation method using ANSYS® software and the phase field modeling with pressure effects, and experiments were also carried out to analyze the microstructure and mechanical properties of the wheel. The differences of mechanical properties at different positions of the wheel and under different pressures were analyzed by the macro- and micro-simulation and experimental results, and the corresponding strengthening mechanism was discussed.

Model and Experiments Description

Mechanical Model for Macro-Simulation

A coupled thermo-mechanical simulation method based on ANSYS® has been established to solve the thermal stress in the squeeze casting solidification process in our previous work [4] and was adopted in this work. A thermal elasto-viscoplastic model for Mg–Al based ternary magnesium alloy was derived from Ref. [8] and shown as below, which was applied to the macro-simulation to calculate items associated with stress.

$$\dot{\varepsilon} = 2.8405 \times 10^{12}[\sinh(0.013 \cdot \sigma_P)]^{5.578} \times \exp\left[\frac{-175.82 \times 10^3}{RT}\right] \quad (1)$$

Phase Field Model for Micro-Simulation

The free energy density of a phase in a multicomponent system consisting of $(n + 1)$ components with n solutes in solid and liquid phases is illustrated as follows,

$$f^m = (c_{1m}, c_{2m}, \ldots, c_{im}, \ldots, c_{nm}) \quad (2)$$

where $m = S$ for the solid phase or $m = L$ for the liquid phase. The governing equations for phase field and concentration field of a multicomponent system at ambient pressure are given as follows:

$$\frac{1}{M_\phi}\frac{\partial \phi}{\partial t} = \varepsilon^2 \nabla^2 \phi - wg'(\phi) - h'_p(\phi)\left(f^S - f^L - \sum_{i=1}^{n}(c_{iS} - c_{iL})\tilde{\mu}_i\right). \quad (3)$$

$$\frac{\partial c_i}{\partial t} = \nabla \cdot [1 - h_d(\phi)]\sum_{j=1}^{n} D^L_{ij}\nabla c_{jL} + \nabla \cdot \left(\frac{\varepsilon}{\sqrt{2w}}(c_{iL} - c_{iS})\right)\frac{\partial \phi}{\partial t}\frac{\nabla \phi}{|\nabla \phi|} \quad (4)$$

where ϕ is the phase field, defined as $\phi = 0$ in the bulk liquid, $\phi = 1$ in bulk solid and $0 < \phi < 1$ in the interfacial region between them. The definition of other parameters and functions involved in the controlling equations can been found in Refs. [5, 9].

To take the effects of pressure into account, it is imperative to couple the multi-component phase field model with thermodynamics with pressure effects. Brosh et al. [10, 11] constructed a thermodynamic model with pressure effect based on the integration of available CALPHAD (CALculation of PHAse Diagrams) descriptions at ambient pressure with a composition-dependent equation of state shown as follows:

$$^0G_i^m(T, P) = {}^0G_i^m(T, P_0) + \int_{P_0}^{P} V(T, P')dP' \quad (5)$$

where P_0 is the reference pressure and the second term is the contribution of pressure to Gibbs free energy. The Gibbs free energy associated with the f^m of the pure component and intermetallics at ambient pressure can be obtained from thermodynamic database, specifically PANDAT® database, and the values with pressure effects can be calculated based on the equation. The calculation of the kinetic parameters, i.e. solute diffusion coefficient D_{ij} with pressure effects can be found elsewhere [9].

Experiments

A kind of magnesium alloy, AT72 (Mg–7Al–2Sn) alloy with good combination of strength and ductility [12] was selected for squeeze casting experiments and macro- and micro-simulation in this study. In the squeeze casting process, the initial punch temperature, mold temperature and pouring temperature are 200 °C, 250 °C and 680 °C, respectively, which were also used as the initial condition in macro-simulation. The pressure was set as 47 and 85 MPa to investigate how the increased pressure influences the microstructure and mechanical properties of the alloy. X-ray inspection has been employed to detect the internal defects in castings, and microstructural characteristics was examined by scanning electron microscopy (SEM). Moreover, the mechanical properties has been obtained by tensile test.

Simulation Results and Discussion

In order to analyze the solidification process in squeeze casting, macro-simulation based on the coupled thermo-mechanical modeling using ANSYS® was carried out. Figure 1a shows the geometry of the squeeze casting wheel of AT72 alloy and liquid metal was filled into the wheel from the bottom riser to the top. Based on the geometry, the distribution of temperature and residual liquid phase was obtained and illustrated in Fig. 1c and d, respectively. It can be seen that the temperature distribution is highly dependent on the location of the wheel which is associated with different heat transfer conditions with lower temperature in thinner parts while higher temperature in thicker parts, as indicated by the hotspot in the junction of spoke and rim shown in Fig. 1c. Another phenomenon is that the temperature of spoke is lower than the liquidus before $t = 4$ s, cutting off the channel through which liquid metal feeds the rim, however, most parts of the rim are not solidified then, as illustrated in Fig. 1d. As a consequence, there is some minor porosity in rim due to insufficient feeding during solidification, as shown in Fig. 1b.

Different from permanent mold casting at ambient pressure, in the process of squeeze casting, in addition to liquid flow feeding, plastic deformation feeding due to the applied pressure is another distinctive feeding where pressure makes the solidified metal plastic deformed and produces casting with high integrity. Figure 2a and b show the distribution of displacement in Z direction under the pressure of 47 MPa and 85 MPa, respectively at $t = 20$ s when all parts of the wheel are solidified. Since the punch moves from bottom to top, the symbol of

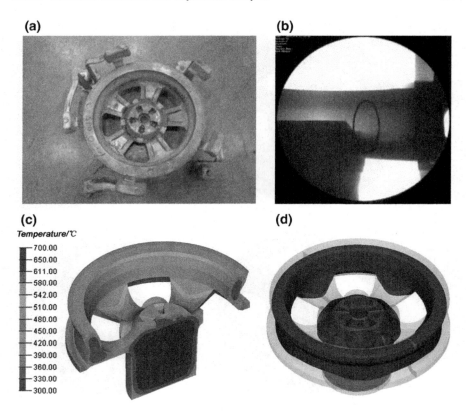

Fig. 1 a Squeeze casting wheel of AT72 alloy, b X-ray inspection result of rim. c and d The temperature and liquid metal distribution at $t = 4$ s, respectively

displacement in Z is negative. With the increase of pressure, the punch moves more upwards, and more material in the riser is punched into the wheel to feed the wheel. Since the degree of plastic deformation feeding can be represented by plastic strain, Fig. 2c and d show the distribution of plastic strain at $t = 20$ s under different pressures. Firstly, the plastic strain happens almost at the location of spoke not rim. The plastic deformation emerges at where the liquid metal has already solidified, such as the spoke shown in Fig. 1d, however the pressure cannot be transferred into the rim to make it plastic deformed since the transfer channel is cut off owing to the early solidified part in spoke. Secondly, the average plastic strain in spoke increases with pressure, indicating that the plastic deformation feeding in spoke is enhanced by increased pressure.

In conclusion, the final average forced feeding to the whole wheel, including liquid flow feeding and plastic deformation feeding, rises with pressure, specially 38% and 51% under pressure of 47 MPa and 85 MPa, respectively at $t = 20$ s, as indicated in Fig. 3. More liquid flow feeding happens in spoke than that in rim and

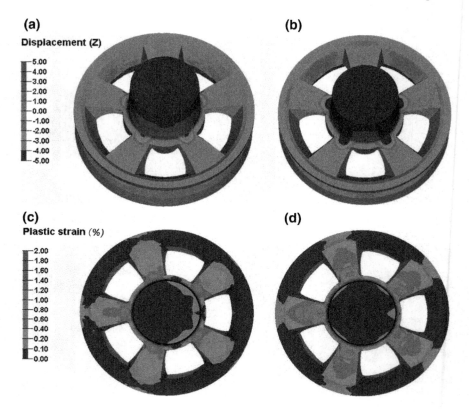

Fig. 2 a and **b** The distribution of displacement in Z direction under pressure of 47 MPa and 85 MPa, respectively, **c** and **d** The distribution of plastic strain under pressure of 47 MPa and 85 MPa, respectively

almost all of the plastic deformation feeding happens in spoke not rim due to the obstructed feeding channel for the rim mentioned above.

The micro-simulation using the phase field model described above, accompanied with some experiments, was executed to analyze the microstructure characteristics under different pressures. For multi-grain growth in simulation, it is necessary to take nucleation into consideration, approximately 9 crystal nuclei and 12 crystal nuclei were seeded in the simulation domain under pressure of 47 MPa and 85 MPa, respectively according to our previous work [9]. It can be seen from Fig. 4a and b that the dendrite grown under pressure of 85 MPa is different from that under pressure of 47 MPa, with more developed secondary dendrite arms and smaller average dendrite size. The reason for it may come in two folds. For one thing, tip growth velocity rises with increasing pressure, leading to more developed dendrite under higher pressure. For another, more nuclei are formed under higher

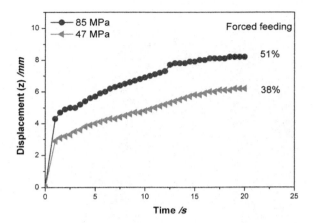

Fig. 3 Forced feeding of the wheel under different pressures

pressure, smaller grain size is obtained due to less growth space under higher pressure, which is consistent with the dendrite morphology by SEM as shown in Fig. 4c and d. Furthermore, as illustrated in Table 1, the average concentration of solute Al and Sn in the α-Mg dendrite of squeeze casting AT72 alloy increases with pressure, resulting in more solution strengthening under higher pressure.

The mechanical properties of squeeze casting wheel of AT72 alloy illustrated in Table 2 are associated with the simulation and experimental results mentioned above. Obviously the differences in mechanical properties resides in two sides, the one between different positions, i.e. spoke and rim, and the other one under different pressures. Firstly, the mechanical properties in spoke are better than those in rim. Since more forced feeding happens in spoke than that in rim, as indicated by the macro-simulation in Figs. 1 and 2, including more liquid flow feeding and almost all of the plastic deformation feeding in spoke mentioned above, higher integrity and less porosity are achieved in spoke, resulting in better strength and ductility. Secondly, the mechanical properties increase with pressure whatever the position is. On one hand, associated with macro-simulation, with increasing pressure, the average forced feeding is enhanced, as shown in Fig. 3, resulting in cast product with higher integrity under pressure of 85 MPa than that under pressure of 47 MPa. On the other hand, associated with micro-simulation and experiments, the dendrite grown under higher pressure is with more developed secondary dendrite arms and smaller average dendrite size, as shown in Fig. 4, and more solutes in the dendrite. As a consequence, the mechanical properties under higher pressure are better according to the strengthening mechanism, including strengthening associated with high integrity, fine-grain strengthening and solution strengthening, the first two of which play the most significant roles in pressurized solidification.

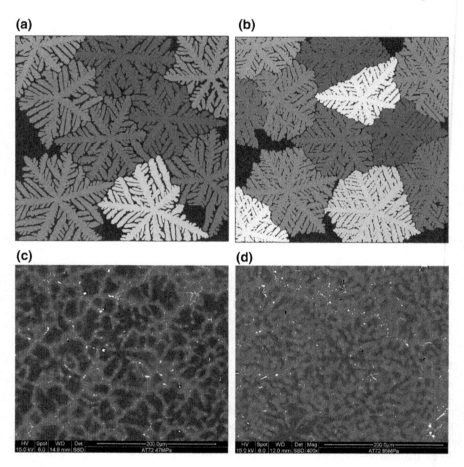

Fig. 4 a and **b** Simulated dendrite morphology under pressure of 47 MPa and 85 MPa, respectively, **c** and **d** The microstructure by SEM under pressure of 47 MPa and 85 MPa, respectively

Table 1 Average concentration of solutes in the dendrite of squeeze casting AT72 alloy under different pressures

Pressure (MPa)	Average concentration in mole fraction	
	Al	Sn
47	0.0250	0.0016
85	0.0298	0.0025

Table 2 Mechanical properties of squeeze casting wheel of AT72 alloy at different positions and under different pressures

Pressure (MPa)	Position	YS (MPa)	UTS (MPa)	EL (%)
47	Spoke	106.0	138.9	1.5
	Rim	116.0	123.2	0.5
85	Spoke	111.4	220.1	6.9
	Rim	98.0	194.4	4.9

Conclusions

In summary, the mechanical properties of the wheel are related to the macro- and micro-simulation and experimental results:

(1) The mechanical properties in spoke are better than those in rim, for the reason that more forced feeding happens in spoke including more liquid flow feeding and almost all of the plastic deformation feeding, resulting in higher integrity in spoke, as shown by macro-simulation results.
(2) The mechanical properties increase with pressure whatever the position is. On one hand, associated with macro-simulation, the average forced feeding is enhanced with increasing pressure, resulting in cast product with higher integrity. On the other hand, associated with micro-simulation and experiments, the dendrites grown under higher pressure are more developed and finer, and with more solutes in dendrite.
(3) The strengthening mechanism that mechanical properties are improved by applied pressure includes strengthening associated with high integrity, fine-grain strengthening and solution strengthening.

Acknowledgements This work was supported by the National Key Research and Development Program of China (No. 2016YFB0701204) and the National Natural Science Foundation of China (No. 51175291). The authors would also like to thank the National Laboratory for Information Science and Technology at Tsinghua University for access to supercomputing facilities.

References

1. M.R. Ghomashchi, A. Vikhrov, Squeeze casting: an overview. J. Mater. Process. Technol. **101**, 1–9 (2000)
2. H.A.A.Y. Henry, Numerical simulation of squeeze cast magnesium alloy AZ91D. Modell. Simul. Mater. Sci. Eng. **10**, 1 (2002)
3. J.H. Lee, H.S. Kim, C.W. Won, B. Cantor, Effect of the gap distance on the cooling behavior and the microstructure of indirect squeeze cast and gravity die cast 5083 wrought Al alloy. Mater. Sci. Eng. A **338**, 182–190 (2002)
4. J. Tang, Z. Han, F. Wang, J. Sun, S. Xu, *A Coupled Thermo-Mechanical Simulation on Squeeze Casting Solidification Process of Three-Dimensional Geometrically Complex Components* (Wiley, 2015), pp. 111–120

5. S.G. Kim, A phase-field model with antitrapping current for multicomponent alloys with arbitrary thermodynamic properties. Acta Mater. **55**, 4391–4399 (2007)
6. J. Allison, Integrated computational materials engineering: a perspective on progress and future steps. JOM **63**, 15–18 (2011)
7. A.A. Luo, Material design and development: from classical thermodynamics to CALPHAD and ICME approaches. Calphad **50**, 6–22 (2015)
8. L. Liu, H. Ding, Study of the plastic flow behaviors of AZ91 magnesium alloy during thermomechanical processes. J. Alloy. Compd. **484**, 949–956 (2009)
9. H. Pan, Z. Han, B. Liu, Study on dendritic growth in pressurized solidification of Mg–Al alloy using phase field simulation. J. Mater. Sci. Technol. **32**, 68–75 (2016)
10. E. Brosh, G. Makov, R.Z. Shneck, Application of CALPHAD to high pressures. Calphad **31**, 173–185 (2007)
11. E. Brosh, R.Z. Shneck, G. Makov, Explicit Gibbs free energy equation of state for solids. J. Phys. Chem. Solids **69**, 1912–1922 (2008)
12. A.A. Luo, P. Fu, L. Peng, X. Kang, Z. Li, T. Zhu, Solidification microstructure and mechanical properties of cast magnesium-aluminum-tin alloys. Metall. Mater. Trans. A **43**, 360–368 (2012)

Solidification Simulation of Fe–Cr–Ni–Mo–C Duplex Stainless Steel Using CALPHAD-Coupled Multi-phase Field Model with Finite Interface Dissipation

Sukeharu Nomoto, Kazuki Mori, Masahito Segawa and Akinori Yamanaka

Abstract A multi-phase field (MPF) model with finite interface dissipation proposed by Steinbach et al. is applied to simulate the dendritic solidification in Fe–Cr–Ni–Mo–C duplex stainless steel. This MPF model does not require an equal diffusion potential assumption and can take into account a substantial non-equilibrium interfacial condition. We develop the MPF code to couple with the CALPHAD thermodynamic database to simulate two-dimensional microstructure evolutions in multi-component alloys using the TQ-interface of Thermo-Calc. The message passing interface parallelization technique is adapted to the program code development to reduce computational elapse time. Solidification calculations were performed in two cases of quinary compositions: Fe–16Cr–2Mo–10Ni–0.08C and Fe–17Cr–2Mo–9Ni–0.08C. We confirm that the developed MPF method can be highly applicable to microstructure evolution simulation of the engineering metal alloy solidification.

Keywords Multi-phase field model · CALPHAD · Solidification · Stainless steel · Quinary system

---

S. Nomoto (✉) · K. Mori
ITOCHU Techno-Solutions Corporation, 3-2-5, Kasumigaseki, Chiyoda-ku, Tokyo 100-6080, Japan
e-mail: sukeharu.nomoto@ctc-g.co.jp

M. Segawa
Department of Mechanical Systems Engineering, Graduate School of Engineering, Tokyo University of Agriculture and Technology, 2-24-16, Naka-cho, Koganei-shi, Tokyo 184-8588, Japan

A. Yamanaka
Division of Advanced Mechanical Systems Engineering, Institute of Engineering, Tokyo University of Agriculture and Technology, 2-24-16, Naka-cho, Koganei-shi, Tokyo 184-8588, Japan

Introduction

The multi-phase field method proposed by Steinbach approximately 20 years ago has been applied to many numerical simulation fields [1]. In particular, the CALPHAD database coupling method that uses the TQ-Interface of Thermo-Calc provides us a means to simulate the microstructure evolution of many engineering metal alloy processes [2]. For example, MICRESS is a typical software tool for general commercial application [3]. The MPF method coupled with solute diffusion is based on the diffuse-interface model in which equal diffusion potential is assumed in the interface region. This model is also generally called the Kim–Kim–Suzuki model [4]. However, in practical calculation using the CALPHAD database coupling, much time is required to obtain quasi-equilibrium concentration at all grid points in the interface region. Recently, Steinbach and Zhang have proposed a finite interface dissipation model, which is based on non-equal diffusion potential [5, 6]. It is called the non-equilibrium MPF model (NEMPFM) in this paper. This method has another advantage in that it avoids iteration calculation to obtain a quasi-equilibrium composition in the interface region. We introduce this model to develop the CALPHAD database coupled with the MPF program for practical engineering steel system.

The evolution of the solidification microstructure of stainless steel is sensitive to the solute composition values, i.e., the ratio of the body-centered cubic (BCC) and face-centered cubic (FCC) phase-stabilized elements Cr and Ni, respectively. In many stainless steel cases, Mo is usually doped, together with C, to obtain hardness. Thus, the quinary system Fe–C–Cr–Mo–Ni, is selected to conform to the CALPHAD database coupled with NEMPFM calculations in this study. We select two cases of a composition in which the primary phase is δ-ferrite, representing ferrite–austenite (FA)-mode solidification. In the FA-mode solidification of stainless steel, austenite usually nucleates in the interface between liquid and δ-ferrite. A numerical method for austenite nucleation is implemented in the NEMPFM program by adapting a heterogeneous crystal nucleation method [7]. This program is parallelized using the open message passing interface (MPI) technique to reduce calculation time because the numerical estimation of the driving force in the interface region incurs much time even if the iteration calculation for the quasi-equilibrium convergence is unnecessary. Calculations of the solidification microstructure evolutions between the two initial compositions of Fe(bulk)–16%Cr–2%Mo–10%Ni–0.08%C and Fe(bulk)–17%Cr–2%Mo–9%Ni–0.08%C are performed. The reliability of the NEMPFM calculation for the multi-component solidification is examined by comparing the two calculation results.

NEMPFM Model

In the finite interface dissipation model, reference volume (RV) is defined as a small area in the phase field interface region. In the practical calculation of the discretized numerical method, e.g., finite difference method, its grid is selected as RV, as shown in Fig. 1. We assume that diffusion among RVs is not considered. The concentration partition rate is assumed to be a function of the total free-energy variation due to the concentration.

$$\dot{c}_\alpha^i \equiv -\sum_{j=1}^{n-1} \tilde{P}_\alpha^{ij} \frac{\delta G}{\delta c_\alpha^j}, \tag{1}$$

where c_α^i is the concentration of the i ($i =$ C, Cr, Ni, Mo) atoms in the α phases ($\alpha =$ liquid, FCC, BCC), G is the total energy of the system, and $\tilde{P}_\alpha^{ij}$ is the interface permeability [4]. The other basic equations are the same as the standard MPF method equations. However, a Lagrange multiplier term is added to the total energy by considering the conservation law.

$$\begin{cases} G = \int_V \left[\sum_{\alpha=1}^{N} \sum_{\beta=\alpha+1}^{N} -\frac{1}{2}\varepsilon_{\alpha\beta}^2 \nabla\phi_\alpha \cdot \nabla\phi_\beta + \sum_{\alpha=1}^{N}\sum_{\beta=\alpha+1}^{N} W_{\alpha\beta}\phi_\alpha \cdot \phi_\beta + \sum_{\alpha=1}^{N} \phi_\alpha f_\alpha(\vec{c}_\alpha) + \sum_{i=1}^{n-1} \lambda^i \left[c^i - \sum_{\alpha=1}^{N}(\phi_\alpha c_\alpha^i) \right] \right] dV, \\ c^i \equiv \sum_{\alpha=1}^{N} (\phi_\alpha c_\alpha^i), \end{cases} \tag{2}$$

where ϕ_α is the order parameter of phase α ($\alpha =$ liquid, FCC, BCC). ϕ_α varies between zero and unity. f_α ($\alpha =$ liquid, FCC, BCC) is the chemical free-energy density of the liquid, austenite, and ferrite. Solving for Lagrange multiplier λ^i by substituting Eq. (2) into Eq. (1) considering definitions $\tilde{P}_\alpha^{ij}(\phi_\alpha) \equiv P_\alpha^{ij}/\phi_\alpha$ and $P_\alpha^{ij} \equiv \delta_{ij} P_\alpha^{(i)}$ and conservation $\dot{c}^i = 0$ leads to the solute partition rate equation.

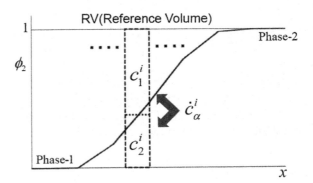

Fig. 1 RV in phase field interface region

$$\dot{c}_\alpha^i = \sum_{\beta=1}^{N} P_{\alpha\beta}^i \phi_\beta \left(\tilde{\mu}_\beta^i - \tilde{\mu}_\alpha^i \right) - \sum_{\beta=1}^{N} \phi_\beta \left(c_\beta^i - c_\alpha^i \right), \qquad (3)$$

where $P_\alpha^{(i)}$ is represented as $P_{\alpha\beta}^i$ because the permeability parameter is defined in the interface between two phases, as shown in Fig. 1. The first and second terms at the right-hand side of Eq. (3) represent the diffusion potential difference and interface moving effect terms, respectively. This partition rate equation is introduced instead of the quasi-equilibrium partitioning calculation process of the standard MPF method.

Parameters $\varepsilon_{\alpha\beta}$ and $W_{\alpha\beta}$ in Eq. (2) represent the gradient energy coefficient and the double-obstacle potential between the αth and βth phases or grains, respectively. These parameters are expressed by the relationship of interface energy $\sigma_{\alpha\beta}$ and interface thickness δ as

$$\sigma_{\alpha\beta} = \frac{\pi}{4\sqrt{2}} \varepsilon_{\alpha\beta} \sqrt{W_{\alpha\beta}}, \qquad (4)$$

$$\delta = \pi \frac{\varepsilon_{\alpha\beta}}{\sqrt{2W_{\alpha\beta}}}. \qquad (5)$$

Substituting Eq. (2) into the time evolution equation (Allen–Cahn equation) yields

$$\frac{\partial \phi_\alpha}{\partial t} = - \sum_{j=1}^{N} \frac{M_{\alpha\beta}}{N} \left(\frac{\delta G}{\delta \phi_\alpha} - \frac{\delta G}{\delta \phi_\beta} \right), \qquad (6)$$

and considering Eqs. (4) and (5) leads to the following NEMPFM equation:

$$\frac{\partial \phi_\alpha}{\partial t} = \sum_{\beta=1}^{N} \frac{M_{\alpha\beta}}{N} \left\{ \sum_{\varsigma=1}^{N} \left[\left(\frac{\pi^2}{\delta^2} \phi_\varsigma + \nabla^2 \phi_\varsigma \right) (\sigma_{\beta\varsigma} - \sigma_{\alpha\varsigma}) \right] + \frac{2\pi}{\delta} \sqrt{\phi_\alpha \phi_\beta} \, \Delta G_{\alpha\beta} \right\}, \qquad (7)$$

where $M_{\alpha\beta}$ is the interface mobility. $\Delta G_{\alpha\beta}$ is the interface driving force.

$$\Delta G_{\alpha\beta} = f_\beta - f_\alpha - \sum_{i=1}^{n-1} \left[\sum_{\varsigma=1}^{N} \phi_\varsigma \tilde{\mu}_\varsigma^i - \frac{1}{P_{\alpha\beta}^i} \sum_{\varsigma=1}^{N} \phi_\varsigma c_\varsigma^i \right] \left(c_\beta^i - c_\alpha^i \right). \qquad (8)$$

Equation (8) is consistent with the driving force representation of the standard MPF model when $P_{\alpha\beta}^i$ becomes infinity.

The diffusion equation is derived by substituting Eq. (2) into the time evolution equation of the conservation type Cahn–Hilliard equation as

$$\frac{\partial c^k}{\partial t} = \nabla \left(\sum_{\alpha=1}^{N} \phi_\alpha \sum_{j=1}^{n-1} D_{kj}^\alpha \nabla c_\alpha^j \right), \quad (9)$$

where D_{kj}^α is the diffusivity whose values are given using Arrhenius-type data in this study. Equation (9) is solved in combination with Eq. (3).

Calculation Conditions

In this study, calculations were performed under a condition with two types of composition, namely, Case 1: Fe–16Cr–2Mo–10Ni–0.08C and Case 2: Fe–17Cr–2Mo–9Ni–0.08C. We selected the Thermo-Calc TCFE8 database. The equilibrium phase fractions of these two cases are shown in Fig. 2. The maximum δ-ferrite fraction of Case 2 is higher than that of Case 1. By comparing the differences between the δ-ferrite and austenite nucleation temperatures in Cases 1 and 2, namely, A and B, respectively, Case 2 B is larger than Case 1 A, as shown in Fig. 2. Thus, we expect that these differences lead to different microstructure evolution.

The two-dimensional finite difference method with explicit temporal integration was applied for the simulation in which the interval time was 1×10^{-5} s. The calculation conditions for the region size, grid size, number of grids, interface width, cooling rate, temperature gradient, and boundary conditions are shown in Fig. 3. The initial δ-ferrite nucleation was set on the left corner at the bottom. The initial temperature values at the bottom for Cases 1 and 2 were set at 1728 and 1736 K, respectively. The maximum temperature to check the austenite nucleation

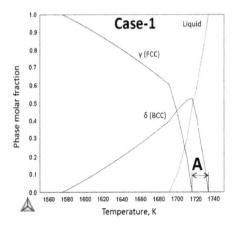
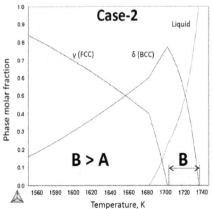

Fig. 2 Equilibrium phase fractions

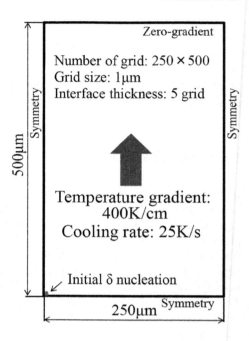

Fig. 3 Calculation-region conditions

ability for Cases 1 and Cases 2 were set at 1711.5 and 1701 K, respectively. The austenite grain was designed to nucleate in the interface region between the liquid and δ-ferrite phases. The checking interval time and average distance of the austenite nucleation were set at 0.025 s and 11 μm, respectively. The physical parameter values, interface energy and mobility, and interface permeability and diffusivity of the Arrhenius plot data are listed in Tables 1, 2, and 3, respectively. The interface energy values were obtained from Ref. [8]. The interface mobility and

Table 1 Interface energy and mobility [8]

Interface	Energy J/m²	Mobility m⁴/J/s
δ–Liquid	0.2	1.5×10^{-10}
δ–γ	0.6	2×10^{-11}
Liquid–γ	0.2	3×10^{-11}

Table 2 Interface permeability

Interface	m³/J/s			
	Cr	Mo	Ni	C
δ–Liquid	2.5×10^{-7}	1.0×10^{-7}	1.0×10^{-7}	2.5×10^{-8}
δ–γ				
Liquid–γ				

Table 3 Diffusivity of solute element (Arrhenius plot type) [9]

Phase	Element	Prefactor (m²/s)	Activation (J/mol)
Liquid	Cr	2×10^{-9}	0
	Mo	2×10^{-9}	0
	Ni	2×10^{-9}	0
	C	2×10^{-9}	0
γ	Cr	1.930×10^{-4}	2.885×10^5
	Mo	3.633×10^{-6}	2.404×10^5
	Ni	1.010×10^{-4}	2.922×10^5
	C	2.200×10^{-4}	1.747×10^5
δ	Cr	0.340×10^{-4}	2.185×10^5
	Mo	0.847×10^{-4}	2.308×10^5
	Ni	1.410×10^{-4}	2.321×10^5
	C	6.449×10^{-7}	7.699×10^5

interface permeability values were determined by numerical examinations. For example, the interface permeability was calibrated to obtain a numerically stable condition by decreasing the parameter value from a large value that caused numerical instability. The Arrhenius-type formulation data of the diffusivity were obtained from the diffusion mobility database MOBFE3 of the Thermo-Calc software [9].

Results and Discussion

Figure 4 shows the phase and C molar-fraction distributions of the two cases. Initially, we can see that the dendrite of the δ-ferrite increases in both cases. However, the second dendrite arm in Case 1 increases much more than that in Case 2. The austenite grain in Case 2 nucleates very late compared with that in Case 1. This difference qualitatively agrees with the equilibrium phase fraction calculated by Thermo-Calc, as shown in Fig. 3. As interstitial element C very highly diffuses in solid, concentration partitioning occurs not only in the liquid and solid interface but also in the δ-ferrite and austenite interface, as shown in Fig. 4.

Figure 5 shows the substitutional element Cr concentration distributions. Because Cr diffuses slower than C in the solid phase, the FA transformation in this system is considered to be close to a negligible partitioning local equilibrium mode. Thus, we can see that a small concentration spike appears in the δ-FA interface, which is also confirmed in the other solute elements Ni and Mo. The temporal concentration distributions are also well calculated to show the same trend as Cr.

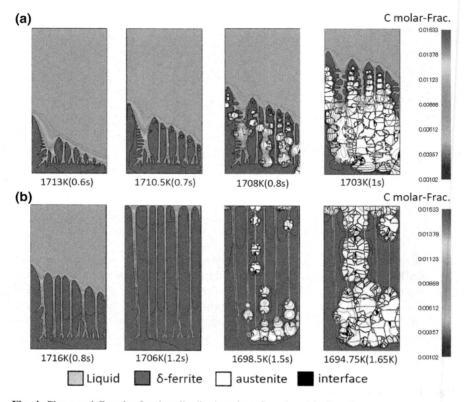

Fig. 4 Phase and C molar-fraction distributions in **a** Case 1 and **b** Case 2

Figure 6 shows the MPI parallel efficiency. Higher acceleration can be obtained with increasing number of CPU cores. We confirm that the MPI parallel program using the TQ-interface library is successfully developed.

Conclusions

An MPI-parallelized program of the non-equilibrium MPF method coupled with CALPHAD database for Fe–Cr–Ni–Mo–C duplex stainless steel composition has been developed by adapting the heterogeneous crystal nucleation feature. The solidification microstructure calculations were successfully performed under the condition of two different compositions. We confirmed that this method have high applicability in microstructure evolution simulation of engineering metal alloys.

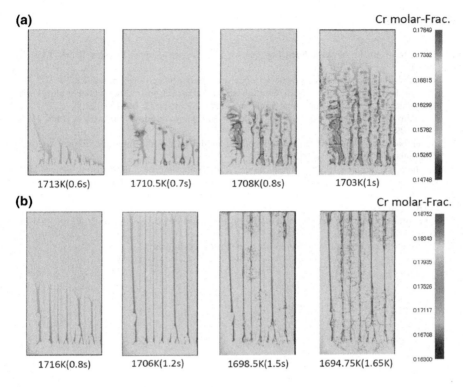

Fig. 5 Cr molar-fraction distributions. **a** Case 1. **b** Case 2

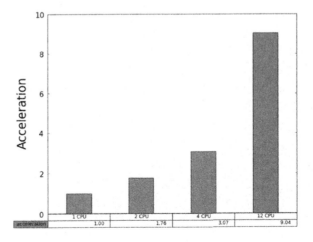

Fig. 6 MPI parallel efficiency

References

1. I. Steinbach et al., A phase field concept for multiphase systems. Physica D **94**, 135–147 (1996)
2. http://www.thermocalc.com/products-services/software/software-development-kits/
3. J. Eiken, B. Boettiger, I. Steinbach, Multiphase-field approach for multicomponent alloys with extrapolation scheme for numerical application. Phys. Rev. E **73**, 066122 (2006)
4. S.G. Kim, W.T. Kim, T. Suzuki, Phase-field model for binary alloys. Phys. Rev. E **60**, 7186–7197 (1999)
5. I. Steinbach, L. Zhang, M. Plapp, Phase-field model with finite interface dissipation. Acta Materialia **60**, 2689–2701 (2012)
6. L. Zhang, I. Steinbach, Phase-field model with finite interface dissipation: extension to multi-component multi-phase alloys. Acta Materialia **60**, 2702–2710 (2012)
7. L. Granasy et al., Phase field theory of heterogeneous crystal nucleation. Phys. Rev. Lett. **98**, 035703 (2007)
8. S. Fukumoto et al., Prediction of sigma phase formation in Fe–Cr–Ni–Mo–N alloys. ISIJ Int. **50**, 445–449 (2010)
9. http://www.thermocalc.com/products-services/databases/mobility/

Phase-Field Modeling of θ′ Precipitation Kinetics in W319 Alloys

Yanzhou Ji, Bita Ghaffari, Mei Li and Long-Qing Chen

Abstract Understanding and predicting the morphology, kinetics and hardening effects of precipitates are critical in improving the mechanical properties of Al-Cu-based alloys through controlling the temperature and duration of the heat treatment process. In this work, we present a comprehensive phase-field framework for simulating the kinetics of θ′ precipitates in W319 alloys, integrating the thermodynamic and diffusion mobility databases of the system, the key precipitate anisotropic energy contributions from literature and first-principles calculations, as well as a nucleation model based on the classical nucleation theory. By systematically performing phase-field simulations, assuming the precipitate peak number densities determined from experiments, we optimize the model parameters to obtain the best possible match to the average diameters, thicknesses and volume fractions of precipitates from experimental measurements at 190, 230 and 260 °C. With these parameters available, the phase-field simulations can be performed at other aging temperatures. The possible extensions of the current phase-field model for more accurate prediction of the precipitate behaviors in W319 alloys will also be discussed.

Keywords Precipitation kinetics · Phase-field model · θ′ (Al_2Cu) · Al-Cu-based alloys

Introduction

Aluminum alloys, due to their light weight, have been widely used in automotive industries. To improve the strength of Al alloys, alloying elements such as Cu are usually incorporated, which enables precipitate hardening. During isothermal aging

---

Y. Ji (✉) · L.-Q. Chen
Department of Materials Science and Engineering, The Pennsylvania State University, University Park, Park, PA 16802, USA
e-mail: yxj135@psu.edu

B. Ghaffari · M. Li
Ford Research and Innovation Center, Ford Motor Company, Dearborn, MI 48124, USA

of Al-Cu-based alloys, the θ' (Al$_2$Cu) precipitate, with a tetragonal crystal structure and partial coherency with the Al-matrix, usually appears at the peak aging point, and is therefore the key strengthening precipitate in these alloys. Over the past decades, the fundamental aspects of the θ' precipitate, including its crystal structure [1], thermodynamic properties [2–4], interface structures [5–8], precipitate morphologies [9, 10], and growth kinetics [7, 11, 12] have been thoroughly investigated through either experimental observations or theoretical calculations. As a result, the θ' precipitate has been included in various undergraduate and graduate-level materials science textbooks as the typical example for precipitate hardening. The knowledge learned from these studies has provided useful guidance for the heat treatment conditions to tune the strength of commercial Al-Cu-based alloys.

In order to accelerate the alloy design process, modeling and prediction of the precipitate hardening effects as functions of processing parameters have been considered as promising alternatives to minimize the development time needed for trial-and-error-based experiments, which is consistent with the integrated computational materials engineering (ICME) initiative. The estimation of precipitate hardening effects largely relies on the prediction of precipitate morphology, volume fraction and spatial distribution during the aging process, which requires meso-scale simulations of the precipitate behavior. There have been a series of numerical models developed for θ' precipitates for this purpose [4, 8, 13–17]. Among these models, the phase-field approach [18], which avoids the explicit tracking of interfaces through the diffuse-interface description, shows significant advantages in accurately and efficiently predicting the precipitate morphology and kinetics in a physics-based manner. Specifically, Vaithyanathan et al. [4, 8] incorporated physical parameters including free energies, lattice parameters, elastic constants and interfacial energies calculated from atomic-scale calculations and predicted the θ' morphology using phase-field simulations. Hu et al. [15] improved the phase-field model's accuracy in predicting the θ' precipitation kinetics. However, these existing works primarily focused on the growth kinetics of already-nucleated θ' particles rather than also considering the nucleation kinetics. Moreover, the quantitative comparisons with experimental measurements on θ' morphologies were lacking.

The primary goal of the current research is to develop a phase-field model capable of predicting the precipitate morphology and kinetics of θ' precipitates in a commercial 319 Al-alloy (Al-3.5 wt%Cu-6.0 wt%Si) at different aging temperatures. The simulated information, including the diameters, thicknesses and volume fractions of the θ' precipitates will be compared with that measured from experiments, in order to adjust certain parameters in the phase-field model, further extend the model for predicting the precipitate kinetics under other aging temperatures, and use the predicted precipitate information for yield strength predictions.

Phase-Field Model

The phase-field model in this study is extended from Hu et al. [15]. Since Cu is the major alloying elements for θ′ precipitates in W319 alloys, we initially simplify the alloy into an Al-Cu binary system with the overall Cu composition of 1.5 at.%. To describe the precipitate morphologies, we use two sets of order parameters: X_{Cu} for the Cu composition and $\{\eta_i\}$ (i = 1, 2, 3) for the three structural variants of θ′. The total free energy within the system is:

$$F = \int_V \left(f_{local}(X_{Cu}, \{\eta_i\}, T) + \frac{1}{2}\sum_{i=1}^{3} \kappa(\theta_i)^2 |\nabla \eta_i|^2 + e_{el} \right) dV \quad (1)$$

The first term $f_{local}(X_{Cu}, \{\eta_i\}, T)$ describes the bulk free energy density of the system:

$$f_{local}(X_{Cu}, \{\eta_i\}, T) = f^{\alpha}(X_{Cu}^{\alpha}, T) \cdot (1 - h(\{\eta_i\})) + f^{\theta'}\left(X_{Cu}^{\theta'}, T\right) \cdot h(\{\eta_i\}) + w \cdot g(\{\eta_i\}) \quad (2)$$

where $f^{\alpha}(X_{Cu}^{\alpha}, T)$ and $f^{\theta'}\left(X_{Cu}^{\theta'}, T\right)$ are molar Gibbs free energies of the α-Al solid solution and the θ′ phase, respectively, which are taken from existing thermodynamic databases [2, 15]; $h(\{\eta_i\}) = \sum_{i=1}^{3}(3\eta_i^2 - 2\eta_i^3)$ is an interpolation function and $g(\{\eta_i\})$ is a double-well type function. The Kim-Kim-Suzuki model is applied here to remove the extra potential at interfaces, as described in [15].

The second term in Eq. (1) describes the contribution of the inhomogeneous distribution of the order parameters to the interfacial energy. The gradient coefficients $\kappa(\theta_i)$, together with the double-well height w, can be calculated from the precipitate interfacial energy $\sigma(\theta_i)$ and interface thickness $2\lambda(\theta_i)$, according to the one-dimensional analytical solution [15]:

$$\kappa(\theta_i)^2 = \frac{3}{2.2}\sigma(\theta_i) \cdot 2\lambda(\theta_i) \quad (3a)$$

$$w = 13.2 \frac{\sigma(\theta_i)}{2\lambda(\theta_i)} \quad (3b)$$

Since the coherent and semi-coherent α-Al/θ′ interfaces have different interfacial energies, we consider the angle-dependent interfacial energy formulation $\sigma(\theta_i)$ [15]:

$$\sigma(\theta_i) = \frac{\sigma_0}{1+\gamma} \begin{cases} 1 + \frac{\gamma}{\sin\phi_0} + \frac{\gamma\cos\phi_0}{\sin\phi_0}\sin\theta_i, & -\frac{\pi}{2} \leq \theta_i \leq -\frac{\pi}{2} + \phi_0 \\ 1 + \gamma\cos\theta_i, & -\frac{\pi}{2} + \phi_0 \leq \theta_i \leq \frac{\pi}{2} - \phi_0 \\ 1 + \frac{\gamma}{\sin\phi_0} - \frac{\gamma\cos\phi_0}{\sin\phi_0}\sin\theta_i, & \frac{\pi}{2} - \phi_0 \leq \theta_i \leq \frac{\pi}{2} \end{cases} \quad (4)$$

where $\theta_i = \arccos\left(\frac{\partial \eta_i/\partial x}{\sqrt{(\partial \eta_i/\partial x)^2 + (\partial \eta_i/\partial y)^2 + (\partial \eta_i/\partial z)^2}}\right) - \frac{\pi}{2}$, $\phi_0 = \frac{\pi}{10000}$ is a small regularization angle to describe the two cusps in the γ-plot due to Wulff construction, $\sigma_0 = 0.49\,\text{J/m}^2$ and $\frac{\sigma_0}{1+\gamma} = 0.24\,\text{J/m}^2$ are the interfacial energies of the semi-coherent and coherent Al/θ' interfacial energies, respectively, which are obtained from first-principles calculations.

The third term in Eq. (1) is the elastic strain energy density, which can be evaluated based on the microelasticity theory. The misfit strain between the Al-matrix and the θ' phase, as well as the elastic constants, are taken from [4]. The system is assumed elastically homogeneous and stress equilibrium is assumed to be much faster than the phase transformation process.

The evolution of θ' precipitates are governed by the Cahn-Hilliard equation (for composition) and Allen-Cahn equation (for order parameters):

$$\frac{\partial X_{Cu}}{\partial t} = \nabla \cdot \left(\frac{D_{Cu}}{\partial^2 f_{local}/\partial X_{Cu}^2} \nabla \frac{\delta F}{\delta X_{Cu}}\right) \tag{5a}$$

$$\frac{\partial \eta_i}{\partial t} = -L(\theta_i)\frac{\delta F}{\delta \eta_i} \tag{5b}$$

where D_{Cu} is the impurity diffusivity of Cu in Al from literature [19], $L(\theta_i)$ is the kinetic coefficient for interface mobility. We assume $L(\theta_i)$ has the same angle-dependence as $\sigma(\theta_i)$, but with a different anisotropy factor β (c.f. γ for $\sigma(\theta_i)$). The equations are solved using semi-implicit Fourier spectral method.

To account for the nucleation kinetics, we incorporate a modified classical nucleation model:

$$j = ZN_0\beta^*\exp\left(-\frac{\Delta G^*_{het}}{k_B T}\right) \tag{6}$$

where j is the nucleation rate, Z is the Zeldovich factor, N_0 is the number of atoms per unit volume, β* is the atomic attachment rate, ΔG^*_{het} is the activation energy for heterogeneous nucleation and k_B is the Boltzmann constant. These parameters are calculated based on [20]. Notably, we take $\Delta G^*_{het} = B \cdot \Delta G^*_{hom}$ where ΔG^*_{hom} is the activation energy for homogeneous nucleation of a plate-shaped particle with interfacial energy anisotropy; the phenomenological parameter B represents the heterogeneity of the nucleation behavior. Since the calculated nucleation rate is very sensitive to the B values, rather than assuming a constant B for all simulation temperatures, we fit different B values for different aging temperatures so that the peak number density in phase-field simulations agrees with the average number density at the same aging temperature as from experimental observations of the W319 alloy.

Results and Discussion

Parameterization

The Nucleation Model

The experimentally measured θ' number density values during different aging times at different aging temperatures are shown in Fig. 1. The mean number densities at different aging temperatures, as well as the fitted B values, are shown in Table 1. Based on the data, B can be fitted into a linear function of the aging temperature: $B(T) = 0.2724 - 4.5 \times 10^{-5} T$. The high number densities and B values at lower temperatures (190 °C) indicates the larger nucleation driving force and the higher heterogeneity of nucleation behavior at lower temperatures.

The phase-field model

In this work, 3D phase-field simulations are performed at the three aging temperatures, 190, 230 and 260 °C, in a 160 × 160 × 160 Δx ($\Delta x = 1$ nm) region. Based on the nucleation model, the numbers of nuclei in the system for the three aging temperatures are 9, 5, and 3, respectively. To make a direct comparison with the precipitation kinetics from experiments, the time step size in the simulation (Δt^*) is related to the real time interval (Δt) through $\Delta t = (\Delta x)^2 / D_{Cu} \cdot \Delta t^*$. Since the θ' precipitates have very large diameter-to-thickness aspect ratios, a large anisotropy factor $\beta = 1000$ is used for the angle-dependent interface mobility coefficients for all three temperatures. This large growth anisotropy can be related to the different atomic-scale growth kinetics along the diameter and thickness directions [7]. Interface thicknesses of 2 nm for the semi-coherent interface and 1 nm for the coherent interface are used for all three temperatures.

Phase-Field Simulations

Figure 2 shows the phase-field simulation results of θ' precipitate morphologies at 190 °C for 10 h, in comparison with the experimental TEM images.

For more quantitative comparisons, we measure the average precipitate diameters, thicknesses and volume fractions from the available TEM images of the W319 alloy, and compare with the corresponding values from the phase-field simulations. Figures 3, 4 and 5 show this comparison, which is the best agreement we achieved using the model parameters mentioned above. Specifically, for all three aging temperatures, the simulated precipitate diameters agree quite well with the experimental measurements. The simulated precipitate volume fractions agree well with the experimental measurements at 230 °C, while for 190 and 260 °C, the predicted

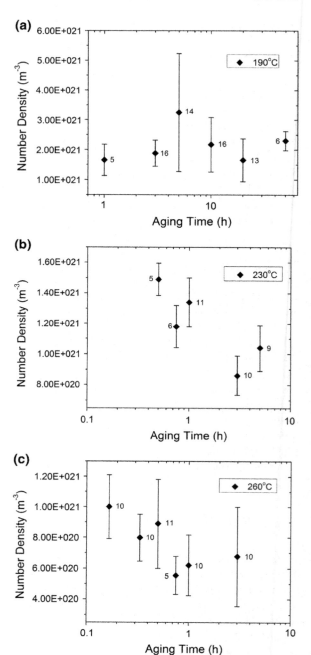

Fig. 1 Experimentally measured number density values at different aging temperatures: **a** 190 °C, **b** 230 °C, and **c** 260 °C. The data points in the figure show the average number density at each aging time. The number beside each error bar indicates the number of measurements for the data point

Table 1 Mean number density of precipitates and fitted B values	Temperature (°C)	Mean number density (/m³)	B
	190	2.2e21	0.00644
	230	1.15e21	0.00456
	260	7.77e20	0.0033

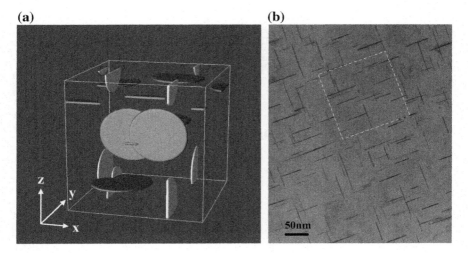

Fig. 2 Simulated θ' morphology and comparison with experimental observation at 190 °C after 10 h of aging. **a** Simulation results with a system size of 160 × 160 × 160 nm³; the different colors represent different θ' variants; **b** Experimental TEM image; the dashed square region has the same size as that in (**a**)

volume fractions are higher and lower than the experimental values, respectively. In contrast, the thickness simulations still need further improvement.

Further Improvements

The deviations between the phase-field simulations and experimental measurements on θ' precipitate kinetics includes two aspects: precipitate thicknesses and volume fractions. For the thicknesses, due to the large aspect ratio, it is numerically expensive to resolve the thickening kinetics by using smaller grid sizes. In the current simulations, the thickening effect is largely eliminated through numerical pinning by the low interface thickness value (1 nm) of the coherent interface, resulting in the constant precipitate thickness of 3 nm. Further improvement of the thickening simulations would have to rely on the development of an efficient numerical method. The thickness issue also has an effect on the volume fraction predictions. However, in addition to the thickness effect, we notice that the experimentally measured precipitate volume fractions finally reach a similar level

Fig. 3 Phase-field simulation results and comparison with experimental values at 190 °C. **a** Mean precipitate diameters. **b** Precipitate volume fractions. **c** Mean precipitate thicknesses

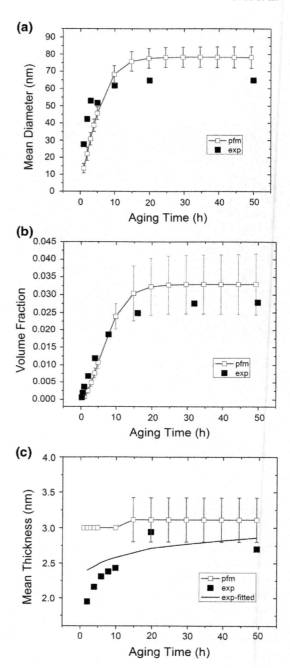

Fig. 4 Phase-field simulation results and comparison with experimental values at 230 °C. **a** Mean precipitate diameters. **b** Precipitate volume fractions. **c** Mean precipitate thicknesses

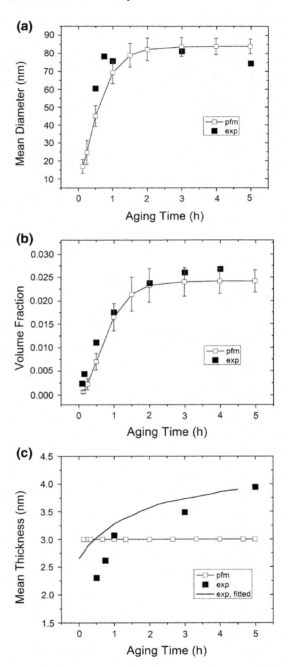

Fig. 5 Phase-field simulation results and comparison with experimental values at 260 °C. **a** Mean precipitate diameters. **b** Precipitate volume fractions. **c** Mean precipitate thicknesses

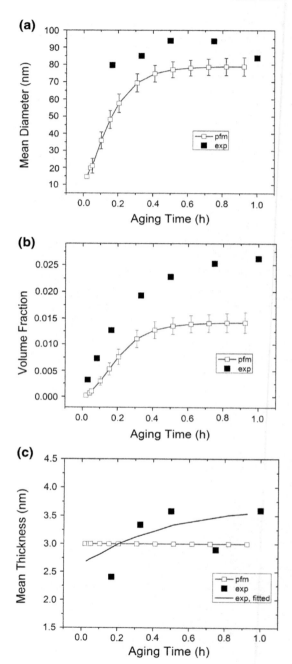

(about 2.7%) for all the three aging temperatures, which is different from the prediction of equilibrium precipitation volume fraction from the thermodynamic database using the lever rule. Such discrepancy can potentially be alleviated through the improvement of the existing thermodynamic database, especially for multi-component, multi-phase Al-databases with better accuracy.

Conclusion

In this study, we developed a 3D phase-field framework for θ' precipitates in W319 alloys, which incorporates the thermodynamic and kinetic information of the system and considers the anisotropic energy and kinetic contributions for the θ' morphology, and can predict both the nucleation and growth kinetics of the precipitate. By adjusting the phenomenological parameter B, we obtained nucleation kinetics that are consistent with the experimental observations. By performing phase-field simulations, we obtained acceptable agreements in precipitate lengthening kinetics and partial agreement in precipitate volume fractions with experimental measurements. Further improvement of the observed agreement would rely on improvement of the current numerical methods and thermodynamic databases. The obtained precipitate kinetics can be further used for mechanical property predictions. The framework can be further extended to predict the precipitate kinetics for other aging temperatures and for other similar alloy systems.

Acknowledgements The authors acknowledge the financial support from the URP program of the Ford Motor Company. The authors are also grateful for Dr. Shannon Weakley-Bollin and Dr. John Allison for providing the TEM images and the useful discussions.

References

1. C. Laird, H.I. Aaronson, Structures and migration kinetics of α-θ' boundaries in Al-4%Cu: I. Interfacial structures. Trans. Metall. Soc. AIME **242**(7), 1393 (1968)
2. J.L. Murray, The aluminium-copper system. Int. Met. Rev. **30**(5), 211–233 (1985)
3. C. Ravi, C. Wolverton, V. Ozoliņš, Predicting metastable phase boundaries in Al-Cu alloys from first-principles calculations of free energies: The role of atomic vibrations. Europhys. Lett. (EPL) **73**(5), 719–725 (2006)
4. V. Vaithyanathan, C. Wolverton, L.Q. Chen, Multiscale modeling of θ' precipitation in Al-Cu binary alloys. Acta Mater. **52**(10), 2973–2987 (2004)
5. A. Biswas, D.J. Siegel, D.N. Seidman, Simultaneous segregation at coherent and semicoherent heterophase interfaces. Phys. Rev. Lett. **105**(7), 076102 (2010)
6. L. Bourgeois, C. Dwyer, M. Weyland, J.-F. Nie, B.C. Muddle, Structure and energetics of the coherent interface between the θ' precipitate phase and aluminium in Al–Cu. Acta Mater. **59**(18), 7043–7050 (2011)
7. L. Bourgeois, N.V. Medhekar, A.E. Smith, M. Weyland, J.F. Nie, C. Dwyer, Efficient atomic-scale kinetics through a complex heterophase interface. Phys. Rev. Lett. **111**(4), 046102 (2013)

8. V. Vaithyanathan, C. Wolverton, L.Q. Chen, Multiscale modeling of precipitate microstructure evolution. Phys. Rev. Lett. **88**(12), 125503 (2002)
9. J.Y. Hwang, R. Banerjee, H.W. Doty, M.J. Kaufman, The effect of Mg on the structure and properties of Type 319 aluminum casting alloys. Acta Mater. **57**(4), 1308–1317 (2009)
10. W.M. Stobbs, G.R. Purdy, The elastic accommodation of semi-coherent-θ′ in Al-4wt.%Cu alloy. Acta Metall. **26**, 1069–1081 (1978)
11. A. Biswas, D. Sen, S.K. Sarkar, Sarita, S. Mazumder, D.N. Seidman, Temporal evolution of coherent precipitates in an aluminum alloy W319: A correlative anisotropic small angle X-ray scattering, transmission electron microscopy and atom-probe tomography study. Acta Mater. **116**, 219–230 (2016)
12. D. Mitlin, V. Radmilovic, J.W. Morris Jr., Catalyzed Precipitation in Al-Cu-Si. Metall. Mater. Trans. A **31A**, 2697–2711 (2000)
13. Abaqus. Providence, RI, USA (2013)
14. S. Hu, M. Baskes, M. Stan, L. Chen, Atomistic calculations of interfacial energies, nucleus shape and size of θ′ precipitates in Al-Cu alloys. Acta Mater. **54**(18), 4699–4707 (2006)
15. S.Y. Hu, J. Murray, H. Weiland, Z.K. Liu, L.Q. Chen, Thermodynamic description and growth kinetics of stoichiometric precipitates in the phase-field approach. Calphad **31**(2), 303–312 (2007)
16. R. Martinez, D. Larouche, G. Cailletaud, I. Guillot, D. Massinon, Simulation of the concomitant process of nucleation-growth-coarsening of Al_2Cu particles in a 319 foundry aluminum alloy. Modell. Simul. Mater. Sci. Eng. **23**(4), 045012 (2015)
17. W. Wang, J.L. Murray, S.Y. Hu, L.Q. Chen, H. Weiland, Modeling of Plate-like Precipitates in Aluminum Alloys-Comparison between Phase Field and Cellular Automaton Methods. J. Phase Equilib. Diffus. **28**(3), 258–264 (2007)
18. L.-Q. Chen, Phase-field Models for Microstructure Evolution. Annu. Rev. Mater. Res. **32**, 113–140 (2002)
19. Y. Du, Y.A. Chang, B. Huang, W. Gong, Z. Jin, H. Xu, Z. Yuan, Y. Liu, Y. He, F.Y. Xie, Diffusion coefficients of some solutes in fcc and liquid Al: critical evaluation and correlation. Mater. Sci. Eng. A **363**(1–2), 140–151 (2003)
20. E. Kozeschnik, I. Holzer, B. Sonderegger, On the Potential for Improving Equilibrium Thermodynamic Databases with Kinetic Simulations. J. Phase Equilib. Diffus. **28**(1), 64–71 (2007)

Part V
Mechanical Performance Using Multi-scale Modeling

Hybrid Hierarchical Model for Damage and Fracture Analysis in Heterogeneous Material

Alex V. Vasenkov

Abstract Complex materials with heterogeneities and discontinuities is a focus of current research in materials science and engineering because such materials promise to have superior properties including enhanced mechanical strength and fatigue resistance. Predictive damage and fracture analysis of complex materials is grueling since it involves the modeling of highly-coupled nonlinear processes at disparate scales. A logical approach taken by many researchers in tackling this challenge is to employ a framework that couples Molecular Dynamic (MD) and Finite Element (FE) modeling in some manner to capture damage processes occurring at different time and length scales. Unfortunately, such coupling typically suffers from the lack of thermodynamic consistency between the MD and FE models and causes pathological wave reflection that commonly occurs at the interface between the MD and FE simulation regions. This work endeavors to circumvent those problems by introducing a Hybrid Hierarchical Model (HHM) that consists of an MD module with an ab initio based force field and a peridynamic continuum mesoscale module. The HHM framework was applied to perform the fracture modeling of a silicon carbide slab with pre-crack and the high-cycle fatigue damage analysis of a turbine blade.

Keywords Computer-aided engineering · Damage · Fracture · Fatigue · Heterogeneous material · Molecular dynamics · Peridynamics

Introduction

Predicting fracture, damage, and failure of materials is one of the central themes of materials science and engineering. Computer-aided Engineering (CAE) software applies the Finite Element Method (FEM) and their more advanced counterparts

---

A.V. Vasenkov (✉)
Sunergolab Inc, Lexington, KY 40511, USA
e-mail: avv@sunergolab.com

such as the extended FEM models [1, 2] and the meshless/mesh-free models [3] in conjunction with the multi-scale computational homogenization techniques [4, 5] to capture the macroscopic stress-strain response [6, 7]. However, the characterization of novel aerospace and automotive materials with heterogeneities and discontinuities requires a different approach capable of providing clear linkage to material microstructure [8]. Fundamental difficulty with implementation of this approach is that damage in complex materials involves highly-coupled nonlinear processes occurring at disparate scales of microstructure, which constitute many known challenges to existing CAE models [6, 7, 9]. CAE models can be classified either as the bottom-up or the top-down types. The top-down models typically apply the FEM to solve the equations of classical continuum mechanics [10]. They can be refined by incorporating a desirable level of detail necessary to account for features of engineering tests. Recently, peridynamic theory, a nonlocal extension of classical continuum mechanics, received considerable attention [11–14]. Peridynamic models have a promise of obtaining the paths and speed of growing cracks without enforcing special kinetic criteria, but rather as part of the solution of motion equations with incorporated bond-level damage law. Peridynamics account for growth and propagation of cracks and damage in an autonomous way that is critical when crack paths and nature or origin of material failure are not known a priori. Bottom-up CAE simulators build detailed models of atomic and molecular processes by means of the Density Functional Theory (DFT), the Molecular Dynamics (MD) method, or the Monte Carlo (MC) method [6, 15]. These models can provide valuable insights to initiation and propagation of crack path in materials with interfaces. The deficiency of the bottom-up models is that they can typically handle intervals of lengths that are several orders of magnitude below the typical characteristic size of microstructure. This precluded many reported MD simulations to correctly predict important features of dynamic fracture [16]. Such problems with the MD simulations are likely because the events of crack formation, propagation, and branching are controlled by the interactions and reflections of waves from the boundaries, which are difficult to correctly account in MD modeling because of severe limitations on the size of simulation domain [17]. The Hybrid Hierarchical Model (HHM) reported here correctly accounts for the boundary effects by updating the boundary conditions for MD simulation using on the fly peridynamic modeling results.

Brief description of the HHM is given in section "Hybrid Hierarchical Model" Fracture modeling results for a silicon carbide slab with pre-crack are discussed in section "Fracture Modeling of a SiC Slab". Application of the HHM to high-cycle fatigue damage analysis of a turbine blade is presented in section "High-Cycle Fatigue Damage Analysis of Gas Turbine Blade". Concluding remarks are in section "Summary".

Hybrid Hierarchical Model

The HHM is a three-dimensional solid mechanics model for massively-parallel multi-physics simulations, which consists of the Peridigm peridynamics module and the Large-scale Atomic/Molecular Massively Parallel Simulator (LAMMPS) module. Both Peridigm and LAMMPS were originally developed by the Sandia National Laboratories. Peridynamics is a nonlocal theory based on direct interactions between points [11–13]. Displacements rather than displacement derivatives are used in its formulation, which give peridynamics advantages over the FEM for modeling of complex materials with heterogeneities and discontinuities. The peridynamic equation of deformation is given by [11, 18]

$$\rho(\vec{x}, t)\ddot{u}(\vec{x}, t) = \int \left\{ \underline{T}(\vec{x}, t)\langle \vec{x'} - \vec{x} \rangle - \underline{T}\left(\vec{x'}, t\right)\langle \vec{x} - \vec{x'} \rangle \right\} dV_{x'} + \vec{b}(\vec{x}, t), \quad (1)$$

where T is the nonlocal force, x is a material point in the reference configuration of a body B, u is the deformation, ρ is the density field, and b is the prescribed external body force density. The subregion H_x of integration in (1) is a spherical neighborhood in B centered at x, and e_x is the unit vector along the bond (x′, x). Since the integral in (1) sums up forces on x′ from all neighbors, the peridynamic model can be thought of as a continuum version of MD. Equation (1) can be recovered from the Cauchy momentum equation by replacing divergence of stress with integral of nonlocal forces [19]. The maximum interaction distance between points, to be called horizon, provides a length scale for peridynamic material model. Currently, the Peridigm peridynamic module supports two damage models: a critical stretch model and a fatigue damage model. The critical stretch model uses critical stretch bond breaking criterion with critical stretch parameters determining when each individual bond is broken [20]. The critical stretch parameters can be estimated based on atomistic data or calibrated using experimentally measured energy release rates. The fatigue damage model follows Silling's et al. approach that does not require any pre-defined criterion in its formulation [12]. Here, the associated remaining life D at any given time t for each bond is determined by

$$\frac{dD}{dt} = \frac{-|1 - R|A \exp(t/\tau)}{\tau} |s|^m, \quad (2)$$

where $\tau = \Delta t / \ln(\text{RPM } \Delta t)$, D is the cyclic bond strain, Δt is the time step of peridynamic solver, R is the loading ratio, A is the fitting coefficient, m is the fitting degree, and RPM is the revolutions per minute.

Atomic-scale LAMMPS geometries and a Peridigm geometry are required to perform the HHM modeling. Each LAMMPS geometry is used around the periphery of pre-crack tip and it constitutes a tiny fraction of Peridigm geometry. Sigma convergence criterion is used in the HHM which forces the ratio of horizon to node spacing in Peridigm to be equal to the ratio of cut-off distance to bond length in LAMMPS. The mesoscale mechanical properties such as displacement, spatial deformation, forces, and force densities are calculated in Peridigm. These properties are sent at the end of each iterative step to the LAMMPS module and used as boundary conditions to model atomistic-scale crack dynamics. The ab initio based force field ReaxFF is employed in LAMMPS. Each LAMMPS simulation is typically performed for a fraction of picoseconds. In contrast, peridynamic simulation are performed using both explicit and implicit schemes. Consequently, LAMMPS simulation is considerably shorter than peridynamic simulation. This can be justified by the fact that there is a significant gap between the timescale of thermal vibration and the timescale of atomic motion [21]. Thermal motions of the atoms in the bulk become quickly enslaved to the atomic motion prescribed at the boundaries. The atomistic crack growth rate and bond strength are calculated in LAMMPS and can be used in Peridigm for updating on the fly bond breaking criterion. The Peridigm and LAMMPS modules are iterated until a required simulation time is achieved.

Fracture Modeling of a SiC Slab

Silicon carbide (SiC) is an attractive material for many technologies that require high performance in extreme environments. However, its utilization is often limited by susceptibility to fracture. The HHM was used to model fracture in SiC. The Peridigm and LAMMPS geometries were slabs with dimensions of 2×0.2 0.002 μm and $200 \times 200 \times 20$ Å, respectively. A constant external tensile force was prescribed along the Peridigm boundary in the y-direction. Strain rate at the LAMMPS boundary in the y-direction was updated on-the-fly based on the Peridigm results. An elastic correspondence model with a Verlet integration scheme was used in Peridigm. LAMMPS simulation was performed for 3C-SiC single-crystal using a velocity Verlet integration scheme and a time step of 0.25 fs. ReaxFF force field was borrowed from [22].

Time-dependent strain profile computed by Peridigm is given in Fig. 1 for an external force of 8×10^{14} dyn. The presented results were obtained by spatially averaging strain over the Peridigm computational cells located along the LAMMPS boundaries in the y-direction. It was observed that strain was harmonically oscillating after an instantaneous application of external force. Frequency of oscillations was determined by the sound speed and the distance between the boundaries to which external force was applied. From this consideration, the computational sound of speed can be estimated as 6×10^3 m/s. Figure 2 presents 2-D spatial profile of displacement calculated at two moments of time corresponding to the highest and lowest strains given in Fig. 1. Displacement was changed by about an order of magnitude within an oscillation cycle. Strains computed in Peridigm were used in LAMMPS for modeling dynamic fracture. Final snapshots of the crack in SiC slab at two external forces of 8×10^{14} dyn and 4×10^{14} dyn obtained using the HHM or the authentic LAMMPS simulation are given in Fig. 3. Both HHM and LAMMPS modeling started from the same initial configuration of a 3C-SiC slab

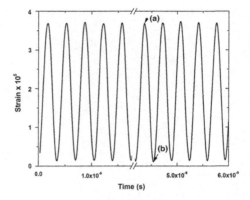

Fig. 1 Time-dependent strain profile computed in Peridigm for an external force of 8×10^{14} dyn, **a** and **b** designates the highest and the lowest strains, respectively

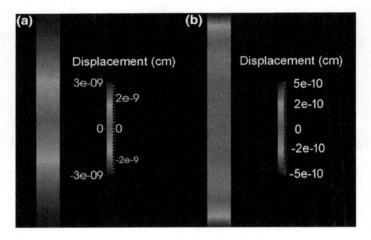

Fig. 2 2-D spatial profile of displacement at two moments of time corresponding to the highest **a** and the lowest **b** strains presented in Fig. 1

with pre-crack. Applied external force created a rough fracture surface, disordering atomic packing within a few atomic planes. Disordering occurred very differently during the HHM and the authentic LAMMPS modeling. The HHM modeling predicted a linear crack propagation, while the authentic LAMMPS simulation yielded crack branching with an angle greater than 90° whereas experiments showed much smaller crack branching angles [23]. This disagreement is likely because the crack branching events are controlled by the interactions and reflections of waves from the boundaries, which are difficult to correctly account in the authentic LAMMPS modeling because of limitations on the size of simulation domain [17]. Also, considering that the modeling system and applying boundary

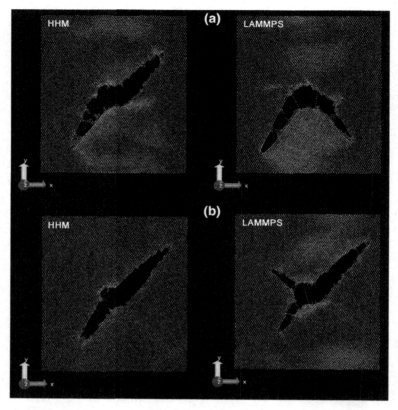

Fig. 3 Final snapshots of crack in SiC slab for two different external forces. The X; Y and Z directions correspond respectively to the [100] directions of the SiC crystal. **a** and **b** designates the results obtained using an external force of 8×10^{14} dyn and 4×10^{14} dyn, respectively

conditions were fully symmetric, one should expect that final snapshots of crack must be close to symmetric. Only the HHM results comply with that requirement.

High-Cycle Fatigue Damage Analysis of Gas Turbine Blade

The peridynamic calculations are almost always done for mesh with nearly constant node spacing [13]. This is a serious disadvantage of many existing peridynamic codes compared to the FEM codes. In contrast, Peridigm module of the HHM can handle a highly non-uniform mesh provided in Exodus/Genesis mesh input. This feature is demonstrated by performing the fracture analysis of a blade of gas-turbine engine. Such an engine generates thrust by jet propulsion. Turbine blades of aircraft engine are responsible for 50–75% of thrust. The high-cycle fatigue (HCF) of the turbine blades and other engine components have traditionally been analyzed with

Goodman diagram. However, increasing incidence of late of the HCF-related engine failures calls for new approaches such as peridynamics.

Procedure for preparing a geometry for peridynamic simulation is illustrated in Fig. 4. Here, the Computer-aided design (CAD) geometry of blade (A) is, first, meshed with an unstructured grid (B) using tetrahedral elements, and, second, translated to the peridynamic geometry (C), where each tetrahedral node serves as a peridynamic point. The peridynamic geometry was used to model dynamics of the blade made from a Ti–6Al–4 V alloy. Temperature was prescribed as a function of height and varied from 400 °C at the bottom of the blade to 500 °C at the top of the blade. RPM was equal to 15,000. Time harmonic external force of a magnitude of 10^3 dyn acting on the pressure side of the blade was prescribed. A fitting coefficient of 3×10^{-8} required for the remaining life estimation in (2) was obtained by developing an analytical fit to the strain ratio versus cycle dependence reported in [24] and, then, extrapolating this fit to N = 1. A fitting degree of 2 in (2) was taken from Paris' law equation for Ti–6Al–4 V alloy, given in [25]. Verlet explicit method was used for modeling blade's dynamics during a single revolution and provided an initial guess to the implicit method. The time harmonic external force caused deformation of the blades as shown in Fig. 5. The displacement was low near the interface between disk and blade. It increased with distance from the interface and peaked at the top free surface. The force density presented in Fig. 6 peaked at the blade/disk interface due to large stresses resulting from the contact between blade and disk. The force density was uniform over the pressure side of the blade where it reached about 10^3 dyn/cm^3. The damage resulting from fatigue of the blade had a strong dependence on the number of cycles as illustrated in Fig. 7. The damage peaked at the bottom of the trailing edge of the blade. It is likely that cracks would form in this area at a larger number of cycles. There was also some damage accumulated at the blade/disk interface where the force density peaked.

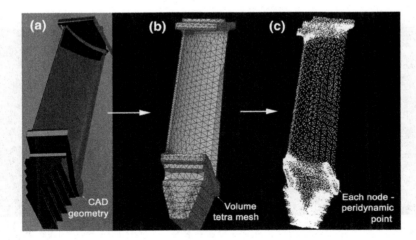

Fig. 4 Illustration of the procedure for preparing a geometry for simulation

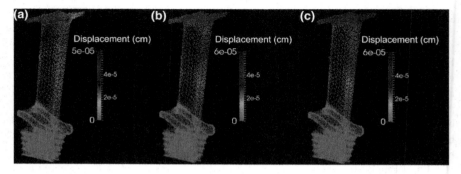

Fig. 5 Displacement versus number of cycles (N): **a** N = 1, **b** N = 100, and **c** N = 10,000. Displacement peaks at the free top surface, far from the blade/disk interface

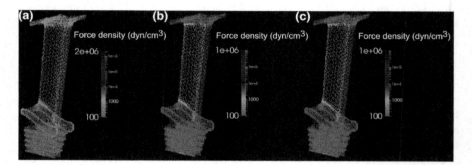

Fig. 6 Force density versus number of cycles: **a** N = 1, **b** N = 100, and **c** N = 10,000. Force density peaks at the blade/disk interface due to large stresses resulting from the contact between blade and disk

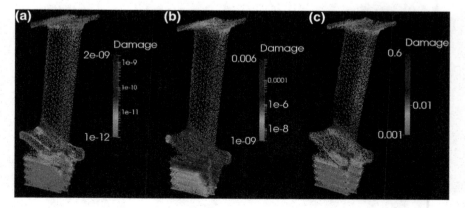

Fig. 7 Damage versus number of cycles: **a** N = 1, **b** N = 1,000, and **c** N = 30,000. Damage peaks at the bottom of the trailing edge of the blade

Summary

The Hybrid Hierarchical Model consisting of the LAMMPS module with ab initio based ReaxFF force field and the Peridigm peridynamic module was presented. The model was applied to the crack nucleation modeling for SiC slab and the high-cycle fatigue damage analysis of turbine blade. LAMMPS was used for the first principle modeling of the crack region where the mechanical strains were the greatest and the most detailed resolution and the highest accuracy were required to predict the crack path. Peridynamic simulations were performed with increasing scale to provide more smeared and continuum-like solution. Link between Peridigm and LAMMPS enabled accurate atomistic-based description of damage without contamination with pathological wave reflection from the MD boundaries. The high-cycle fatigue modeling was performed for the real geometry of Pegasus blade. The use of non-uniform node spacing made the peridynamic simulations practical. Verlet integration method for a single revolution and implicit method for up to 30,000 revolutions were employed. The external forces applied to the pressure side of the blade caused deformation resulting in fatigue damage. Locations of expected crack formation were depicted based on the analysis of peridynamic modeling results obtained without predefined criterion.

Acknowledgements The author would like to thank Prof. Adri van Duin for providing the ReaxFF SiC force field.

References

1. H. Cheng, X.P. Zhou, A multi-dimensional space method for dynamic cracks problems using implicit time scheme in the framework of the extended finite element method. Int. J. Damage Mech **24**(6), 859–890 (2015)
2. G. Vigueras et al., An XFEM/CZM implementation for massively parallel simulations of composites fracture. Compos. Struct. **125**, 542–557 (2015)
3. S. Wu et al., A front tracking algorithm for hypervelocity impact problems with crack growth, large deformations and high strain rates. Int. J. Impact Eng. **74**, 145–156 (2014)
4. A. Shojaei et al., Multi-scale constitutive modeling of ceramic matrix composites by continuum damage mechanics. Int. J. Solids Struct. **51**(23–24), 4068–4081 (2014)
5. J. Fish, Q. Yu, Multiscale damage modelling for composite materials: theory and computational framework. Int. J. Numer. Meth. Eng. **52**(1–2), 161–191 (2001)
6. B. Cox, Q.D. Yang, In quest of virtual tests for structural composites. Science **314**(5802), 1102–1107 (2006)
7. C. Kassapoglou, *Modeling the Effect of Damage in Composite Structures: Simplified Approaches* (Wiley, 2015)
8. C. Przybyla, CMC behavior and life modeling workshop. AFRL-RX-WP-TP-2011-4396 Summary Report (2011)
9. R.O. Ritchie, In pursuit of damage tolerance in engineering and biological materials. MRS Bull. **39**(10), 880–890 (2014)
10. V. Vavourakis et al., Assessment of remeshing and remapping strategies for large deformation elastoplastic Finite Element analysis. Comput. Struct. **114**, 133–146 (2013)

11. D. Littlewood, S. Silling, P. Seleson, Local-nonlocal coupling for modeling fracture ASME 2014, in *International Mechanical Engineering Congress and Exposition held November 8–13, 2014 in Montreal, Canada* (2014)
12. S.A. Silling, A. Askari, *Peridynamic Model for Fatigue Cracking*. SAND2014-18590 (2014)
13. E. Madenci, E. Oterkus, *Peridynamic Theory and Its Applications* (Springer, New York, 2013)
14. Z. Chen, F. Bobaru, Selecting the kernel in a peridynamic formulation: A study for transient heat diffusion. Comput. Phys. Commun. **197**, 51–60 (2015)
15. K.W.K. Leung, Z.L. Pan, D.H. Warner, Atomistic-based predictions of crack tip behavior in silicon carbide across a range of temperatures and strain rates. Acta Mater. **77**, 324–334 (2014)
16. S.J. Zhou et al., Dynamic crack processes via molecular dynamics. Phys. Rev. Lett. **76**(13), 2318–2321 (1996)
17. K. Ravi-Chandar, Dynamic fracture of nominally brittle materials. Int. J. Fract. **90**(1–2), 83–102 (1998)
18. D.J. Littlewood et al., *Strong Local-Nonlocal Coupling for Integrated Fracture Modeling*. SAND2015-7998 Report (2015)
19. S.A. Silling, Reformulation of elasticity theory for discontinuities and long-range forces. J. Mech. Phys. Solids **48**(1), 175–209 (2000)
20. M.L. Parks et al., *Peridigm Users' Guide*. SAND2012-7800 (2012)
21. C.W. Gear, I.G. Kevrekidis, C. Theodoropoulos, "Coarse" integration/bifurcation analysis via microscopic simulators: micro-galerkin methods. Comput. Chem. Eng. **26**, 941 (2002)
22. D.A. Newsome, D. Sengupta, A.C.T. van Duin, High-temperature oxidation of SiC-based composite: rate constant calculation from ReaxFF MD simulations, Part II. J. Phys. Chem. C **117**(10), 5014–5027 (2013)
23. M. Ramulu, A.S. Kobayashi, Mechanics of crack curving and branching—a dynamic fracture analysis, In *Dynamic Fracture*, eds. by M.L. Williams, W.G. Knauss (Springer, Netherlands, 1985) pp. 61–75
24. C.D. Lykins, S. Mall, V. Jain, An evaluation of parameters for predicting fretting fatigue crack initiation. Int. J. Fatigue **22**(8), 703–716 (2000)
25. R.O. Ritchie et al., Thresholds for high-cycle fatigue in a turbine engine Ti–6Al–4 V alloy. Int. J. Fatigue **21**(7), 653–662 (1999)

Fatigue Performance Prediction of Structural Materials by Multi-scale Modeling and Machine Learning

Takayuki Shiraiwa, Fabien Briffod, Yuto Miyazawa and Manabu Enoki

Abstract Structural materials having higher performance in strength, toughness, and fatigue resistance are strongly required. In the conventional materials development, many fatigue tests need to be conducted to validate statistical behavior of fatigue failure. Accordingly the evaluation of fatigue properties with shorter time becomes quite essential. Based on such background, we are developing fatigue prediction methods for wide range of structural materials by multi-scale finite element analysis (FEA) and machine learning in the Materials Integration (MI) system. The multi-scale FEA consists of the following procedures: (i) mechanical and thermal properties are estimated by using commercially available software and database; (ii) temperature field, residual stress and distortion generated during a manufacturing process is calculated on the macroscopic model by thermo-mechanical FEA; (iii) macroscopic stress field under cyclic loading condition is calculated with a hardening constitutive model; (iv) the microscopic stress field is derived by finite element model with the polycrystalline structures and the cycles for a fatigue crack initiation is analyzed by strain energy accumulation on the slip plane; (v) the cycles for fatigue crack propagation is analyzed by extended finite element method (X-FEM) and the total number of cycles to the failure is obtained. The second approach is to use machine learning techniques to obtain empirical prediction formula. The database was prepared from published resources and experiments. Deterministic machine learning techniques such as multivariate linear regression and artificial neural network provided accurate equations to predict fatigue strength from materials and process parameters. Additionally, the concept of model-based machine learning was adopted to incorporate prior knowledge of microstructures and properties, and to account for uncertainty on fatigue life. The results showed that model-based machine learning was a promising tool for predicting fatigue performance in structural materials. The features and limitations of our prediction methods will be discussed.

T. Shiraiwa (✉) · F. Briffod · Y. Miyazawa · M. Enoki
Department of Materials Engineering, School of Engineering, The University of Tokyo, 7-3-1 Hongo, Bunkyo-Ku, Tokyo 113-8656, Japan
e-mail: shiraiwa@rme.mm.t.u-tokyo.ac.jp

Keywords Fatigue life prediction · Multi-scale finite element method · Model-based machine learning · Two-point spatial correlation · Materials integration

Introduction

Fatigue design curves are widely used in design of structures such as bridges, ships, aircraft, power plants and industrial machines. Although many factors affect fatigue properties of welding structures, the geometries have the most influence on fatigue of welding structure. Thus, in the conventional approach, the fatigue design curve is defined by the geometries of the welded joint. On the other hand, several novel materials have been developed to improve the fatigue performance. For example, low transformation temperature (LTT) welding material reduces the residual stress [1], and dual-phase steel improves the fatigue property by controlling the microstructure [2]. Therefore, the evaluation of the effect of residual stress and the microstructure is becoming more important. The objective of the current study is to develop an extensible framework to predict the fatigue performance of various materials using physically-based simulations and machine learning techniques.

Methods

Physically-Based Fatigue Simulation

The first approach is a physically-based simulation using a multi-scale finite element method. The overview of the simulation method is shown in Fig. 1. The simulation consists of the following procedures: (i) mechanical and thermal properties are estimated by using commercially available software and database; (ii) temperature field, residual stress and distortion generated during a welding process is calculated on the macroscopic model by thermo-mechanical FEA; (iii) macroscopic stress field under cyclic loading condition is calculated with a hardening constitutive model; (iv) the microscopic stress field is derived by finite element model with the polycrystalline structures and the cycles for a fatigue crack initiation is analyzed by Tanaka-Mura model [3]; (v) the cycles for fatigue crack propagation is analyzed by extended finite element method (X-FEM) and the total number of cycles to the failure is obtained. The input values in this simulation are chemical composition of base metal and welding material, shape parameters of the joint, initial microstructure, welding conditions, and loading conditions. The output is the S-N curve with the survival probability.

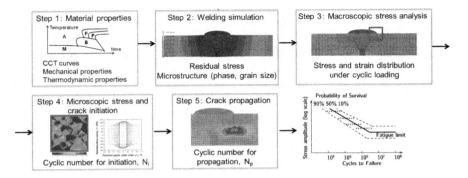

Fig. 1 Overview of proposed method for predicting fatigue life of welded joints

Machine Learning to Predict Fatigue Life

The second approach is to apply machine learning techniques to obtain empirical prediction formula from fatigue database accumulated over the years. In the field of machine learning, numerous learning algorithms have been developed over several decades. In the traditional approach, we must select an appropriate learning algorithm to solve a problem from the set of the numerous algorithms. While this traditional approach has resulted in successful applications in various fields, when trying to apply to the fatigue prediction, there are difficulties in incorporating prior knowledge between microstructure and properties, and handling the scattering of fatigue life. To overcome these limitations, a concept of model-based machine learning [4] was adopted in the current study. This concept is also referred to as physics-informed machine learning [5] in physical problems such as continuum mechanics of fluids and solids. Our interpretation of these methodologies is that all assumptions (called as "model") which obey physical constraints are made explicit in a mathematical form, and all parameters are expressed as random variables. Two approaches for using machine learning techniques to predict fatigue life in the current study are shown in Fig. 2. One is deterministic machine learning based on the traditional methods, and the other is model-based machine learning.

In the deterministic machine learning, a dataset of fatigue life in steels was obtained from National Institute of Material Science (NIMS) fatigue datasheet [6], which includes chemical composition, processing parameters (reduction ratio, heat treatment), inclusion sizes, and fatigue strength at 10^7 cycles. The details of the features in the dataset are given in Table 1. The machine learning was performed using two algorithms: Multivariate Linear Regression (MLR) and Artificial Neural Network (ANN).

In the model-based machine learning, the linkage of process-structure-property-performance is expressed by the conditional probability distributions as shown in Fig. 2. We focused on the linkage between properties and structures ($p(\mathbf{p}|\mathbf{s})$ in Fig. 2) in the current study. Microstructures and stress-strain curves

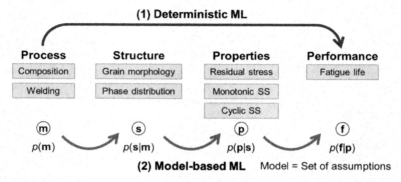

Fig. 2 Two approaches for using machine learning techniques to predict fatigue life

Table 1 Features in the fatigue dataset extracted from NIMS fatigue datasheet

Features	Mean	Max	Min
Fe (wt%)	97.228	99.072	95.108
C (wt%)	0.413	0.63	0.22
Si (wt%)	0.320	2.05	0.16
Mn (wt%)	0.829	1.6	0.21
P (wt%)	0.016	0.031	0.004
S (wt%)	0.014	0.03	0.003
Ni (wt%)	0.506	2.78	0.01
Cr (wt%)	0.537	1.12	0.01
Cu (wt%)	0.064	0.26	0.01
Mo (wt%)	0.069	0.24	0
Al (wt%)	0.001	0.04	0
N (wt%)	3.81×10^{-4}	0.0153	0
Ti (wt%)	8.65×10^{-5}	0.01	0
O (wt%)	8.14×10^{-5}	0.003	0
Non-metallic inclusions dA (%)	0.048	0.13	0
Non-metallic inclusions dB (%)	0.003	0.05	0
Non-metallic inclusions dC (%)	0.008	0.058	0
Reduction ratio	977	5530	289
Tlog (t) for Normalizing	1681	1733	1622
Tlog (t) for Quenching	1610	1681	0
Tlog (t) for Tempering	1503	1695	0
Fatigue strength (MPa)*	499.88	906	225

*10^7 cycles, rotating bending

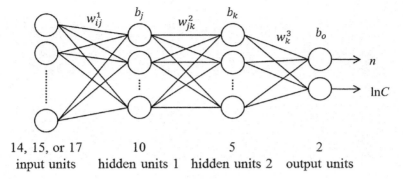

Fig. 3 Artificial neural network used in the model-based approach

in 40 low carbon steels with different chemical composition and heat treatment conditions were prepared to create the learning dataset. The microstructures were obtained from optical microscopy observation and the stress-strain curves were acquired in a wide range by a combination of strain gauge and noncontact extensometer. The optical micrographs were binarized to determine the ferrite phase (phase 1) and the other phases (phase 2). Then two-point spatial correlation (f^{11}) was calculated for each microstructure and principal component analysis (PCA) was performed to reduce the dimension of f^{11}. The first three principal components were added to the input variable in ANN. The structure of ANN is given in Fig. 3. Each stress-strain curve was fitted to a power law, $\sigma = C\varepsilon^{n}$, and the C and m were used as the output variable in the ANN. The machine learning using the ANN was carried out with and without principal components of f^{11} in the input layer.

Results and Discussion

In the first part of the physically-based simulation, a macroscopic simulation was conducted using the standard finite element analysis to predict the microstructure and stress distribution in welded joints. Various material properties were calculated from chemical composition, using commercial software JMatPro. In the second step, welding simulations were carried out using the commercial FEM software SYSWELD. Temperature distribution and volume fraction of each phase were obtained from heat conduction analysis. Residual stress distribution and welding deformation were also recorded in this step. Moreover, the distribution of Vickers hardness was derived from empirical equation. The austenite grain size was also calculated by the incremental equation. Figure 4 shows the distributions of volume fraction of martensite, prior austenite grain size, and residual stress in butt joint. In the third step, the finite element software, Abaqus, was utilized to calculate the stress distribution under the cyclic loading with importing the hardness distribution

Fig. 4 Distribution of **a** volume fraction of martensite, **b** prior austenite grain size, **c** residual stress of butt joint

and residual stress distribution obtained from the previous step. Figure 5 shows the distribution of maximum principal stress when the toe curvature was 0.1 mm and the applied stress was 150 MPa. The maximum principal stress (σ_1) is reached maximum value around the weld toe, and the direction of maximum principal stress was in parallel with the surface. The magnitude of σ_1 and the location of the element were recorded. These results were used in next part. In the fourth step, fatigue crack initiation analysis was conducted with mesoscopic polycrystalline aggregates models [7]. The distribution of Mises stress on the mesoscopic model is given in Fig. 6. The cyclic number for crack initiation was calculated by Tanaka-Mura equation [3].

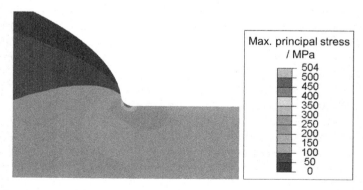

Fig. 5 Distribution of maximum principal stress at last cycle ($\rho = 0.1$ mm, $\sigma_{\mathrm{ap}} = 150$ MPa)

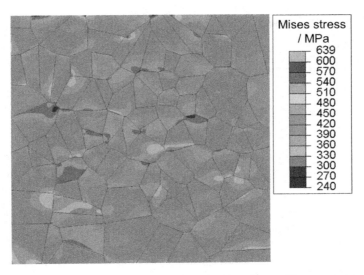

Fig. 6 Mises stress distribution on the mesoscopic polycrystalline aggregates model ($\rho = 0.1$ mm, $\sigma_{ap} = 150$ MPa)

Using developed framework, sensitivity analysis of residual stress, stress concentration, and microstructure were conducted. It showed that the effect of residual stress was not significant in the current conditions. The fatigue life increased with the toe radius, and the scattering in high cycle region appeared due to considering microstructure. Additionally, the predicted SN curve was compared with fatigue design curves proposed by Japanese Society of Steel Construction (JSSC) as shown in Fig. 7. It was demonstrated that the predicted results were consistent with conventional fatigue design recommendations.

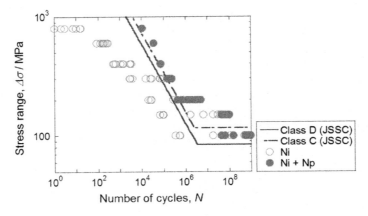

Fig. 7 Predicted SN curve and fatigue design curve ($\rho = 0.1$ mm)

Fig. 8 Results of parameter selection using cross validation method

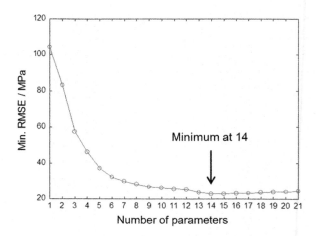

In the machine learning approach, to avoid overfitting, leave-one-out cross validations (CV) were conducted using multivariate linear regression analysis for all (10^{21} −1) combinations of the input parameters. The results of parameter selection using CV are shown in Fig. 8. The best combination of inputs was determined as a set of 14 inputs parameters which includes element contents of C, Si, Mn, P, Ni, Cr, Cu, Mo, Al, N, Ti, O and heat conditions of quenching and tempering. In the learning procedure, the dataset extracted from NIMS fatigue datasheet were divided into two groups. One was training data for learning, and the other was test data for evaluating the root mean squared error (RMSE) of the model. The RMSE value of MLR with the best combination was 23.8 MPa for the training data, and 29.8 MPa for the test data. These RMSE values were lower than a previous study using similar dataset [8]. In ANN model, Levenberg-Marquardt method with Bayesian framework was used as the training algorithm. Comparing several network structures, a neural network model with standardized input and output variables, ramp-type activation function and two hidden layers showed the lowest RMSE for the test data. The prediction results of fatigue strength by the ANN model is shown in Fig. 9. The RMSE of the training data and test data are 13.4, 16.4 MPa, respectively. These results showed that the proposed MLR and ANN models are more accurate than previous models. However, it does not provide any knowledge for the microstructure and properties.

In the model-based machine learning, strain hardening coefficients C and m were predicted from the dataset of 12 element contents, 2 heat conditions and 3 principal components of two-point spatial correlation. Two different datasets were also used for comparison. One is a dataset without spatial correlation (Without SC), and the other is a dataset which contains volume fraction of ferrite instead of f^{11} (Vf). The coefficient C predicted by these three dataset are shown in Fig. 10. The RMSE for the training data in the three dataset (Without SC, Vf, With SC) were 0.110, 0.101 and 0.085 MPa, respectively. Therefore, it was suggested that two-point spatial correlation is effective to predict stress-strain curves.

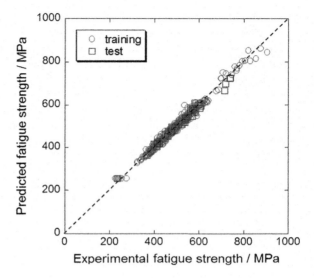

Fig. 9 Fatigue strength predicted by neural network model with standardized input and output variables, ramp activation function and two hidden layers

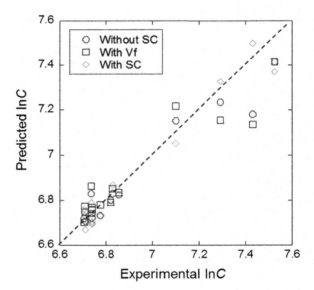

Fig. 10 Strain hardening coefficient predicted by three types of model with and without spatial correlation (SC) and volume fraction of ferrite (Vf)

Conclusions

A framework for prediction of fatigue performance in welded structures was proposed using multi-scale finite element analysis and machine learning techniques. The effects of residual stress, toe radius and microstructure were evaluated. The effect of residual stress was not significant in the current conditions. The fatigue life increased with the toe radius, and the scattering in high cycle region appeared due to considering microstructure. Prediction of total number was close to conventional fatigue design curve. It can be expected to improve this framework by implementing various models for microstructure and performance prediction. Additionally, machine learning techniques such CV, MLR and ANN were applied to fatigue datasets. These results showed that the proposed MLR and ANN models provide more accurate prediction of fatigue strength than the previous results. The potential benefits of model-based machine learning were considered that it allowed for incorporation of prior knowledge of structure and property, and it can account for uncertainty such as scattering of fatigue life.

Acknowledgements This study was partially supported by the Cross-ministerial Strategic Innovation Promotion Program (SIP)—Structural Materials for Innovation—unit D62 operated by The Cabinet Office, Japan. The authors would like to thank Dr. T. Kasuya (The University of Tokyo) who generously provided the data of microstructure and stress-strain curve used in the current study.

References

1. C. Shiga, H. Murakawa, K. Hiraoka, N. Osawa, S. Tsutsumi, H. Yajima, T. Tanino, Fatigue Improvement of Steel Welded Joints Using Elongated-Bead Weld Method for Low Temperature Transformation Welding Materials. Jpn. Weld. Soc. (2016), F-23-27
2. J.A. Wasynczuk, R.O. Ritchie, G. Thomas, Effects of microstructure on fatigue crack-growth in duplex ferrite martensite steels. Mater. Sci. Eng. **62**, 79–92 (1984)
3. K. Tanaka, T. Mura, A dislocation model for fatigue crack initiation. J. Appl. Mech. **48**, 97–103 (1981)
4. C.M. Bishop, Model-based machine learning. Trans. R. Soc. A **371**, 1–17 (2013)
5. H. Xiao, J.L. Wu, J.X. Wang, R. Sun, C.J. Roy, Quantifying and reducing model-form uncertainties in reynolds averaged navier-stokes equations: an data-driven, physics-based bayesian approach. J. Comput. Phys. 115–136 (2016)
6. National Institute of Materials Science Fatigue Datasheet, http://smds.nims.go.jp/fatigue/index.html
7. Fabien Briffod, Takayuki Shiraiwa, Manabu Enoki, Fatigue crack initiation simulation in pure iron polycrystalline aggregate. Mater. Trans. **57**, 1–6 (2016)
8. A. Agrawal, P.D. Deshpande, A. Cecen, G.P. Basavarsu, A.N. Choudhary, S.R. Kalidindi, Exploration of data science techniques to predict fatigue strength of steel from composition and processing parameters. Integrating Mater. Manuf. Innov. **3**, 1–19 (2014)

Nano Simulation Study of Mechanical Property Parameter for Microstructure-Based Multiscale Simulation

K. Mori, M. Oba, S. Nomoto and A. Yamanaka

Abstract We proposed a microstructure-based multiscale simulation of duplex stainless steel by using the multi-phase field method and finite element method software. Use of an accurate elastic constant is the key to the success of these simulations. However, it is difficult to obtain the elastic constant for the each constituent phase in multicomponent steels from a database and datebook. Herein, the elastic constant of each constituent phase of duplex stainless steel (Fe–Cr–Ni alloy) was calculated by first-principles and molecular dynamics (MD) simulation. The commercial software VASP was used to estimate the elastic constants of the bcc structure. On the other hand, the open-source MD software LAMMPS was used to estimate the elastic constants of the fcc structure. Calculations were performed using 10,000 models in which the Cr and Ni atoms were repositioned at random by MD simulation. The elastic constant volumes showed a Gaussian-like distribution, and this was explained using a radial distribution function. The calculated elastic constants C11, C12, and C44 were good agreement with experimental values.

Keywords Microstructure-based multiscale simulation · First-principles calculation · Molecular dynamics simulation · Duplex stainless steel

---

K. Mori (✉) · M. Oba · S. Nomoto
ITOCHU Techno-Solutions Corporation (CTC), 3-2-5 Kasumigaseki,
Chiyoda-ku, Tokyo 100-6080, Japan
e-mail: kazuki.mori.013@ctc-g.co.jp

A. Yamanaka
Division of Advanced Mechanical Systems Engineering,
Institute of Engineering, Tokyo University of Agriculture and Technology,
2-24-16 Naka-cho, Koganei-shi, Tokyo 184-8588, Japan

© The Minerals, Metals & Materials Society 2017
P. Mason et al. (eds.), *Proceedings of the 4th World Congress on Integrated Computational Materials Engineering (ICME 2017)*,
The Minerals, Metals & Materials Series, DOI 10.1007/978-3-319-57864-4_30

Introduction

Development of multiscale simulation techniques has long been pursued [1]. We have previously proposed a microstructure-based multiscale simulation framework using various types of commercial simulation software for analyzing the hot rolling of duplex stainless steel [2]. In the framework various commercial software, not only multi-phase field method and homogenization method but also atomistic simulation and finite element method was bridged. Among them, the asymptotic homogenization method was used to identify the parameters of the homogenized macro material from its microstructure and material properties of constituent phases [3]. As the homogenization method is a method for only averaging the properties, its accuracy strongly depends on that of each phase. Therefore, accurate material properties of constituent phases in microstructure is required for a microstructure-based multiscale simulation. First-principles calculations that is one of atomistic simulation can compensate for the lack of experimental crystal mechanical data. However, the size of the systems modeled based on the conventional first principles methods is often limited to a few hundred atoms because of the high calculation costs and long calculation time associated with the development of a high accuracy model. Therefore, it is very difficult to calculate the elastic constant of duplex stainless steel because the simulation model needs to include a large number of atoms for accurate estimation of the material properties and elastic constant. Past studies aimed at calculating the elastic constant of simple alloys using first-principles methods [4, 5] did not take into account the effect of atom distribution on the dispersion of the elastic constant. In this work, we calculated the elastic constant of the Fe–Cr–Ni ternary system by considering atom distribution. The commercial software VASP (package version 5.3.5) [6–8] is used to estimate the elastic constants of a bcc structure. On the other hand, the open-source molecular dynamics (MD) software LAMMPS [9] is used to estimate the elastic constants of the fcc structure, because a sufficient number of accurate interatomic potential parameters are available for the fcc structure in the Fe–Cr–Ni ternary system. Further, the LAMMPS (package version 16Feb2016) entails a shorter calculation time than VASP. In addition, we considered the atom distribution, then we conformed the dispersion of the elastic constant of fcc structure in Fe–Cr–Ni ternary system. This reason that elastic constant has the dispersion was observed by considering Radial Distribution Function (RDF).

Methodology

In the first-principles calculation for the bcc structure using VASP, the electron wave functions were described based on the Blochl method in the implementation by Kresse and Joubert [10, 11]. The generalized gradient approximation was used for treating the exchange and correlation effects [12]. The kinetic energy cutoff was set as 400 eV. The specific k points were selected using $4 \times 4 \times 4$ Monckhorst-Pack grids for integration in the Brillouin zone. The convergence of the calculations was investigated by suppressing the energy difference to less than 1e-6 eV in the repetition of the self-consistent loop. The lattice parameters were obtained from the corresponding energy minimization at constant volumes and by fitting the Birch-Murnaghan equation of state to the resulting energy-volume data [13, 14]. The total number of atoms for the first principles calculation was set as 54. Calculations of 11 different atom distribution cases were additionally performed. The elastic constant value was obtained from the average of the 11 cases. The elastic constants of the cubic cell were calculated in reference to a method proposed by Nishimatsu et al. [15].

Next, we first calculated the elastic constants of the fcc structure in the Fe–Cr–Ni ternary alloy at 0 K using LAMMPS with the embedded atom potential proposed by Bonny et al. [16]. To construct the initial atomic configuration, the lattice parameter was calculated by energy minimization with the conjugate gradient algorithm. The total number of Fe, Cr, and Ni atoms was 4000. These atoms were randomly distributed in the Fe–18%Cr–8%Ni (wt%) composition. To obtain statistically meaningful results, the elastic constants were estimated for 10,000 different atom distribution cases, because the calculated internal energy was dependent on the initial structure. The atom distribution was investigated by using the RDF expressed as

$$g(r) = \frac{n(r)}{\rho 4\pi r^2 \Delta r} \quad (1)$$

where $g(r)$ is the RDF, $n(r)$ is the mean number of atoms in a shell of Δr at distance r, and ρ is the mean atomic density.

Results and Discussion

First, the elastic constants C11, C12, and C44 of the bcc structure at 0 K were calculated as 239.2, 121.1, and 116.1 GPa, respectively, using VASP. Since the first-principles calculations were performed using a very small model comprising 54 atoms, the arrangement of the Cr and Ni atoms had no obvious effect on the obtained elastic constants. In other words, the number of Cr and Ni atoms in the first-principles model was too small for estimating the effect of atomic positions on the results. However, the first-principles calculation allows for the estimation of material properties without the need for any potential parameter, unlike the MD simulation.

In a previous work, the averaged values of the elastic constants C11, C12, and C44 for the fcc structure at 0 K were obtained as 229.7, 154.3 and 126.9 GPa, respectively, by using LAMMPS [2]. The deviation of the calculated elastic constant distribution in 11 cases resembled a Gaussian distribution. The maximum and

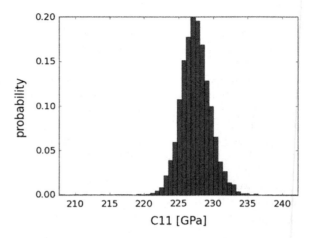

Fig. 1 Distribution of elastic constants C11 in the fcc structure of Fe–18%Cr–12%Ni

minimum C11 of the fcc structure were 231.5 GPa and 227.6 GPa, respectively. The maximum and minimum C12 were 154.8 GPa and 153.4 GPa, respectively, while the corresponding values for C44 were 127.7 and 126.4 GPa. Therefore, the elastic constants of the fcc structure were simply the arithmetic average of the values in the 11 cases. The calculated elastic constants C11, C12, and C44 for the Fe–18%Cr–12%Ni (wt%) composition were 215.9, 144.6, and 128.9 GPa, respectively, which were in good agreement with the experimental data [17].

In this study, the distribution of the calculated elastic constant of bcc structure was estimated over 10,000 cases. Figure 1 shows the distribution of the elastic constants C11. The distributions of C12 and C44, which resembled a Gaussian distribution, were identical to the case of C11. The standard deviation volume of the Gaussian distribution of the calculated elastic constants C11, C12, and C44 was determined as 78.3%, 87.6%, and 72.9%, respectively. The deviation of the C12 distribution was the smallest, while that of the C44 distribution was the largest. C44 was easily influenced by atom distribution, even if the calculation was performed using 10,000 cases in which Cr and Ni atoms were randomly repositioned. The RDF plot with regard to the Ni–Ni distance ($g(r)_{Ni-Ni}$) is shown in Fig. 2, with the $g(r)_{Ni-Ni}$ for the maximum and minimum C11 indicated by red and blue lines, respectively. The first peak of $g(r)_{Ni-Ni}$ was observed at 2.5 Å in both structures. However, the first peak intensity of the maximum C11 was twice that of the minimum C11. The second peak of $g(r)_{Ni-Ni}$ was observed at 3.6 Angstrom in both structures. However, the second peak intensity of the maximum C11 was less than that for the minimum C11. The RDF between two Cr atoms ($g(r)_{Cr-Cr}$) is shown in Fig. 3, with the $g(r)_{Cr-Cr}$ of the maximum and minimum C11 depicted as red and blue lines, respectively. The peak volumes for both cases were approximately the same. These observations indicated that the Cr atom distribution did not affect the distribution of the elastic constant in the Fe–Cr–Ni ternary system.

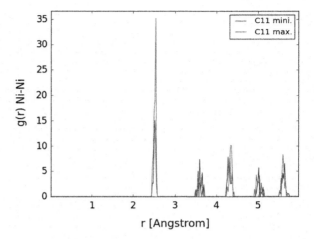

Fig. 2 Radial distribution function (RDF) versus Ni–Ni distance in the fcc structure of Fe–18%Cr–12%Ni (wt%) for two different structures: *red* maximum C11; *blue* minimum C11 (color online)

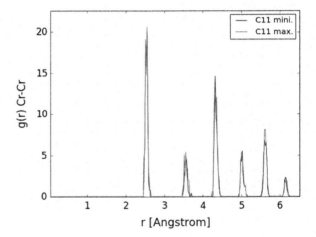

Fig. 3 Radial distribution function (RDF) versus Cr–Cr distance in the fcc structure of Fe–18%Cr–12%Ni (wt%) for two different structures: *red* RDF of the maximum C11 structure; *blue* RDF of the minimum C11 structure (color online)

Conclusion

We report for the first time the use of first-principles calculations and MD simulation to obtain the elastic constant for an Fe–Cr–Ni ternary alloy system. The calculated elastic constant of this ternary system was in good agreement with the experimental data and showed a Gaussian distribution. The elastic constant of the fcc structure increased with the first peak intensity corresponding to the RDF for the Ni–Ni distance. Estimation of the elastic constant by MD simulation would be useful if is a sufficient number of accurate interatomic potential parameters.

References

1. A. Jaramillo-Botero et al., First-principles-based multiscale, multiparadigm molecular mechanics and dynamics methods for describing complex chemical processes. Top. Curr. Chem. **307**, 1–42 (2012)
2. S. Nomoto et al., Microstructure-based multiscale analysis of hot rolling of duplex stainless steel using various simulation software. Integr. Mater. Manuf. Innov. **6**, 69–82 (2017)
3. G. Laschet, Homogenization of the thermal properties of transpiration cooled multi-layer plates. Comput. Method Appl. Mech. **191**(41–42), 4535–4554 (2002)
4. H. Zhang et al., Single-crystal elastic constants of ferromagnetic bcc Fe-based random alloys from first-principles theory. Phys. Rev. B **81**, 184105–184105–184105–184114 (2010)
5. S. Reeh, Elastic properties of fcc Fe–Mn–X (X = Cr Co, Ni, Cu) alloys from first-principles calculations. Phys. Rev. B **87**, 224103 (2013)
6. https://www.vasp.at/
7. G. Kresse, J. Hafner, Ab initio molecular dynamics for liquid metals. Phys. Rev. B **47**, 558–561 (1993)
8. G. Kresse, J. Furthmuller, Efficient iterative schemes for ab initio total-energy calculations using a plane-wave basis set. Phys. Rev. B **54**, 11169–11186 (1996)
9. http://lammps.sandia.gov
10. P.E. Blochl, Projector augmented-wave method. Phys. Rev. B **50**, 17953–17979 (1994)
11. G. Kresse, G.D. Joubert, From ultrasoft pseudopotentials to the projector augmented-wave method. Phys. Rev. B **59**, 1758–1775 (1999)
12. J.P. Perdew et al., Generalized gradient approximation made simple. Phys. Rev. Lett. **77**, 3865–3868 (1996)
13. F. Birch, The effect of pressure upon the elastic parameters of isotropic solids, according to Murnaghan's theory of finite strain. J. Appl. Phys. **9**, 279–288 (1938)
14. F. Birch, Finite elastic strain of cubic crystals. Phys. Rev. **71**, 809–824 (1947)
15. T. Nishimatsu et al., First principles accurate total-energy surfaces for polar structural distortions of BaTiO3, PbTiO3 and SrTiO3: consequences to structural transition temperatures. Phys. Rev. B **82**, 134106–134115 (2010)
16. G. Bonny et al., Interatomic potential for studying ageing under irradiation in stainless steels: the FeNiCr model alloy. Model. Simul. Mater. Sci. Eng. **21**, 085004–085016 (2013)
17. G. Bradfield, Comparison of the elastic anisotropy of two austenitic steels. J. Iron Steel Inst. **202**, 616 (1964)

Part VI
ICME Success Stories and Applications

Multiscale, Coupled Chemo-mechanical Modeling of Bainitic Transformation During Press Hardening

Ulrich Prahl, Mingxuan Lin, Marc Weikamp, Claas Hueter, Diego Schicchi, Martin Hunkel and Robert Spatschek

Abstract We present our recent developments in multiscale and multiphysics modeling of bainite formation under mechanical loads in the press hardening process. The full field description of the bainitic microstructure by a multi-phase field model is linked to a crystal plasticity model at micrometer scale to simulate the mutual interaction between the phase transformation and plastic accommodation of the lattice distortion. Homogenized parameters of the phase fields are then coupled with a finite element model at a larger length scale to facilitate the calculation of transformation plasticity. The crystallographic features, transformation strain and texture of the simulated phase structure are discussed with comparison with experimental observations.

Keywords Phase field · Crystal plasticity · Bainite · Press hardening

Introduction

Press hardening offers high strength components of light-weight vehicles. In conventional processes, the martensitic microstructure is obtained through in-tool quenching of austenite. With pre-heated tools, this process can produce bainitic microstructure with improved strength ductility balance [2, 11, 17]. Modeling the phase transformation during this process is a challenging problem because of two reasons. First, there is a strong coupling between mechanical load, internal stresses, chemical composition and phase transformation kinetics. Due to the displacive nature of the bainitic transformation, a large eigenstrain is expected for the bainitic ferrite,

---

U. Prahl (✉) · M. Lin
Department of Ferrous Metallurgy, RWTH Aachen University, Aachen, Germany
e-mail: ulrich.prahl@iehk.rwth-aachen.de

M. Weikamp · C. Hueter · R. Spatschek
Forschungszentrum Juelich, Jüelich, Germany

D. Schicchi · M. Hunkel
Stiftung Institut Fuer Werkstofftechnik, Bremen, Germany

© The Minerals, Metals & Materials Society 2017
P. Mason et al. (eds.), *Proceedings of the 4th World Congress on Integrated Computational Materials Engineering (ICME 2017)*,
The Minerals, Metals & Materials Series, DOI 10.1007/978-3-319-57864-4_31

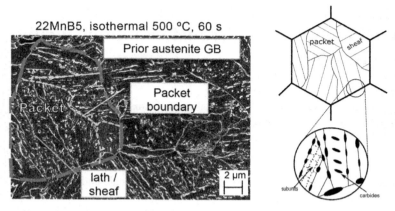

Fig. 1 Microstructure of bainite formed isothermally under 500 °C

which results in elastic/plastic deformation in the microstructure. Second, Bainite is a hierarchical structure spanning various length scales [8]. We model the coupled thermo-chemical and thermo-mechanical processes under various length scales. At the mesoscale, the anisotropic migration of phase boundaries is modeled by a multi-phase field (MPF) method coupled with elastic and crystal plasticity (CP) models. The activation of different slip systems in both BCC bainitic ferrite and FCC austenite and the rotation of local orientation are emphasized in the crystal plasticity model. At the macroscale, the transformation plasticity is modeled by a finite element method. The evolution of bainite at each FE integration point is described by the phase field model from the lower length scale. Finally numerical results are compared with tests on sheet specimens.

The morphology and transformation kinetics of bainite are distinct depending on the C and Si content [3]. In this paper our scope is limited to the bainites with low to medium C. High C steels are rarely used for press hardening in automobile manufacturing due to their poor weldability. Figure 1 shows a bainitic microstructure of 22MnB5 after 60 s isothermal transformation at 500 °C.

Plastic Deformation Around a Sub-unit Array

One of the most significant factors that influence the motion of the $\gamma-\alpha$ phase boundary is the plastic deformation in both the BCC and FCC crystals. At the transformation temperature of low or medium carbon bainite, dislocation slip is the dominant plastic deformation mechanism. In this session we use a coupled crystal-plasticity and phase-field model to capture the interaction between plastic deformation and phase transformation.

The phase field evolution equation reads

$$\dot{\phi}_i = \sum_{j \neq i} m \left\{ \kappa \left[\phi_i \nabla^2 \phi_j - \phi_j \nabla^2 \phi_i + \frac{\pi^2}{2\eta^2}(\phi_i - \phi_j) \right] - \frac{\pi}{\eta} \sqrt{\phi_i \phi_j} \Delta g_{ij} \right\} \quad (1)$$

where m, κ and η are the interface mobility, interface tension and interface width, respectively. We only consider the γ–α interface so they are constants. The driving force term Δg_{ij} is given by

$$\Delta g_{ij} = \frac{\Delta G_{ij}^m}{V^m} - \frac{1}{2} S : (F_{ij}^{*T} F_{ij}^* - I) \quad (2)$$

in which ΔG^m is the molar chemical driving force and V^m is the molar volume. The deformation gradient tensor $F_{ij}^* = F_i^* \mathrm{inv}(F_j^*)$ is the relative eigenstrain which contains the lattice mismatch and crystal orientation. The stress field S is coupled with the crystal plasticity model via the Hooke's equation,

$$S = \frac{1}{2} \mathbb{C} : (F_e^T F_e - I) \quad (3)$$

$$F_e = F F_p^{-1} \left(\sum_i F_i^* \phi_i \right)^{-1} \quad (4)$$

The subscripts e and p denote elastic and plastic, respectively. The plastic deformation gradient F_p can be decomposed into the contribution of 12 slip systems

$$\dot{F}_p = \left(\sum_{\alpha=1}^{12} \dot{\gamma}_\alpha(\tau_\alpha) M_\alpha^{\mathrm{Schmid}} \right) F_p \quad (5)$$

$$\tau_\alpha = S : M_\alpha^{\mathrm{Schmid}} \quad (6)$$

where $M_\alpha^{\mathrm{Schmid}}$ is the Schmid tensor, defined as the tensor product of the Burger's vector and the norm of slip plane ($\vec{b}_\alpha \otimes \vec{n}_\alpha$). The evolution of the shear strain γ_α of each slip system is given by the phenomenological equations of Roters et al. [13].

$$\dot{\gamma}_\alpha = \dot{\gamma}_0 \left| \frac{\tau_\alpha}{\tau_\alpha^{\mathrm{CRSS}}} \right|^n \mathrm{sgn}(\tau_\alpha) \quad (7)$$

$$\dot{\tau}_\alpha^{\mathrm{CRSS}} = h_0 \left(1 - \frac{\tau_\alpha^{\mathrm{CRSS}}}{\tau_{\mathrm{sat}}} \right)^w \sum_\beta h_{\alpha\beta} \dot{\gamma}_\beta \quad (8)$$

where τ^{CRSS} is the critical resolved shear stress (CRSS).

The aforementioned crystal plasticity model has been implemented in an open source software package, DAMASK [12]. Its spectral solver [5, 6] is used to solve the static stress equilibrium equation

$$\nabla \cdot \mathbf{P} \equiv \nabla \cdot (\mathbf{FS}) = \vec{0} \tag{9}$$

in the Fourier space. By the nature of the spectral method, the solution $\mathbf{F}(\vec{x})$ is restricted to a periodic field. The mechanical loads from the upper length scale are applied as the zero-frequency component in the Fourier space (equivalent to the average of the field). The applied strain and stress tensors are called periodic boundary conditions (PBC). In the current simulation a small tensile strain is imposed along the X direction ($\dot{\bar{F}}_{11} = 0.003 \text{ s}^{-1}$) and the stress at Z direction is kept zero ($\bar{P}_{33} = 0$). We only consider the <110>{111} slip systems for the FCC phase and the <111>{110} slip systems for the BCC phase.

The phase field evolution Eq. (1) is solved with a finite difference solver implemented in OpenPhase [1]. The phase-, temperature- and composition fields are also subjected to the PBC. The temperature and composition fields only affect the chemical driving force in Eq. (2). They are not considered in the current analysis so a constant chemical driving force is assumed. The input parameters are summarized in Table 1. The orientation relationship and lattice difference between austenite and bainitic ferrite are taken from the measurement of [9]. The eigenstrain can be inter-

Table 1 Model parameters for the austenite to ferrite transformation

Symbol	Description	Value for ferrite	Value for austenite
$\varphi_1, \Phi, \varphi_2$	Euler angles (degree)	0.366, 8.318, 6.756	−45, 0, 0
$\dot{\gamma}_0$	Referential glide velocity	0.001	0.001
$\tau^{CRSS}(t=0)$	Initial CRSS	55 MPa	60 MPa
τ_{sat}	Saturation CRSS	205 MPa	210 MPa
h_0	Hardening parameter	312 MPa	150 MPa
h_{ij}	Cross hardening (same slip plane)	1	1
h_{ij}	Cross hardening (other slip plane)	1.4	1.4
n	Exponent resolved shear stress	20	20
w	Exponent hardening	1.0	2.25
$\mathbb{C}_{11}$	Elastic constant	223.3 GPa	185.0 GPa
$\mathbb{C}_{12}$	Elastic constant	135.5 GPa	104.4 GPa
$\mathbb{C}_{44}$	Elastic constant	118.0 GPa	49.0 GPa
G_i	Bulk free energy	−56.4 J/cm^3	0
F^*	Eigenstrain tensor	$\begin{pmatrix} 1.1173 & -0.0331 & 0.0973 \\ 0.0176 & 1.1189 & 0.0893 \\ -0.1404 & -0.1232 & 0.7853 \end{pmatrix}$	

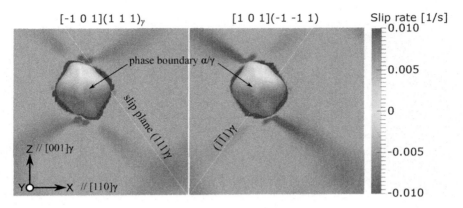

Fig. 2 Shear strain rate of two of the active slip systems in the austenite (γ) matrix (figure reproduced from [10]). The square is the cross section of the simulation domain cutting through the center of the bainitic subunit. Common crystallographic notation is used: The vector in *square bracket* represents the slip direction (parallel to Burger's vector) and the *round bracket* represents the slip plane

preted as contraction along the $[100]_\gamma$ crystallographic direction and expansion along the other two axes [4]. In the initial configuration the $[100]_\gamma$ is aligned parallel to the z-axis. The CRSS values are fitted from room- and high temperature tensile curves. The chemical driving force is estimated from the TCFE steel database of Thermo-Calc.

The growth of a single orientation variant of bainite has been simulated. An aggregate of bainitic ferrite grains (sub-units) with identical orientation is called a sheaf which typically has a plate- or lath-like shape. We consider the sheaf as an array of small ferritic grains and simulate their growth. With the periodic BC, the problem can be simplified to a single grain. We use a cubic simulation domain with edge length 6.4 μm and a regular mesh of $64 \times 64 \times 64$. Initially a spherical nucleus of ferrite (α) of 0.3 μm radius is placed in the matrix of austenite (γ). The initial stress equilibrium is obtained by gradually increasing ϕ_α from 0 to 1 inside the nucleus. The results after the growth of 1 s is presented in the following figures. The shape of the subunit and the shear rate of the active FCC slip systems in the austenite matrix are shown in Fig. 2. The subunits preferentially grow at a small angle to the (111) plane of the FCC crystal. It qualitatively reproduces a well known phenomenon observed in bainite [3], that is, the habit plane of the subunit forms a small angle with one of the 111 planes of austenite. Common habit planes of bainite are {223} and {557}. The slip bands in the matrix are affected by the neighboring subunits which lie with an equal distance at the $\pm X$, $\pm Y$ and $\pm Z$ directions. The stacking directions of the subunit arrays are imposed by the periodic BC and therefore the interaction between neighbors does not capture the reality. Inside the bainitic ferrite, only 8 of the 12 BCC slip systems are active. Figure 3 shows the shear rate of these 8 slip systems

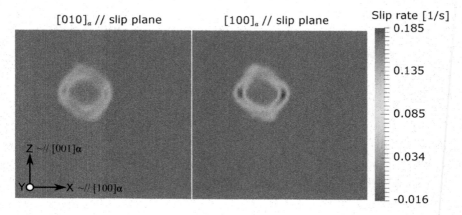

Fig. 3 Shear strain rate of two groups of active slip systems in the bainitic ferrite (α) subunit (figure reproduced from [10]). The displayed values are average of 4 slip systems of a Bain group (the slip planes within one Bain group are parallel to one of the <001> directions of the BCC crystal)

grouped according to their relation with the <100> directions. The distribution of slip rate inside the BCC crystal can account for the in-sheaf misorientation (deviation to the Kurdiumov-Sachs or Nishiyama-Wasserman OR) and the morphology of the subunit.

Variant Selection and Transformation Plasticity

Homogenized results from the mesoscopic phase field (PF) simulations shall be efficiently linked to the material description of a macroscopic finite element model (FEM) in the respect of transformation plasticity. In this session, isothermal bainitic transformation under constant tensile stress is simulated using a previously proposed PF-FE model [15], where the transformation plasticity at each integration point (IP) is evaluated from an individual phase-field simulation with a mesoscopic representative volume element (RVE). As shown in Eqs. (1) and (2), the phase field evolution depends on the local temperature and stress, which is incrementally taken from the thermal- and mechanical states of the IP. Then the transformation strain is estimated from the overall microstructure parameters of the phase fields. Thermal expansion and normal plasticity are described by classical mean field models in our previous work [8, 15].

The width of a bainitic sheaf or subunit is smaller than the austenite grain size by 1 order of magnitude. Thus, it is difficult to simulate a bainitic RVE while keeping the spacial resolution of single sheaf/subunit. The sheaves with similar orientation are aggregated into packets, whose size is comparable to the austenite grain size. Therefore, we can use a single phase field ϕ to represent one packet and only simulate the motion of packet boundaries in the phase field model. By neglecting the small angle grain boundaries inside the packet, we can construct a bainite RVE which contains a sufficient number of prior austenite grains to properly capture the variant selection effect [3, 9]. The RVE can be used to bridge a mesoscopic phase-field model and a macroscopic finite-element model.

The initial austenite grain structure is constructed in 2 steps. Firstly, a voxel-based tessellation technique implemented in DREAM3D [14] is used. Random orientations are assigned to each grain. Subsequently, phase field profile across grain boundaries is stabilized by a virtual relaxation process with Eq. (1) ($\Delta g = 0$). RVE data are stored and transmitted in the Hierarchical Data Format (HDF) according to the ICME recommendations [16].

The transformation plasticity is related to the overall eigenstrain of variants, whose nucleation and growth show strong dependency on the stress state. In order to capture the variant selection effect, we use a stress-dependent nucleation probability p,

$$\boldsymbol{R}_{\alpha,k} = \boldsymbol{R}_\gamma \boldsymbol{R}_{\mathrm{CS},k} \boldsymbol{R}_{\mathrm{OR}} \tag{10}$$

$$p_k = p_0 T \tanh \left[\frac{\Delta G^{\mathrm{m}} + \boldsymbol{S} : (\boldsymbol{R}_{\alpha,k} \boldsymbol{\varepsilon}^* \boldsymbol{R}_{\alpha,k}^{-1})}{RT} \right] \tag{11}$$

where $\boldsymbol{\varepsilon}^*$ is the eigenstrain; $\boldsymbol{R}$ denotes the rotation matrix for austenite grain (γ) and bainite nucleus (α) of variant k. The orientation relationship between austenite and bainite, $\boldsymbol{R}_{\mathrm{OR}}$, is given by [9]. $\boldsymbol{R}_{\mathrm{CS},k}, (k = 1 \ldots 24)$ represents the 24 symmetric operations on cubic lattice.

Figure 4 shows the behavior of an RVE of 57 equiaxed austenite grains under various tensile stress. Constant force along the longitudinal direction is applied to the top nodes of the hexagonal FEM element and their displacement is plotted. The number of nucleation sites, packet boundary mobility and ground nucleation probability are

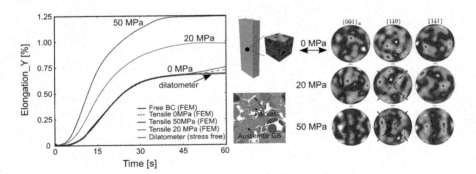

Fig. 4 Influence of applied static tensile stress on the transformation strain (*left*); calculated orientation distribution (texture) of bainite after complete transformation (*right*). The *dashed lines* indicate the peaks that agree with the experimental measurements of Hase et al. [7]

calibrated from dilatometer measurements (red curve in Fig. 4) on a tensile sample of 22MnB5 [15]. The influence of stress on the texture of bainitic ferrite is shown in the pole figures. It was reported [7] that the angle between stress axis and sheaf habit planes (close to {110} plane of BCC and {111} of FCC) is preferentially 45°, which is marked by dash lines in the {110} pole figures of the stress affected bainite.

Summary

A mutually interacted phase-field and crystal plasticity model and a coupled phase-field and finite-element model have been demonstrated. At the length scale of sub-unit (sheaf), the activity of slip systems, morphology and crystallographic features of the bainitic ferrite are qualitatively captured. At a larger length scale the overall transformation plasticity and texture can be predicted by the scale bridging PF-FE model.

Acknowledgements This work has been supported by the Deutsche Forschungsgemeinschaft (DFG) under the priority program SPP 1713, with project M7—"Modelling bainitic transformations during press hardening". The authors gratefully acknowledge the computing time granted by the JARA-HPC Vergabegremium and provided on the JARA-HPC Partition part of the supercomputer at RWTH Aachen University.

References

1. www.openphase.de
2. A. Abdollahpoor, X. Chen, M.P. Pereira, N. Xiao, B.F. Rolfe, Sensitivity of the final properties of tailored hot stamping components to the process and material parameters. J. Mater. Process. Technol. (2015)
3. H.K.D.H. Bhadeshia, *Bainite in steels: Transformations, microstructure and properties*, 2nd edn. (IOM Communications, London, 2001)
4. H.K.D.H. Bhadeshia, *Worked Examples in the Geometry of Crystals*, 2nd edn. (Institute of Metals, Brookfield, VT, USA, 2001)
5. M. Diehl, A spectral method using fast Fourier transform to solve elastoviscoplastic mechanical boundary value problems. Master's thesis, Technische Universitaet Muenchen, Munich, 2010
6. P. Eisenlohr, M. Diehl, R.A. Lebensohn, F. Roters, A spectral method solution to crystal elasto-viscoplasticity at finite strains. Int. J. Plast. **46**, 37–53 (2013)
7. K. Hase, C. Garcia-Mateo, H.K.D.H. Bhadeshia, Bainite formation influenced by large stress. Mater. Sci. Technol. **20**(12), 1499–1505 (2004)
8. C. Huter, M. Lin, D. Schicchi, M. Hunkel, U. Prahl, R. Spatschek, A multiscale perspective on the kinetics of solid state transformations with application to bainite formation. AIMS Mater. Sci. **2**(4), 319–345 (2015)
9. S. Kundu, K. Hase, H.K.D.H. Bhadeshia, Crystallographic texture of stress-affected bainite. Proc. R. Soc. A: Math. Phys. Eng. Sci. **463**(2085), 2309–2328 (2007)
10. M. Lin, U. Prahl, A parallelized model for coupled phase field and crystal plasticity simulation. Computer Methods in Materials Science. **16**(3), 156–162 (2016)

11. M. Naderi, M. Abbasi, A. Saeed-Akbari, Enhanced mechanical properties of a hot-stamped advanced high-strength steel via tempering treatment. Metall. Mater. Trans. A **44**(4), 1852–1861 (2013)
12. F. Roters, P. Eisenlohr, C. Kords, D.D. Tjahjanto, M. Diehl, D. Raabe, DAMASK: the Düsseldorf Advanced Material Simulation Kit for studying crystal plasticity using an FE based or a spectral numerical solver. Procedia IUTAM **3**, 3–10 (2012)
13. F. Roters, P. Eisenlohr, T.R. Bieler, D. Raabe, Crystal plasticity finite element methods, in *Materials Science and Engineering* (Wiley, Somerset, 2011)
14. D.M. Saylor, J. Fridy, B.S. El-Dasher, K.-Y. Jung, A.D. Rollett, Statistically representative three-dimensional microstructures based on orthogonal observation sections. Metall. Mater. Trans. A **35**(7), 1969–1979 (2004)
15. D.S. Schicchi, M. Lin, U. Prahl, M. Hunkel, A combined finite element—phase field model approach on the bainitic transformation, in *Proceedings of European Conference on Heat Treatment* (2016)
16. G.J. Schmitz, A. Engstrom, R. Bernhardt, U. Prahl, L. Adam, J. Seyfarth, M. Apel, C. Agelet de Saracibar, P. Korzhavyi, J. Ågren, B. Patzak, Software solutions for ICME. JOM **68**(1), 70–76 (2016)
17. A. Turetta, S. Bruschi, A. Ghiotti, Investigation of 22MnB5 formability in hot stamping operations. J. Mater. Process. Technol. **177**(1–3), 396–400 (2006)

Development of Microstructure-Based Multiscale Simulation Process for Hot Rolling of Duplex Stainless Steel

Mototeru Oba, Sukeharu Nomoto, Kazuki Mori and Akinori Yamanaka

Abstract Recent improvement of multi-phase field method enables us to simulate microstructure formed by various material processes and homogenization method attracts attention as the way of bridging microstructure and macro homogenized material properties. We have proposed microstructure-based multiscale simulation framework and it was applied to the simulation of hot rolling process of duplex stainless steel. In the framework various commercial software, not only multi-phase field method and homogenizaiton method but also nanoscale molecular dynamics simulation and finite element method was bridged. Multi-phase field method coupled with CALPHAD database was used to simulate microstructure evolution by columnar and equiaxed solidifications during continuous casting. Elastic property for the constituent phases in the duplex stainless steel was calculated by molecular dynamics simulation and first principles calculation. Plastic property was obtained by nano-indentation tests. Homogenization calculation gave macro elastic property from microstructure and property of each phase and virtual material test performed by finite element method served homogenized plastic property. With the material properties hot rolling process was simulated by dynamic explicit simulation of finite element method. Recrystallization by hot rolling process was performed by multi-phase field method. In this paper, the results are discussed to reveal the usefulness and problem for performing microstructure based multiscale analysis. Further discussion is given for the framework here: the method for obtaining material property of each micro phase, anisotropy of homogenized elastic constants, three-dimensional recrystallization calculation. Through these discussions, our simulation framework becomes more reliable.

M. Oba (✉) · S. Nomoto · K. Mori
ITOCHU Techno-Solutions Corporation, 3-2-5, Kasumigaseki, Chiyoda-ku, Tokyo 100-6080, Japan
e-mail: mototeru.ooba@ctc-g.co.jp

A. Yamanaka
Division of Advanced Mechanical Systems Engineering, Institute of Engineering, Tokyo University of Agriculture and Technology, 2-24-16, Naka-cho, Koganei-shi, Tokyo 184-8588, Japan

Keywords Multi-phase field method · Multiscale analysis · Homogenization method · Molecular dynamics · Finite element method

Introduction

Over several decades since finite element method started to be applied to practical analyses, material properties had been obtained by experimental tests. In this case, microstructure of a metal was ignored and simple homogenized material model was assumed. Its parameters, for example, young's modulus and Poisson's ratio for elastic property and yield stress and strain hardening exponent for plastic property, were identified with the various experimental tests. On the other hand, asymptotic homogenization method was studied to identify the parameters of the homogenized macro material from its microstructure numerically [1]. The method could be applied to composite materials whose microstructure was known. But idealized virtual microstructure was assumed for the study of the homogenized method because of no method for obtaining the microstructures of practical metals [2]. Thus it is hard to obtain material properties of metals based on these microstructure.

However, recent improvement of multi-phase field (MPF) method [3] can simulate the process to generate the microstructure. With the MPF method and the homogenization method, predicting material properties of metals based on its

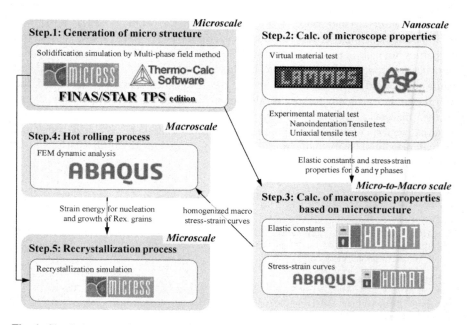

Fig. 1 Simulation procedure and software used in this procedure [6–10, 14, 15]

microstructure becomes enabled. In our previous study [4], we proposed a multi-scale simulation framework shown in Fig. 1 and applied to hot rolling processes along with solidification during continuous casting of duplex stainless steel comprising two stable phases: δ-ferrite and γ-austenite phases [5]. We bridged various commercial simulation software for nano-, micro-, and macroscopic length scales and built the simulation process for evaluating the mechanical material properties based on its microstructures.

Concept of Bridging Multi-scale Software

In our previous study, we proposed a multiscale simulation framework using various commercial simulation software for nano-, micro-, and macroscopic length scales [4]. The concept of our multiscale and software-bridging simulation framework is illustrated in Fig. 1. The framework were applied to hot rolling processes along with solidification during continuous casting of duplex stainless steel comprising two stable phases: δ-ferrite and γ-austenite phases [5].

Small hot rolling experiment in laboratory was assumed. The dimension of a test piece is shown as Fig. 2. Solute elements of the stainless steel, chromium, nickel and carbon (Fe-18 wt%Cr-8 wt%Ni-0.08 wt%C) was selected and the average composition was defined as same as values of SUS304. During the hot rolling process, temperature of steel was assumed to be kept 900 °C.

As a first step, microstructure evaluations of solidification process were performed by using MPF method coupled with CALPHAD database. The columnar and equiaxed solidification of duplex stainless steel were simulated separately. We used MICRESS [6] and Thermo-Calc [7] for the microstructure evolution, and also used FINAS/STAR TPS Edition [7] for predicting cooling profile of each phase.

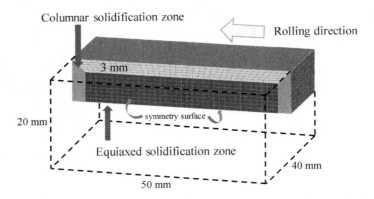

Fig. 2 Dimensions of slab specimen for small hot rolling experiment in laboratory

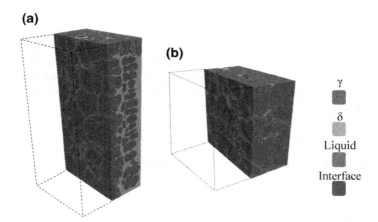

Fig. 3 Center cross section of columnar and equiaxed microstructure calculated by MPF method **a** columnar, **b** equiaxed

Center cross section of columnar and equiaxed microstructure calculated by MPF method is shown in Fig. 3.

The duplex stainless steel consists of δ-ferrite and γ-austenite phase, material properties of each phase were required to perform homogenization method. Nano-simulation was used for elastic properties. By using molecular dynamics software LAMMPS [9] and first principles software VASP [10], elastic constants at 0 K were calculated and according to the temperature dependency [11–13], elastic constants at 900 °C were calculated. The results are shown in Table 1. In order to obtain plastic stress-strain curves a nano-indentation test was performed. Single indenter with Berkovich triangular pyramid indenter was used and parameters were identified according to the reverse algorithm by Dao et al. [16]. We obtain yield stress and strength hardening exponent of δ-ferrite phase at room temperature. By using experimental data of stress–strain curve of SUS304 stainless steel at both room temperature and 900 °C [17], we identified stress-strain curves of δ-ferrite and γ-austenite at 900 °C.

With the properties of each phase, homogenization method and virtual tensile test were applied to the microstructure to obtain homogenized macro material properties, Homogenization software HOMAT [14] and non-linear FEM software ABAQUS/Standard [15] were used. Homogenized elastic constants for columnar and equiaxed microstructure is shown in Table 2.

Hot rolling simulation was performed by ABAQUS/Explicit solver. Hot rolling reduction is 40%. Obtained macroscopic stress-strain property values in previous

Table 1 Elastic constants of each phase at 900 °C

(GPa)	C11	C12	C44
γ-austenite	171.4	107.4	95.1
δ-ferrite	180.3	104.3	94.3

Table 2 Homogenized elastic constants for columnar and equiaxed microstructure

(GPa)	C11	C22	C33	C12	C13	C23	C44	C55	C66
Columnar	205.3	205.3	172.4	74.1	107.1	107.1	53.6	95.1	95.1
Equiaxed	219.6	211.2	205.6	80.6	86.2	94.7	60.3	66.1	70.8

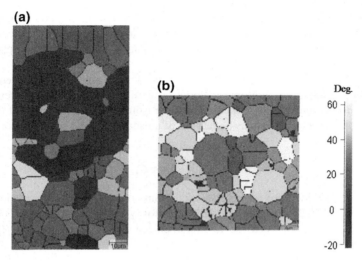

Fig. 4 Recrystallized structures by adding strain energy density generated in hot rolling process **a** columnar, **b** equiaxed

step were used. After hot rolling, maximum plastic strain energy density value of 195.50 MJ/mm^3 for surface columnar area and 116.73 MJ/mm^3 for inner equiaxed area were obtained.

The last step was recrystallization process. Two-dimensional calculation is performed for vertical center-cross-section using MICRESS. It is assumed that recrystallization time is the same as roller reduction time and before starting recrystallization, residual δ phase was transformed to γ phase. Temporal crystalline direction distributions of the results are shown as Fig. 4. Grain growth competition with unified crystal direction can be confirmed. Furthermore, this unidirectional growth competition in the columnar case seemed to be stronger than equiaxed one. This is reasonable from experimental knowledge.

Discussion

In order to identify plastic material property from nano-indentation experiment result, the reverse algorithm by Dao et al. was used. The algorithm can be covered not only plastic property but also elastic property, but Young's modulus calculated by the algorithm was much different from our nano-simulation results. To avoid the sensitivity Young's modulus was given by the nano-simulation result and applied the algorithm for only plastic property. This problem in Dao's algorithm has already known: the algorithm is too sensitive with small experimental errors [18, 19]. In the previous study appropriate parameters seemed to be identified but the error might be large because of the sensitivity. As a solution of the sensitivity dual nano-indentation method was proposed [18, 19]. Though the method requires two experiment results of indenters with different apex angles, it can be identified plastic property robustly. In the study dual nano-indentation method will be applied to identify the plastic property of each phase.

The obtained homogenized elastic constants in Table 2 for the equiaxed microstructure are the orthotropic material properties. However if the microstructure was truly equiaxed, the property had to be isotropic. Bhattacharyya et al. used the following simple formula as a measure of anisotropy in the plane of axes 1 and 2 [20]:

$$\alpha = 2C_{44} - C_{11} + C_{12} \quad (1)$$

Applying this measure to the elastic constants of each single phase showed in Table 1, $\alpha_\gamma = 126.2$ GPa can be calculated for γ-austenite. $\alpha_\delta = 112.6$ GPa for δ-ferrite. To apply the measure to homogenized equiaxed property in Table 2, components are simply averaged as follows:

$$\begin{cases} \overline{C}_{11} = \frac{1}{3}(C_{11} + C_{22} + C_{33}) = 212.1 \text{ GPa} \\ \overline{C}_{12} = \frac{1}{3}(C_{12} + C_{13} + C_{23}) = 87.2 \text{ GPa} \\ \overline{C}_{44} = \frac{1}{3}(C_{44} + C_{55} + C_{66}) = 65.7 \text{ GPa} \end{cases} \quad (2)$$

Then the measure can be applied:

$$\alpha_e = 2\overline{C}_{44} - C_{11} + C_{12} = 6.5 \text{ GPa} \quad (3)$$

It means anisotropy measure of homogenized equiaxed microstructure α_e is much smaller than that of constituent phases α_γ and α_δ. The result is natural because equiaxed microstructure should be isotropic. However the result also is not natural because elastic constants in Table 2 looks far from isotropic: for example, C_{11} value of 219.6 GPa and C_{33} value of 205.6 GPa are much different. To compare the anisotropy before/after homogenization, building the measure of anisotropy for general elastic constants are required.

As already shown in Fig. 4, the recrystallization calculations in previous study were performed in two-dimensional space by cutting the vertical center-cross-section out of three-dimensional solidification simulation results. Because of the issues of uncertain interfacial mobility and anisotropic models in multi-phase junction energy [21], we selected two-dimensional analysis even if the solidification simulations were executed in three dimensions. However, the recrystallization calculation should be simulated for whole three-dimensional solidification simulation results, not two-dimensional cutting cross-section. Three-dimensional recrystallization calculation will be challenged.

Conclusion

We discussed some topics for the multiscale simulation framework using various simulation software. In order to identify plastic material property, nano-indentation test by single indenter have problem of low stability for experiment error. For our study, dual nano-indentation experiments is better for our study. The obtained homogenized elastic constants for equiaxed microstructure expected to be isotropic, but it remains anisotropy of constitutive phases. The simple measure of anisotropy by Bhattacharyya et al. was not able to detect the anisotropy and more general method to measure anisotropy is required. Recrystallization process was simulated using two-dimensional cross-section of three-dimensional solidification simulation results. Though three-dimensional simulation has some problems, three-dimensional recrystallization simulation will be challenged.

References

1. G. Laschet, Homogenization of the thermal properties of transpiration cooled multi-layer plates. Comput. Method Appl. M. **191**(41–42), 4535–4554 (2002)
2. I. Watanabe, K. Terada, M. Akiyama, Two-scale analysis for deformation-induced anisotropy of polycrystalline metals. Comput. Mater. Sci. **32**(2), 240–250 (2005)
3. I. Steinbach et al., A phase field concept for multiphase systems. Phys. D **94**, 135–147 (1996)
4. S. Nomoto et al., Microstructure-based multiscale analysis of hot rolling of duplex stainless steel using various simulation software. Integr. Mater. Manuf. Innov. **6**, 69–82 (2017)
5. G. Laschet et al., Asymptotic homogenization of 3-D microstructures simulated either by the multi-phase field method or by cellular automata, in *1st International Workshop on Software Solutions for ICME* (2014)
6. MICRESS, http://web.micress.de/
7. Thermo-Calc Software, http://www.thermocalc.com/
8. FINAS/STAR TPS Edition, http://www.engineering-eye.com/FINAS_TPS/
9. LAMMPS Molecular Dynamics Simulator, http://lammps.sandia.gov/
10. VASP, https://www.vasp.at/
11. A.R. Wazzan et al., Temperature dependence of the singlecrystal elastic constants of Corich Co-Fe alloys. J. Appl. Phys. **44**, 2018–2024 (1973)
12. X. Sha, R.E. Cohen, First-principles themoelasticity of bcc iron under pressure. Phys. Rev. B **74**, 214111 (2006)
13. P. Renaud, S.G. Steinemann, High temperature elastic constants of fcc Fe-Ni invar alloys. Phys. B **161**, 75–78 (1990)
14. HOMAT, http://web.micress.de/
15. Abaqus Unified FEA, http://www.3ds.com/products-services/simulia/products/abaqus/
16. M. Dao et al., Computational modeling of the forward and reverse problems in instrumented sharp indentation. Acta Mater. **49**, 3899–3918 (2001)

17. Y. Tsuchida et al., High and ultra-high temperature tensile tests on SUS304, SUS321 and 2.25Cr-1Mo steels. *JAEA report*, PNC-TN941 (1985), pp. 85–128
18. J.L. Bucaille et al., Determination of plastic properties of metals by instrumented indentation using different sharp indenters. Acta Mater. **51**, 1663–1678 (2003)
19. N. Chollacoop, M. Dao, S. Suresh, Depth-sensing instrumented indentation with dual sharp indenters. Acta Mater. **51**, 3713–3729 (2003)
20. S. Bhattacharyya et al., A phase-field model of stress effect on grain boundary migration. Modell. Simul. Mater. Sci. Eng. **19**, 035002 (2011)
21. E. Miyoshi, T. Takaki, Extended higher-order multi-phase-field model for three-dimensional anisotropic-grain-growth simulations. Comput. Mater. Sci. **120**, 77–83 (2016)

A Decision-Based Design Method to Explore the Solution Space for Microstructure After Cooling Stage to Realize the End Mechanical Properties of Hot Rolled Product

Anand Balu Nellippallil, Vignesh Rangaraj, Janet K. Allen, Farrokh Mistree, B.P. Gautham and Amarendra K. Singh

Abstract Manufacturing a product involves a host of unit operations and the end properties of the manufactured product depends on the processing steps carried out in each of these unit operations. In order to couple the material processing-structure-property-performance spaces, both systems-based materials design and multiscale modeling of unit operations are required followed by integration of these models at different length scales (vertical integration). This facilitates the flow of information from one unit operation to another thereby establishing the integration of manufacturing processes to realize the end product (horizontal integration). In this paper, we present a goal-oriented inverse, decision-based design method using the compromise Decision Support Problem construct to achieve the vertical and horizontal integration of models by identifying the design set points for hot rod rolling process chain. We illustrate the efficacy of the method by exploring the design space for the microstructure after cooling stage that satisfies the requirements identified for the end mechanical properties of a hot rolled product. Specific requirements like managing the banded microstructure to avoid distortion in forged gear blanks are considered for the problem. The method is goal-oriented as the design solutions after exploration of microstructure space are passed in an inverse manner to cooling and rolling stages to identify the design set points in order to

---

A.B. Nellippallil · V. Rangaraj · J.K. Allen (✉) · F. Mistree
Systems Realization Laboratory @ OU, University of Oklahoma,
Norman, OK 73019, USA
e-mail: janet.allen@ou.edu

B.P. Gautham
Tata Consultancy Services, Pune 411013, India

A.K. Singh
Department of Materials Science and Engineering, Indian Institute of Technology Kanpur, Kanpur 208016, India

realize the end product. The method is generic and has the potential to be used for exploring the design space of manufacturing stages that are connected to achieve the integrated decision-based design of the product and processes.

Keywords Decision-based design · Microstructure-property correlations · Hot rod rolling and cooling

Frame of Reference

The widespread popularity of steel as an engineering material in manufacturing industries is due to the fact that diverse range sets of mechanical properties and microstructures are possible by carefully managing the materials processing resulting in improved performances of products. The defining players for the properties of a steel product that is rolled are the chemical composition of material, the deformation history during the rolling process and the thermal history during subsequent cooling operation. Large number of plant trials are needed to produce a new grade of steel product mix having specific target properties and performances. In plant set-up, these trials are usually expensive and time consuming. The alternative is to exploit the advancements in computational modeling tools and frameworks to carry out simulation-based design exploration of different manufacturing processes involved in order to identify ranged set of solutions that satisfies the requirements identified for the process as well as the end product.

In the model based realization of complex systems, we have to deal with models that are typically incomplete, inaccurate and not of equal fidelity. We believe that the fundamental role of a human designer is to make decisions given the uncertainties associated with the system [1]. Thus, we try to find robust *satisficing* solutions that are relatively insensitive to change rather than optimum solutions that perform poorly when the conditions are changed. The compromise Decision Support Problem (cDSP) construct is proposed by Mistree and co-authors for robust design under multiple goals [2]. Using the cDSP, several solutions are identified which are further explored to identify solutions that best satisfy specific requirements. In this paper, we present an inverse decision-based design method using the cDSP construct and solution space exploration to determine the set points of the hot rod rolling and cooling stages to realize the microstructure and mechanical properties of the end product. Allen and co-authors describe the foundational problem that we are addressing in [3]. Nellippallil and co-authors describe the goal-oriented inverse design method, the cDSP construct and illustrate the utility of the same for roll pass design in [4, 5]. Information on the mathematical models we use to achieve integration of different processes in hot rod rolling process chain and the framework that we use to formulate the cDSPs is presented by Nellippallil and co-authors in [6]. Hence, we are not providing the details of the method, the cDSP construct and the mathematical models used for the problem of hot rod rolling process chain in this paper.

In this paper, we explore the solution space for the microstructure after cooling stage that satisfies the goals identified for the mechanical properties of end product. We identify the influence of different fractions of ferritic and pearlitic microstructures on end mechanical properties like yield strength, tensile strength, toughness (impact transition temperature) and hardness. We demonstrate the efficacy of the method and solution space exploration by designing the microstructure after cooling to realize the end product mechanical properties. In Sect. 2, we describe the problem and the proposed goal-oriented inverse decision-based design method. In Sect. 3, we highlight the results obtained. We close the paper with our remarks in Sect. 4.

Problem Definition and Solution Strategy

There has been an increasing trend in developing algorithms for predicting the behavior of materials during complex manufacturing processes like the hot rod rolling, as the final properties of end steel product produced depends on its processing route [7–11]. One of the major issue during the hot rod rolling process is the segregation of alloying elements such as manganese (Mn) during the progress of solidification in casting and affects the entire downstream processing as well as the mechanical properties of the end product [12]. These segregates, known as microsegregates, are typically of the size of grains and are formed due to limited solid solubility of these solutes. During the hot rolling process, the concentration profile changes due to the deformation of these structures. During the subsequent cooling process, austenite to ferrite phase transformation occurs. Supposing the steel is of hypo eutectoid composition the ferrite phase will form in regions with low content of austenite stabilizing solute content and the rest of the phase will be pearlite. Thus due to the alternate layers of low and high solute regions induced during hot rolling, we will see a banded microstructure formation having both ferrite and pearlite [12]. These banded microstructures are a major factor for distortions in gear blanks after forging process. Thus, managing the factors associated with banding will indirectly affect the final mechanical properties of the product. To predict the final mechanical properties of the product as a function of the composition variables, rolling and cooling parameters, there is a need for series of modeling integrations, both vertical and horizontal. We define the integration of multiple length scale models within a process as vertical integration and the integration of the different stages or processes ensuring information flow as horizontal integration. More information on the specific problem addressed and the vertical and horizontal integration of models are provided in Ref. [6].

Goal-Oriented, Inverse Decision-Based Design Method

The goal-oriented, inverse decision-based design method proposed in this work will be explained using the information flow diagram shown in Fig. 1. The method is goal-oriented because we start with the end goals that needs to be realized for the product as well as process. The decisions that are taken for the end requirements of the product/process are then communicated to the stages that precedes to make logical decisions at those stages that satisfies the requirements identified thereby making it an inverse design process. Brief descriptions of the steps are provided below

Step 1: Establish forward modeling and information flow across models for the problem formulated. In Step 1, the designer makes sure that there is proper flow of information as models are connected across different stages (from rolling to cooling to end product mechanical properties). Mathematical models are either identified or developed in this step.

Step 2: In Step 2, a cDSP for the mechanical properties of the final end product is formulated using the models identified in Step 1. Information, requirements (manage banding) and the correlations of mechanical properties with microstructure after cooling stage (ferrite grain size, interlamellar spacing, phase fractions and

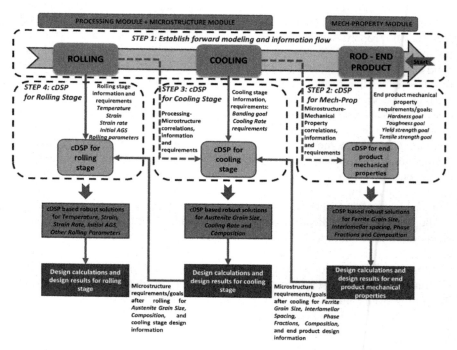

Fig. 1 Goal-oriented inverse decision-based design method. (More details in [6])

composition) are communicated to this cDSP formulated. For the hot rod rolling problem formulated, the end mechanical property goals and requirements for yield strength, tensile strength, hardness and toughness (impact transition temperature) are identified. On exercising the cDSP, the best combinations for ferrite grain size, phase fractions, interlamellar spacing and compositions that satisfy the requirements for properties are identified.

Step 3: In Step 3, a similar process followed in Step 2 is carried out to formulate the cDSP for cooling stage. This cDSP will have goals and requirements for ferrite grain size, phase fraction and composition that are based on the solutions obtained from first cDSP. Also, information from cooling stage like banding requirements, cooling rate requirements are included into the cDSP. The information, requirements and correlations of variables after the end of rolling (austenite grain size, composition) with the cooling stage parameters are communicated to the cDSP formulated. The goals for this cDSP are target ferrite grain size and target phase fractions subjected to constraints. On exercising the cDSP, the best combinations for austenite grain size, cooling rate, composition elements like carbon and manganese that satisfy the requirements are identified.

Step 4: In Step 4, we follow a similar procedure as in Steps 2 and 3 to formulate the cDSP for rolling considering the information generated from cooling side and the rolling stage information and requirements identified. This cDSP will have grain growth module, static, dynamic and meta dynamic recrystallization modules and hot deformation module. These modules may require individual cDSPs that needs to be integrated following a similar pattern.

In this paper, we illustrate the efficacy of the goal-oriented method by formulating the cDSP for the mechanical properties of end product (Steps 1 and 2). The details of the method, models used, response surface models developed and cDSP formulated for different stages is explained in Ref. [6].

Results and Discussion

In this paper, we address the following inverse problem: Given the end mechanical properties of a new steel product mix, what should be the microstructure after cooling that satisfies the requirements? We carry out exploration of microstructure solution space. In Table 1, we list the requirements for the end product as well as the requirements after the cooling stage. The end product mechanical properties and their target values/ranges are defined. The requirements from the cooling stage are to have a high ferrite fraction (≥ 0.8) and to achieve a minimum ferrite grain size after cooling. The ferrite fraction is defined as a goal in the cDSP to manage banded microstructure. A very high ferrite fraction denotes a less banded structure with pearlite phase. This is true in case of a very high pearlite fraction too as there will be less ferrite leading to less banded structure.

Table 1 Target values and design preferences for the requirements identified

Requirements/goals	Target ranges/values	Design preferences
Yield strength goal (YS)	220–400 MPa	Maximum possible
Tensile strength goal (TS)	500–780 MPa	Maximum possible
Ferrite fraction goal (X_f)	≥ 0.8 (Min Banded Microstructure)	Close to target
Impact transition temperature (ITT) requirement	−90 to −30 °C	Minimum possible
Hardness requirement (HV)	150–250	Maximum possible
Ferrite grain size requirement	5–10 μm	Minimum possible

Table 2 System variables and their ranges

System variables	Ranges defined
Ferrite grain size (FGS) (μm)	5–25
Ferrite fraction (X_f)	0.1–1.0
Pearlite interlamellar spacing (S_0) (μm)	0.15–0.25
Chemical composition of Silicon (%)	0.18–0.3
Chemical composition of Nitrogen (%)	0.007–0.009

In Table 2, we list the system variables and their corresponding ranges. More information on the dependence of the system variables on the final mechanical properties are available in [6].

The cDSP is exercised for different scenarios by assigning different weights to the goals associated. Ternary plots are created using the design and operating set points generated after exercising the cDSPs. We use these ternary plots to determine the appropriate weights for the goals and predict the required design set points, Figs. 2, 3, 4, 5 and 6. For Goal 1, the process designer is interested in maximizing the yield strength of the end product to a target value of 400 MPa. On analyzing Fig. 2, we see that the values in the dark red contour region demarcated by the blue dashed line achieve the maximum yield strength of around 329 MPa. For Goal 2, the process designer is interested in maximizing the tensile strength of the end product to a target value of 780 MPa. On analyzing Fig. 3, we see that the values in the dark red contour region identified by the green dashed lines achieves the maximum tensile strength of 759 MPa. For Goal 3, the process designer is interested to manage the banded microstructure by identifying high ferrite fraction regions. On analyzing Fig. 4, we see that region in the red contour identified by the violet dashed line has ferrite fraction from 0.7 to 0.99609 with maximum being at the same region where yield strength is seen to have the highest value. Similarly, we see in Fig. 4 that the blue region identified with the violet dashed lines has the

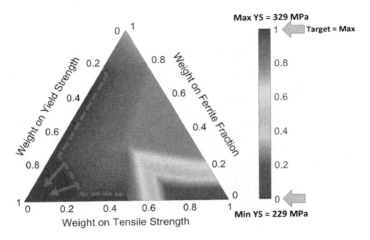

Fig. 2 Ternary plot—yield strength

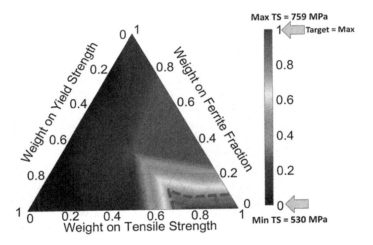

Fig. 3 Ternary plot—tensile strength

lowest ferrite fraction (0.3 to 0.100049) leading to a high pearlite fraction. This region corresponds with the region where tensile strength has the maximum value. The region between these two dashed violet lines has the highest banded microstructure of ferrite and pearlite. Thus, it is clear from the ternary analysis that the ferrite fraction plays a major role in defining the yield strength and tensile strength of the end product. A high ferrite fraction improves the yield strength of the product while compromising the tensile strength and a high pearlite fraction improves the tensile strength of the product while compromising the yield strength of the product. Another requirement identified is to have a minimum impact transition temperature (ITT) for the product. We plot the solution space for ITT in

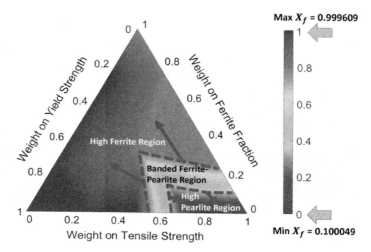

Fig. 4 Ternary plot—ferrite fraction

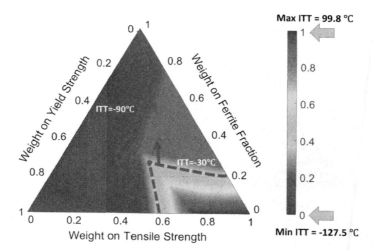

Fig. 5 Ternary plot—ITT solution space

Fig. 5. We see from Fig. 5 that the ITT drops in those regions with high ferrite fraction. The target regions for ITT of −30 and −90 °C are identified by the red dashed lines in Fig. 5.

Now, since the designer's interest is in identifying regions that satisfy all the conflicting requirements, there is a need to visualize all the design spaces in one ternary plot. Therefore, we plot the superimposed ternary plot shown in Fig. 6. If there is a common region that satisfies all the requirements identified, then we select

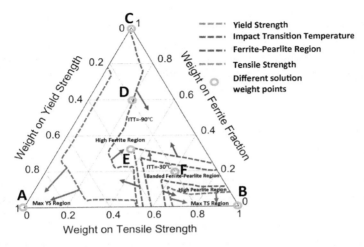

Fig. 6 Ternary plot—superimposed ternary plot (color online)

Table 3 Solution points for microstructure after cooling and end mechanical properties

Sol. Pts	FGS (μm)	X_f	S_0 (μm)	%Si	%N	YS (MPa)	TS (MPa)	ITT (°C)	HV
A	5	0.996	0.15	0.3	0.009	328.7	541.8	−120.7	166.1
B	5	0.1	0.15	0.299	0.0089	229	759.35	99.81	242.65
C	5	0.8	0.15	0.18	0.007	306.9	589.05	−72.5	182.7
D	5	0.87	0.15	0.299	0.0089	314.76	572.17	−90	176.76
E	5	0.7997	0.15	0.299	0.0089	306.9	589.05	−72.5	182.7
F	5	0.55	0.15	0.299	0.0089	279.13	649.46	−10.95	203.97

solutions from that region. If not, we identify compromised solutions that satisfy our requirements to the best possible. From the superimposed ternary plot, we are analyzing 6 solution points A, B, C, D, E and F in Table 3.

On analyzing the solution points in Table 3, we see that ferrite grain size (FGS) and pearlite interlamellar spacing (S_0) is low for all the solution points. Thus a smaller ferrite grain size and smaller interlamellar spacing is preferred to enhance the end mechanical properties of the product. This is important information that must be communicated as the goal for the preceding rolling and cooling stages that produces the end product. Solution point A with the highest ferrite fraction has the highest YS and lowest ITT while achieving a TS and HV that is acceptable. Solution point B with highest the pearlite fraction has the highest TS and HV while falling short in YS and ITT leading to rejection of the point. Solution points C, D, E with ferrite fractions around 0.8 and above achieve acceptable targets for YS, TS, ITT and HV. Solution point F with a ferrite fraction of 0.55 achieves better TS and HV than points C, D and E but the values drops for YS and ITT respectively. This point is rejected due to the highly banded microstructure generated. Based on this analysis, we pick solution point D that generates the best combination of values of the system variables that satisfies the requirements of YS, TS, ITT and HV. The goal of the designer next is to formulate the cDSP for cooling and rolling stages to identify design set points for cooling that will generate the identified microstructure space solution point D by means of the proposed goal-oriented inverse design method (Steps 3 and 4 of method).

Closing Remarks

In this paper, we demonstrate the utility of the proposed goal-oriented inverse decision-based design method by exploring the solution space for the microstructure after cooling stage that satisfies the requirements for the end mechanical properties of the product. We study the effect of ferrite fraction, ferrite grain size, pearlite interlamellar spacing and composition in defining the mechanical properties like yield strength, tensile strength, toughness and hardness. We illustrate the efficacy of ternary plots to explore microstructure space solutions that satisfies the conflicting mechanical property goals in the best possible manner by carrying out design trade-offs. The results for microstructure space obtained will be studied further to design (identify design and operating set points) preceding manufacturing stages like rolling and cooling by following the proposed design method in order to realize the end product. The proposed inverse decision-based design method is generic and supports the integrated decision-based design exploration of manufacturing stages that are connected. The primary advantage of the proposed method is to enable a process designer to rapidly explore the design space for manufacturing processes using simulation models and reduce the need for expensive plant trials resulting cost involved in the production of a new grade of product mix.

Acknowledgements The authors thank TRDDC, Tata Consultancy Services, Pune for supporting this work (Grant No. 105-373200). J.K. Allen and F. Mistree gratefully acknowledge financial support from the NSF Grant CMMI 1258439 and the L.A. Comp and John and Mary Moore Chairs at the University of Oklahoma.

References

1. D.L. McDowell, J. Panchal, H.-J. Choi, C. Seepersad, J. K. Allen, F. Mistree, *Integrated Design of Multiscale, Multifunctional Materials and Products* (Elsevier, New York, 2010)
2. F. Mistree, O.F. Hughes, B. Bras, Compromise decision support problem and the adaptive linear programming algorithm. Prog. Astronaut. Aeronaut. **150**, 251 (1993)
3. J.K. Allen, F. Mistree, J.P. Panchal, B.P. Gautham, A.K. Singh, S. Reddy, P. Kumar, Integrated realization of engineered materials and products: a foundational problem, in *Proceedings of the 2nd World Congress on Integrated Computational Materials Engineering* (The Minerals, Metals & Materials Society, Warrendale, PA, 2013) pp. 279–284

4. A.B. Nellippallil, K.N. Song, C.-H. Goh, P. Zagade, B.P. Gautham, J.K. Allen, F. Mistree, A goal-oriented, sequential, inverse design method for the horizontal integration of a multi-stage hot rod rolling system. J. Mech. Des. **139**(3), 031403–031403-16. doi:10.1115/1.4035555
5. A.B. Nellippallil, K.N. Song, C.-H. Goh, P. Zagade, B.P. Gautham, J.K. Allen, F. Mistree, A goal oriented, sequential process design of a multi-stage hot rod rolling system, in *ASME Design Automation Conference*, Paper No. DETC2016-59402 (2016)
6. A.B. Nellippallil, V. Rangaraj, B.P. Gautham, A.K. Singh, J.K. Allen, F. Mistree, A goal-oriented, inverse decision-based design method to achieve the vertical and horizontal integration of models in a hot-rod rolling process chain, in *ASME Design Automation Conference*, Cleveland, Ohio, USA (Accepted) (2017)
7. S. Phadke, P. Pauskar, R. Shivpuri, Computational modeling of phase transformations and mechanical properties during the cooling of hot rolled rod. J. Mater. Process. Technol. **150**(1), 107–115 (2004)
8. P. Hodgson, R. Gibbs, A mathematical model to predict the mechanical properties of hot rolled C-Mn and microalloyed steels. ISIJ Int. **32**(12), 1329–1338 (1992)
9. J. Majta, R. Kuziak, M. Pietrzyk, H. Krzton, Use of the computer simulation to predict mechanical properties of C-Mn steel, after thermomechanical processing. J. Mater. Process. Technol. **60**(1–4), 581–588 (1996)
10. R. Kuziak, Y.W. Cheng, M. Glowacki, M. Pietrzyk, Modeling of the microstructure and mechanical properties of steels during thermomechanical processing. NIST Tech. Note (USA) **1393**, 72 (1997)
11. A.B. Nellippallil, A. Gupta, S. Goyal, A.K. Singh, Hot rolling of a non-heat treatable aluminum alloy: thermo-mechanical and microstructure evolution model. Trans. Indian Inst. Metals 1–12 (2016). doi:10.1007/s12666-016-0935-3
12. E. Jägle, Modelling of microstructural banding during transformations in steel. Master of Philosophy thesis (Department of Materials Science & Metallurgy, University of Cambridge, 2007)

Influence of Computational Grid and Deposit Volume on Residual Stress and Distortion Prediction Accuracy for Additive Manufacturing Modeling

O. Desmaison, P.-A. Pires, G. Levesque, A. Peralta, S. Sundarraj, A. Makinde, V. Jagdale and M. Megahed

Abstract Powder Bed Additive Manufacturing offers unique advantages in terms of cost, lot size and manufacturability of complex products. The energy used however leads to distortions during the process. The distortion of single layers can be comparable with the powder layer thickness. The contact between the coater blade and the deposited material could terminate the build process. Furthermore, accumulated residual stresses can lead to deviations of the final shape from the design. This work focusses on the accuracy of quick residual stress and distortion models that will both provide layer by layer distortion data as well as the final work piece residual stress and shape. The residual stress and distortion models are implemented in an ICME platform that takes powder size distribution as well as the heat source powder interaction into account. Lower scale models are briefly introduced and data required for the residual stress analysis are documented prior to the analysis of some large components assessing manufacturability and final work piece shape.

Keywords Metal additive manufacturing · Powder bed · Process modeling · Residual stress · Distortion

---

O. Desmaison (✉) · P.-A. Pires · M. Megahed
ESI Group, Paris, France
e-mail: Olivier.Desmaison@esi-group.com

G. Levesque · A. Peralta · S. Sundarraj
Honeywell Aerospace, Phoenix, USA

A. Makinde
GE Global Research, Niskayuna, USA

V. Jagdale
UTRC, East Hartford, USA

Introduction

Qualifying powder bed fusion processes for performance critical components is a challenge currently encountered by massive experimental campaigns to identify a suitable process window, optimize the build strategy and fine tune final material properties. Physics based modelling was suggested as a promising tool for rapid qualification of additive manufacturing [1].

The heat energy used for powder melting leads to workpiece distortions during the build process. The main mechanical phenomenon is thermal shrinkage. It occurs once the metal is deposited and the laser moves to process other regions [2]. The distortion of single layers can be comparable with the powder layer thickness. If a contact between the coater blade and the deposited material occurs, it could terminate the build process. In addition to possible machine damage, it means the whole process has to be repeated, significantly increasing the production cost. Furthermore, the accumulated residual stresses can lead to deviations of the final shape from the design. The expected dimensions might not be reached and the part ruled out.

Although simulation of Powder Bed Additive Manufacturing (PB-AM) is a relatively new field, it can rely on the large experience of the welding community [3, 4]. Indeed, the physics are similar: fusion of a supplied material and its fast solidification driven by internal conduction. Nevertheless, the complexity of PB-AM process lies in the multi-physics and multi-scale coupling required to resolve the process.

ICME Platform

Modelling additive manufacturing processes involves the resolution of very disparate length scales. The powder particles are in the order of microns in diameter, the work piece can be several cubic meter in volume. The heat source travels several hundreds of meters to deposit the material. The model should accordingly resolve length scales that differ by 8–9 orders of magnitude. The time during which the heat source interacts with the powder is as short as a few micro-seconds. The build time can be a few hours and may span several days. The time grid therefore also spans 8–9 orders of magnitude.

Resolving the details of the physical phenomena taking place while modelling the production of the complete work piece is a challenge that cannot be solved with currently available computational resources. The problem is therefore subdivided into distinct models that focus on different aspects of the additive manufacturing process: Powder spreading, powder melting and distortion/residual stress modelling (detailed below). Each solver communicates with a meso scale module providing material properties and metallurgical phase transformations as needed [5–7].

Distortion Model for Powder Bed Additive Manufacturing

The coupling between different scales dictates the level of effort when modelling distortions and residual stresses. The use of assumptions and process simplifications makes the simulations easier to handle, quicker to run and more efficient for industrial problems. The described finite element model has been developed in this sense.

Considering the distortion evolution of a part is induced by thermal strains, one may conclude a thermo-mechanical analysis is compulsory to achieve an accurate prediction of both stresses and strains. The unsolidified powder is almost adiabatic [8, 9] (i.e. it is an insulator due to reduced contact area and the air trapped between the particles), the temperature of the manufactured part is quite stable and its cooling is only driven by its own conductivity. Consequently, the final thermal strain occurring on each deposited layer is determined by the thermo-mechanical material properties. This simplification is true especially as the larger thermal strains occur at a low temperature range, i.e. when the part has reached a global cooling/warming. From this observation, it is possible to numerically deposit a complete layer using a single mechanical solution, where the applied load would be equivalent to the total thermal strain load supplied by the process. The behavior law of both baseplate and workpiece are considered as elastoplastic.

Based on this approach, the successive deposition of layers is modeled using multi-state elements activation method defined in a Lagrangian framework. Considering the current layer being processed, all elements of the already processed layers are activated (including the current layer). The elements which compose the next layer are in a "quiet" state.

The speed of any finite element model depends on the non-linearity of the mechanical problem (e.g. material behavior laws or contact conditions) and on the mesh resolution (i.e. number of nodes). The slicing of the initial geometry into structured layers has encouraged the PB-AM simulation community to use hexahedral 3D meshes. The described model is based on a similar mesh coupled to a remeshing tool. In order to reduce the size of the model during the build process, the mesh of the already processed layers is coarsened. The high resolution mesh is concentrated around the layer being deposited.

Validation Test Cases

The validation of the ICME platform has been carried out through a close partnership with Honeywell Aerospace. The qualitative and quantitative aspects of the distortion prediction are evaluated based on both "academic" and "industrial" test cases. An EOS M290 and Inconel 718Plus® are used. The table displacement is 25 µm. The thermo-mechanical properties are extracted from Sysweld database [10].

Academic Test Case: Beam on Five Supports

The geometry corresponds to a beam on five supports (cf. Fig. 1). The beam is 40 mm long, 5 mm wide and 2 mm thick. The supports are 5 mm high, 5 mm wide and 2 mm thick. The part is subdivided into 280 layers of 25 μm thick each. The mesh is composed of 455,343 nodes and 408,720 3D elements for the workpiece and 21,799 nodes and 14,484 3D elements for the baseplate. Once the build process is simulated, two additional steps are modeled: the release of the baseplate from the machine and the removal of the part from the baseplate.

Figure 2b shows a three dimensional scan of the built component. It is interesting to note that whereas the intended bar is flat, we obtain an arch structure between the supports. Experimental data could not clarify whether the arching was due to lack of material (recoating problems) or whether it is distortion effect. The numerical simulations of distortion were performed assuming ideal recoating conditions predicting. The final workpiece shape is shown in Fig. 2a. It is interesting to note that the solver actually changed the grid topology to a distortion very close to that observed experimentally. As shown on Fig. 3a, the first deposited layers of the beam, while connecting the supports, tend to strongly distort upwards along their free edges. Indeed, the effects of the thermal shrinkage are maximal on these areas since their bottom surfaces are not supported. This behavior continues for several layers until a steady state is reached. It can be therefore concluded that the recoating process is not the root cause of the geometric changes but rather the thermo-mechanical behavior of the material. On Fig. 3b, a table compares several lengths measures on both numerical and experimental parts. The numerical prediction error is lower that 2%.

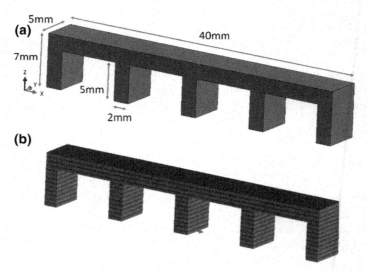

Fig. 1 a Academic test case: beam on five supports. **b** Initial mesh and deposition sequence

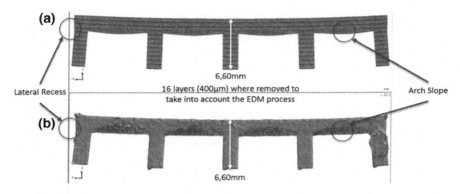

Fig. 2 Comparison of final shape numerically (**a**) and experimentally (**b**) obtained

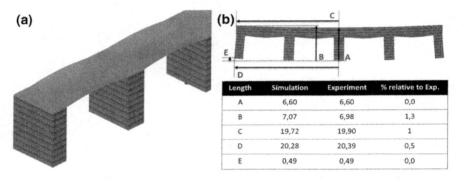

Fig. 3 **a** Zoom on the first deposited layers of the beam and the arch slope growth. **b** Comparison of characteristic dimensions between the experiment and the simulation

Sensitivity Study on Deposition Volume and Grid Resolution

One of the challenges of the PB-AM simulation is to provide a correct prediction of the final built shape within a reasonable computational time. Considering the non-linearity of the mechanical problem being difficult to reduce, it is expected to decrease the computational time by increasing the deposit volume (lumping) or by coarsening the mesh. The first option consists in lumping several layers altogether into a single deposit (or activation) step for all of them. A lumping strategy ×2 means there will be 140 fictive layers to deposit rather than 280, each one with a thickness of 50 μm (instead of 25 μm). The second option is using a coarser mesh. It consists in merging the adjacent elements along the z-axis (if the lumping method is applied) or along the X-Y plane (by increasing the aspect ratio of each element—

Table 1 Sensitivity study of the model to volume deposition and grid fineness. Green case for good results (deviation <5% or elapse time <1 h and arch slope correctly modeled) (colour online)

Model identification	Elements Number	Lumping Strategy	Rel. deviation to experiment (%)	Elapse time	Arch Slope Modeling
0	N	X1	0.5	12h	Yes
1	N	X5	4.6	3h	Yes
2	N	X10	5.3	2h 20 min	No
3	N/2	X2	4.1	2h 42 min	Yes
4	N/2	X5	5.8	1h 20 min	No
5	N/8	X2	4.5	40 min	Yes
6	N/8	X5	6.2	20 min	No
7	N/16	X2	5.4	17 min	Yes
8	N/16	X5	10.6	11 min	No

thickness over surface). Both options were studied, the results are summarized in Table 1. The relative deviation from experiments is based on dimension A, B, C, D and E of Fig. 3b. Model #0 corresponds to the model described above.

As expected, by lumping layers, the final computational elapse time is reduced by a factor of 4 from no lumping to ×5 lumping (models 0 and 1 respectively). The error increases from 0.5 to 4.6%. Doubling the lumping again (Model 2) does not reduce the computational effort significantly, the reference dimensional accuracy is almost constant but the model fails to resolve the arching. By reducing the number of grid elements (N/2, N/8 and N/16), while using the same lumping ×2—models 3, 5 and 7 respectively, the computational time decreases considerably. The relative deviation from measurements is similar for the three models (around 5%) and arching is always predicted correctly. Combining a coarse grid with lumping previously found to be acceptable (i.e. ×5 for instance) leads to loss of dimensional and structural accuracy (models 4, 6 and 8).

In conclusion, given a fine grid with no lumping, the algorithm is able to accurately predict large geometric deviations. Larger deposits (lumping) and coarser grids are alternative approaches to reduce the computational effort allowing quick trend studies, the larger the lumping and the coarser the grid the higher the prediction error.

Industrial Test Case: Thin Wall Blade

A blade shape representative for an industrial turbo application is used. The dimensions are around 50 mm high, 40 mm long and 20 mm wide. The blade wall is approx. 0.4 mm thick. Figure 4a shows the blade geometry without the utilized support structure. The central plate is supported along its bottom surface.

Considering the large part dimensions compared to the walls' thickness, it has been decided to use a lumping factor of 5 and to use remeshing to refine/coarsen the grid as needed. To do so, five submeshes were automatically generated in order to model the complete manufacturing. These meshes are shown on the Fig. 4b. The maximal number of nodes is 1 million nodes. Mapping and projection procedures were set to move from one mesh to another and ensure the continuity of the variables (displacements, strains and stresses). In total 434 layers are deposited during the simulation.

Using 8 cores, the job ran approx. 45 h. This computational cost could be reduced by increasing the lumping factor and increasing the mesh resolution; as demonstrated above. Nevertheless, it has been decided to focus on max. possible accuracy as a preliminary validation step.

Comparison with Experimental Measurements

Figure 5 shows an overlay of a 3D scan of the built blade and the simulated final shape of the build. Each surface is colored by a different color, if one color is visible, then that surface is in front of the other—it is then displayed in a transparent mode, so as to be able to see the surface behind it. The proximity of both surfaces to one another confirm the accuracy of the numerical prediction in a qualitative manner. The main difference can be seen at the top of the blade, where the numerical results tend to predict a slightly higher twisting of the blade.

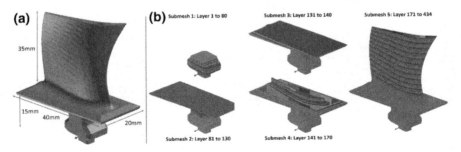

Fig. 4 **a** CAD geometry of the industrial test case. **b** Meshes subdivision

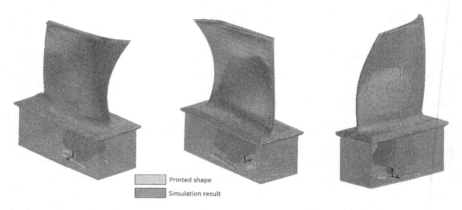

Fig. 5 Comparison of deviation at the end of the process (post-release of the baseplate)

Figure 6 compares the profiles at z = 40 mm. The negative x-axis side shows a very slight deviation; the positive x-axis side is identical. The blade twisting is correctly predicted by the numerical tool.

Figure 7a shows the Von Mises residual stress state once the part has been manufactured and the baseplate has been released from the machine. Maximal stresses are located in the pillar and the central table. The stresses on the blade's walls are low. A quick analysis of the manufacturability of the part is provided in Fig. 7b. The maximal vertical distortion after every 10 layers are deposited is extracted from the numerical solution and is compared to the experimental table displacement. The distortion remains lower than the value of 25 µm except for the 80th layer, where it reaches 35 µm which is still acceptable confirming the manufacturability of the part.

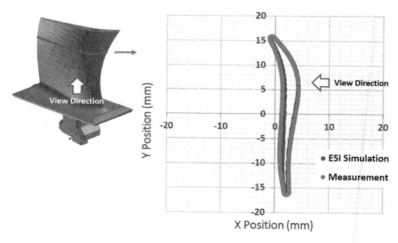

Fig. 6 Comparison of the final Shape of the Blade at z ~ 40 mm (Experiment vs. Simulation)

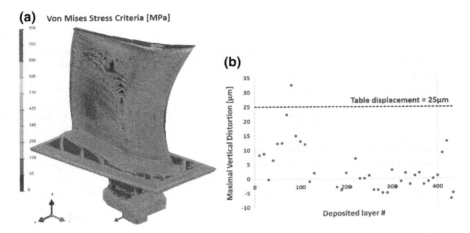

Fig. 7 **a** Von Mises residual stress after release of the baseplate from the machine. **b** Maximal vertical displacement of the last deposited layer

Conclusions

The proposed process simplification leads to accurate distortion predictions. The distortion is resolved for each layer to confirm manufacturability of the component and is accumulated throughout the build process to deliver the final workpiece shape. Predicted residual stresses are qualitatively correct explaining the validated distortions reliably. The computational effort of an industrial workpiece, typically to be pursued using highest possible accuracy, was 45 h. Further reduction of this effort can be achieved by increasing number of cores without losing accuracy. Easy calculation acceleration strategies such as the use of coarse grids and resolution of larger deposits (lumping) do successfully reduce the computational effort at the cost of reduced accuracy. Limiting the maximum allowable error to be lower than 5%, the utilized algorithm and remeshing tools were able to converge within minutes.

Acknowledgements This effort was performed under the America Makes Program entitled 'Development of Distortion Prediction and Compensation Methods for Metal Powder-Bed Additive Manufacturing' and is based on research sponsored by Air Force Research Laboratory under agreement number FA8650-12-2-7230. The U.S. Government is authorized to reproduce and distribute reprints for Governmental purposes notwithstanding any copyright notation thereon.

The views and conclusions contained herein are those of the authors and should not be interpreted as necessarily representing the official policies or endorsements, either expressed or implied, of Air Force Research Laboratory or the U.S. Government.

References

1. A.D. Peralta, M. Enright, M. Megahed, J. Gong, M. Roybal, J. Craig, Towards rapid qualification of powder bed laser additively manufactured parts. Integr. Mater. Manuf. Innov. **5**, 8 (2016)
2. P. Mercelis, J.-P. Kruth, Residual stresses in selective laser sintering and selective laser melting. Rapid Prototyp. J. **12**(5), 256–265 (2006)
3. J. Tejc, J. Kovařík, V. Zedník, A. Sholapurwalla, M. Nannapuraju, H. Porzner, F. Boitout, Using numerical simulation to predict the distortion of large multi-pass welded assemblies, in *Proceedings from the 12th International Conference on Modeling of Casting, Welding, and Advanced Solidification Processes* (2009)
4. J.-M. Bergheau, V. Robin, F. Boitout, Finite element simulation of processes involving moving heat sources. Application to welding and surface treatment, in *Proceedings of 1st International Conference on Thermal Process Modelling and Computer Simulation*, Shanghai (2000)
5. M. Megahed, H.-W. Mindt, N. N'Dri, H.-Z. Duan, O. Desmaison, Metal additive manufacturing process and residual stress modelling. Integr. Mater. Manuf. Innov. **5**, 4 (2016)
6. H.-W. Mindt, M. Megahed, N.P. Lavery, M.A. Holmes, S.R. Brown, Powder bed layer characteristics: the overseen first-order process input. Metall. Mater. Trans. A **47**(8) (2016)
7. H.-W. Mindt, O. Desmaison, M. Megahed, A. Peralta, J. Neumann, Modeling of powder bed manufacturing defects, in *MS&T 16*, Salt Lake City, Utah, USA, 23–27 Oct 2016
8. P. Fischer, V. Romano, H.P. Weber, N.P. Karapatis, E. Boillat, R. Glardon, Sintering of commercially pure titanium powder with a Nd:YAG laser source. Acta Mater. **51**, 1651–1662 (2003)
9. K. Dai, L. Shaw, Thermal and mechanical finite element modeling of laser forming from metal and ceramic powders. Acta Mater. **52**, 69–80 (2004)
10. ESI-Group, Sysweld User Manual (2016)

Author Index

A
Adamovic, Nadja, 79
Ågren, John, 103
Allen, Janet K., 353
Asinari, Pietro, 79

B
Babu, Aravind, 3
Becerra, Andres E., 183
Bernhardt, R., 133
Bernhardt, Ralph, 3
Bhattacharjee, Akash, 15
Borgenstam, Annika, 103
Briffod, Fabien, 317

C
Carroll, Lance, 23
Chen, Chong, 155
Chen, Cong, 169
Cheng, Kaiming, 155
Chen, Kangxin, 193
Chen, Li, 155
Chen, Long-Qing, 293
Chen, Weimin, 169
Christensen, Mikael, 45

D
Das, Prasenjit, 69, 93
Deng, Peng, 155, 169
Desmaison, O., 365
Duan, Zhenhu, 203
Du, Jinglian, 263
Du, Yong, 155, 169

E
Enoki, Manabu, 317
Everhart, Wes, 23

F
Farivar, H., 133
Farivar, Hamidreza, 3
Frankel, Dana, 23
Fröck, Hannes, 237

G
Gautham, B.P., 3, 15, 69, 93, 353
Ghaffari, Bita, 293
Goldbeck, Gerhard, 79
Graser, Matthias, 237
Guo, Zhipeng, 263

H
Han, Zhiqiang, 273
Hashibon, Adham, 79
Hatcher, Nicholas, 23
Hermansson, Kersti, 79
Hristova-Bogaerds, Denka, 79
Hu, Bin, 273
Hueter, Claas, 335
Hunkel, Martin, 335

J
Jagdale, V., 365
Ji, Yanzhou, 293
John, Deepu Mathew, 3

K
Kalariya, Yagnik, 33
Kessler, Olaf, 237
Khan, Danish, 3, 15
Kong, Yi, 155
Koopmans, Rudolf, 79
Kumar, Ranjeet, 3

L

Lechner, Michael, 237
Levesque, G., 365
Li, Han, 155
Li, Juntao, 249
Li, Kai, 155
Li, Mei, 293
Li, Na, 155
Linder, David, 103
Lin, Mingxuan, 335
Liu, Baicheng, 203, 249
Li, Wei, 249
Long, Jianzhan, 155
Luo, Alan A., 273

M

Macioł, Andrzej, 113
Macioł, Piotr, 113
Makinde, A., 365
Malhotra, Chetan, 69, 93
Mantripragada, Vishnu Teja, 145
Megahed, M., 365
Merklein, Marion, 237
Mistree, Farrokh, 353
Miyazawa, Yuto, 317
Mori, K., 327
Mori, Kazuki, 283, 345
Mosbah, Salem, 183, 227

N

Nellippallil, Anand Balu, 353
Nomoto, S., 327
Nomoto, Sukeharu, 283, 345

O

Oba, M., 327
Oba, Mototeru, 345
Olofsson, Jakob, 217
Olson, Gregory B., 23

P

Pan, Yafei, 155
Pathan, Rizwan, 33
Peng, Yingbiao, 155, 169
Peralta, A., 365
Phanikumar, G., 3
Pires, P.-A., 365
Prahl, U., 133
Prahl, Ulrich, 3, 335

Q

Qiu, Yue, 155

R

Rangaraj, Vignesh, 353
Rauch, Łukasz, 113
Reddy, Sreedhar, 69, 93
Reich, Michael, 237
Reith, David, 45
Rothenbucher, G., 133
Rothenbucher, Gerald, 3

S

Salomonsson, Kent, 217
Salvi, Amit, 33
Sarkar, Sabita, 145
Schicchi, Diego, 335
Schmitz, G.J., 57
Schmitz, Georg J., 45
Sebastian, Jason, 23
Segawa, Masahito, 283
Shaik, Azeez, 33
Shamamian, Vasgen A., 183
Shang, Shan, 273
Shen, Houfa, 193, 203
Shiraiwa, Takayuki, 317
Singh, Amarendra Kumar, 353
Skinner, Kwan, 183
Snyder, David, 23
Spatschek, Robert, 335
Srimannarayana, P., 15
Suhane, Ayush, 15
Sundarraj, S., 365
Sun, Weihua, 273

T

Tennyson, Gerald, 15
Tian, Haixia, 155
Tu, Wutao, 193

V

Vale, Sushant, 69
Vasenkov, Alex V., 307
Verbrugge, Tom, 79
Vernon, Greg, 23

W

Wang, Shequan, 155
Wang, Yanbin, 249
Wang, Yaru, 155
Weikamp, Marc, 335
Wen, Guanghua, 155
Wimmer, Erich, 45, 79
Wolf, Walter, 45
Wu, Chang Kai, 183, 227

X

Xie, Wen, 155
Xiong, Shoumei, 263
Xue, Xin, 249
Xu, Qingyan, 249
Xu, Tao, 155

Y

Yamanaka, A., 327
Yamanaka, Akinori, 283, 345
Yang, Jing'an, 203
Yang, Manhong, 263
Yan, Xuewei, 249

Yao, Lei, 249
Yeddula, Raghavendra Reddy, 69
Yun, Gun Jin, 123

Z

Zagade, Pramod, 3, 15
Zhang, Cong, 155, 169
Zhang, Jinfeng, 169
Zhang, Weibin, 155, 169
Zhang, Zhongjian, 155
Zhao, Gang, 249
Zhou, Peng, 155, 169

Subject Index

A
Ab-initio calculations, 270
θ′ (Al_2Cu), 294
Al-Cu-based alloys, 293, 294
Aluminium alloy, 238, 239, 241, 246
Anisotropy function, 252, 263, 264, 266–268, 270
Artificial intelligence neural network, 124
Asymptotic homogenization, 6
Atomic mobility, 169, 171–173, 177

C
CALPHAD, 25, 26, 283, 284, 290
Carburizing process, 133, 137, 141
Case-hardening process, 137
Cast iron, 217, 218
Cellular automaton finite element (CAFE), 203, 204, 209, 211, 213, 214
Cemented carbides, 169, 170, 172, 175, 178
Characterization, 218, 219
Cold-rolling, 16, 17, 19
Composite, 33–35, 38, 40–42
Computer-aided engineering (CAE), 307, 308
Confined hardness
Coupling/Linking, 84, 87, 89

D
Damage, 307–309, 313–315
Data repositories, 82, 85, 86, 89, 91
Decision-based design, 354, 356, 362
Dendritic morphology, 267, 269, 270
Design workflows, 96, 97, 99, 100
DICTRA simulation, 170, 176, 178
Differential Scanning Calorimetry (DSC), 186, 238, 239, 242, 244, 246
Diffusion database, 157, 158, 169, 170, 173, 175, 176
Distortion, 133–135, 137, 139, 140, 142
Domain specific search engine, 69

DP steel, 3–6, 12
Duplex stainless steel, 327, 328

E
Economic impact, 86, 90
e-CUDS
EN AW-6060, 239, 240, 244
European Materials Modelling Council (EMMC), 79–84, 86–91
Eutectic, 183–189, 191

F
Fatigue, 308, 309, 312, 313, 315
Fatigue life prediction, 319, 320
Finite element analysis, 321, 325
Finite element method, 4, 6, 7, 345, 346
First-principles calculation, 328, 329, 331
Flow-charts, 58
Fracture, 307, 308, 310, 312

G
Gear, 3, 4, 7, 8, 12
Glass-ceramic-to-metal seals, 24, 26, 29, 30
Governing equation, 62, 147, 148, 193, 265
Grain structure, 228, 233, 234

H
HDF5, 45, 47–50, 52, 54
Heterogeneous material, 308, 309
Hierarchy, 48
Homogenization, 3, 6, 8, 10, 345, 346, 348
Hot rod rolling and cooling, 354
Hybrid hierarchical model, 307–309, 315

I
Indentation behavior, 103, 105, 106, 109, 110
Industrial deployment, 86
Information extraction, 76
Information retrieval, 76

380

Subject Index

Integrated computational materials engineering (ICME), 24, 26, 29, 32, 41, 42, 46, 93–97, 101, 133, 142
Inter-critical annealing (ICA), 3–5, 7, 12, 16, 19, 20
Interoperability, 60, 62, 65, 68, 82, 85, 86, 88, 89, 91
IT Platforms, 96, 97

K
Knowledge engineering, 94, 101

L
Lagrangian tracking, 228
LMC process, 250, 254

M
Macro- and Micro-simulation, 273, 274, 276, 281
Macroscale simulation, 7, 10
Macroscopic grain structures, 203, 204, 208, 209, 213, 214
Macrosegregation, 193–195, 199, 201
Magnesium alloys, 264, 266–268, 270
Manufacturing, 24, 25, 28, 32
Marketplace, 81, 82, 85, 86, 89, 90
Material Big-Data, 123, 124, 126, 129
Materials design, 24, 25
Materials engineering, 70, 71
Materials integration, 317
Materials modelling, 46, 79–86, 88–91
Meso-scale grain, 251, 254, 258
Metadata, 45, 47, 52, 54, 55
Metadata schema, 85, 88
Micress, 3, 5, 6, 10, 12
Micro dendrite, 251
Micromechanics, 17, 19
Microstructural state, 59, 62, 63, 67
Microstructure, 217–224
Microstructure-based multiscale simulation, 327, 328
Microstructure modelling, 16, 17
Microstructure-property correlations, 155, 156, 164, 166
Model-based machine learning, 317, 319, 324, 326
Model classes, 61, 62
Models, 81–86, 88–91
Molecular dynamics (MD), 308, 309, 345, 348
Molecular dynamics simulation, 327
Multicomponent, 193, 195, 197, 201
Model-driven engineering, 41, 76
Multi-phase field method, 283, 284, 345, 346
Multiphase flows, 153

Multiphase modeling, 195–197, 201
Multiscale analysis, 35, 345
Multi-scale finite element method, 317, 318, 325
Multiscale modeling, 46, 191
Multiscale simulation, 8, 9, 327, 328, 345, 347, 351

N
Navy C-Ring, 137–143
Nonlinear finite element analysis, 125
Numerical simulation, 249

O
Ontology, 10, 69, 71–76, 97, 98, 101, 113, 115–119
Ontology modeling, 72, 76
Open simulation platform, 82, 85, 88
Orientation selection, 264, 266, 269, 270

P
Peridynamics, 308, 309, 313, 315
Phase-field model, 3, 5, 17, 19, 263–265, 273–275, 278, 293–295, 297, 336, 341
Phase field simulation, 161, 264, 268, 270, 293, 294, 296, 297, 299–303, 340
Phase transformation, 19
Plastic deformation, 170, 276–279, 281, 336
Precipitation kinetics, 294, 297
Process and steel alloy design, 134
Process integration, 16–18
Property prediction, 17, 19, 21

Q
Quench factor, 229
Quinary system, 284

R
Reflexive modeling, 71
Repetitive unit cell (RUC), 35, 39, 40

S
Self-learning simulation, 124
SelfSim computational algorithm, 124, 125
Si alloy, 184
Simufact, 4, 7, 8, 10
Simulation, 133, 140–143, 203, 204, 209–211, 213
Slag eye, 146, 150–153
Software, 46
Solidification, 27, 183–185, 188–191, 190, 191, 194, 227–232, 234, 283, 284, 290
Squeeze casting magnesium alloy, 274
Stainless steel, 283, 284, 290

Steel ingot, 193, 194, 197, 199–201, 203, 204, 206, 208, 214
Steel making ladle, 145, 146, 153
Strengthening mechanism, 273, 274, 279, 281

T
Tailor heat treated profiles (THTP), 238, 239, 242, 244–246
TCS PREMAP, 20, 71–72, 75, 76, 93, 96–102
Thermal analysis, 239, 244
Thermocouples, 205–207, 214
Thermodynamic database, 30, 31, 156, 169, 175, 176, 275, 283, 303
Thermo-mechanical analysis, 237, 242, 244, 246, 367

Thermophysical databases, 156, 186
Translation, 82, 86, 88, 90
TiC–Co alloy, 174, 175
Two-point spatial correlation, 321, 324

U
Uncertainty quantification, 25, 32

V
Validated software, 83, 85
Virtual testing, 4, 6–10, 40, 41, 63

X
X-ray tomography, 217, 219

CPSIA information can be obtained
at www.ICGtesting.com
Printed in the USA
BVOW06*1836100517
483751BV00001B/1/P

9 783319 578637